T0136816

THE PHYSICS
OF *THE HEALING*

◆

BOOKS I & II

◆

This publication was funded through the support of the Kuwait Foundation for the Advancement of Sciences

Avicenna

The Physics of *The Healing*

Books I & II

A parallel English-Arabic text
translated, introduced, and annotated by

Jon McGinnis

Brigham Young University Press ◆ Provo, Utah ◆ 2009

Library of Congress Cataloging-in-Publication Data is available.

ISBN: 978–0–8425–2747–7 (alk. paper)

PRINTED IN THE UNITED STATES OF AMERICA.

2 3 4 5 6 7 8 9 17 16 15 14 13 12

First Edition

To

EVERETT ROWSON

and

JAMES ROSS

for teaching me how to read Avicenna's Physics

and how to appreciate what I read

✦

In Memoriam

MICHAEL E. MARMURA

1929–2009

Contents

Volume 1

◆ ◆ ◆

First Book
On the Causes and Principles of Natural Things

◆ ◆ ◆

Second Book

On Motion and That Which Follows It

Volume 2

◆ ◆ ◆

Third Book

Concerning What Belongs to Natural Things Owing to Their Quantity

◆ ◆ ◆

Fourth Book

On the Accidents of These Natural Things and Their Interrelations, as Well as the Things That Are Necessary Concomitants of Their Interrelations

Foreword to the Series

Brigham Young University and its Middle Eastern Texts Initiative are pleased to sponsor and publish the Islamic Translation Series (ITS). We wish to express our appreciation to James L. and Beverley Sorenson of Salt Lake City for their generous support, which made ITS possible, and to the Ashton Family Foundation of Orem, Utah, which kindly provided additional funding so that we might continue.

Islamic civilization represents nearly fourteen centuries of intense intellectual activity, and believers in Islam number in the hundreds of millions. The texts that appear in ITS are among the treasures of this great culture. But they are more than that. They are properly the inheritance of all the peoples of the world. As an institution of The Church of Jesus Christ of Latter-day Saints, Brigham Young University is honored to assist in making these texts available to many for the first time. In doing so, we hope to serve our fellow human beings, of all creeds and cultures. We also follow the admonition of our own tradition, to "seek . . . out of the best books words of wisdom," believing, indeed, that "the glory of God is intelligence."

—Daniel C. Peterson
—D. Morgan Davis

Foreword to the Volume

The Kuwait Foundation for the Advancement of Sciences (KFAS) is a leading institution in the Arab world that provides support for scientific projects and activities. KFAS publishes scientific books that are either originally written in Arabic or are translated into Arabic. These publications provide valuable information to scientists and researchers. In this regard, KFAS concentrates on publishing Arabic and Islamic scientific works, shedding light on the contribution of Islamic scientists and scholars to the development of science and technology throughout the whole world.

In its aim of furthering science and technology, KFAS has always been active in cooperating with other institutions of higher learning, both locally and internationally. The cooperation between KFAS and Brigham Young University to introduce the remarkable scientific achievement of the *Physics* by the famous scholar Ibn Sīnā (Avicenna) is a successful example of such collaboration. By sponsoring the publication of this text, which is a part of Ibn Sīnā's book *The Healing*, KFAS aims to highlight the singular achievements and contributions of Islamic intellectuals to important fields of knowledge such as physics. In doing so, KFAS is actually introducing to scientists and researchers all over the world the views and findings of men of learning who sought knowledge about the natural world during the golden age of Islamic civilization.

This translation of the *Physics* is the first full English translation that has ever been done from the text that was originally written in Arabic by Ibn Sīnā. It primarily concerns the study of natural motion, as opposed to metaphysics or cosmology. Some of his thinking in these areas has been heavily criticized by many Islamic scholars, but at the same time it has also gained support from others. The sponsorship by KFAS of this volume should in no way be viewed as an endorsement of

Ibn Sīnā's metaphysical or philosophical system. It is, rather, a proof of the Foundation's commitment to raise awareness throughout the world of important Islamic contributions to the history of science, as Muslims throughout the ages have sought knowledge and learning.

—Director General
Kuwait Foundation for the Advancement of Sciences

مقدمة

تعد مؤسسة الكويت للتقدم العلمي إحدى المؤسسات الرائدة في العالم العربي في دعم المشاريع العلمية. وهي تعمد في هذا الإطار إلى نشر الكتاب العلمي العربي بهدف دعم المكتبة العربية بالمؤلفات والترجمات، حيث تقدم المؤسسة خدمة جليلة للقارئ والباحث العربي بما يعينه على البحث والعلم والمعرفة. وتولي المؤسسة اهتماما بالغا بنشر كتب التراث العلمي العربي والإسلامي، والتي تلقي الضوء على مساهمات العلماء العرب والمسلمين والتي كان لها أثر بالغ في تطوير العلوم.

والمؤسسة تسعى لتعزيز التعاون بينها وبين العديد من المؤسسات العلمية الأخرى على المستويين المحلي والدولي لتحقيق أهدافها من أجل نشر المعرفة العلمية. وكان من ثمرة هذا التعاون المميز المشروع المشترك بين المؤسسة وجامعة برغهام يونغ، حيث يقدم المشروع عملا مميزا لأحد أعلام الفكر الإسلامي في عهد يعج بالعلماء المسلمين الأفذاذ من أمثال ابن سينا وكتابه «الطبيعيات» الذي هو جزء من كتابه الكبير «الشفاء» ليكون مثالا لجهود علماء المسلمين في ذلك العصر وتأثيرهم في الفكر الإنساني.

والمؤسسة إذ تساهم في هذا الإصدار المميز لابن سينا في مجال الفيزياء «الطبيعيات» فإنها تقدم للباحثين والمهتمين في مختلف أنحاء العالم نتاج أحد أبرز العلماء

خلال عصر النهضة الإسـلامية ، ليكون هذا الإصـدار الأول من نوعه في العالم لترجمة كتاب «الطبيعيات» لابن سـينا إلـى اللغة الانجليزية . وقد اقتصـرت الترجمة على علم الفيزياء عند ابن سـينا من الكتاب و خصوصاً في حركة الأجسام الطبيعية و ذلك لإبراز الجانب العلمي عنده ، دون تبني أو الخوض في تفاصيل آرائه في ما فوق الطبيعة أو فكره الفلسـفي والذي كان له العديد من المعارضين والأنصار سـواء في العالم الإسـلامي أو الأجنبي ، حيث استمر هذا التأثير من تأييد ومعارضة إلى يومنا هذا .

وتأتي مساهمة المؤسسة في هذا الإصدار لإبراز دور العلماء المسلمين في نشر الثقافة والمعرفة العلمية التي كتب العديد منها باللغة العربية في حينها ، وإيمانا من المؤسسة بالدور الذي تلعبه الترجمة في التواصل بين الحضارات وبناء الجسور بينها .

المدير العام

مؤسسة الكويت للتقدم العلمي

Acknowledgments

My love affair with Avicenna's *Physics* began while I was still a graduate student. Thus, my debt to individuals and institutions alike goes back many years. I would first like to thank all those on my dissertation committee who read through the nearly five hundred-page behemoth that was my dissertation and included a translation of 2.10–13 of Avicenna's *Physics.* These include Susan Sauvé Meyer, Gary Hatfield, Everett Rowson, and James Ross. I am particularly thankful to Everett Rowson for the countless hours he spent reading Avicenna's Arabic with me and to James Ross, who, although he does not know Arabic, immensely improved my translation by his constant admonition, "This translation cannot be right; Avicenna's smarter than that!" (Jim, as far as I can tell, was always right.) From my Penn days, I would also like to thank Susan Peppers, Todd Bates, and Shane Duarte for their never-ending willingness to talk about natural philosophy, and philosophy more generally, whether ancient, medieval, or early modern.

Many thanks go out to those who have been willing to read bits and pieces of earlier translations. In particular, I would like to mention Peter Adamson; David Reisman (who read the entire first half of my translation of book 2); Jules Janssens, who not only provided me with his provisional edition of book 3 of the *Avicenna Latinus* but also graciously read my corresponding translation; and Edward Macierowski, who worked through and commented on my entire translation. I also have an "inclination" to thank Alexander Bellamy for his comments on Avicenna's discussion of *mayl.* Their suggestions and acute observations greatly improved my initial efforts. I am also appreciative of the conversations with Thérèse-Anne Druart, Deborah Black, Richard Taylor, and Asad Ahmad about Avicenna and philosophy done in the medieval Arabic-speaking world. My efforts have been greatly enriched by all of them. A very special thanks goes to Paul Roth, who, while having absolutely no training in either Arabic philosophy or the history of physics, always seemed to see what "my guys" were up to more clearly than I and thus put me on the right track. I would also like to say *obrigado* to my friends in Brazil working on the history of natural philosophy, and especially to Tadeu Verza, a fellow scholar of Arabic natural philosophy, whose encouragement and inspiration has been a genuine godsend. Finally, Daniel C. Peterson and D. Morgan Davis of the Middle Eastern Texts Initiative have been saints in putting up with me and my quirks. The overall success of this volume is greatly indebted to their untiring work, aided by Jude Ogzewalla and Elizabeth Watkins.

On the institutional front, I would like, first, to acknowledge the University of Missouri–St. Louis for two summer grants that supported my initial translation of book 2. Additionally, I would like to thank the Center of International Study at UM–St. Louis for the significant travel support they have provided me over the years that has allowed me to present my work on Avicenna's *Physics* at conferences both here and abroad. I am also very grateful for the support offered by two University of Missouri system-wide Research Board awards that allowed me time off to work on the present translation. Beyond the local level, I was blessed with a National Endowment for the Humanities summer stipend and two NEH fellowships, all of which allowed me to focus on Avicenna's *Physics*. One of the NEH Fellowships additionally provided for my expenses while I was a member at the Institute for Advanced Study in Princeton, whose staff and permanent members I also want to thank for the intellectually stimulating environment that they have created, and where I was able to work on parts of books 1 and 4 uninterrupted.

I also want to acknowledge the countless number of undergraduate and graduate students who had to put up with whole lectures on Avicenna's natural philosophy and at times were even forced to read my provisional translations of parts of Avicenna's *Physics*. Here, three students stand out most clearly in my mind for helping me see things in Avicenna's arguments that I had originally missed. These are Cynthia Winter, Josh Hauser, and Stuart Reeves. I would also like to thank the departmental administrative assistant, Nora Hendren, who has graciously initiated me into the esoteric secrets of Microsoft Word, particularly its graphic functions. Finally, I want to thank my family: my wife for seeing that I got the time to work on this project, and my boys for seeing that I got enough play time away from this project.

Translator's Introduction

Unlike Avicenna's metaphysics, philosophical psychology, and even logic, his natural philosophy—that is his general physics—has received relatively little attention.[1] One indication of the discrepancy in scholarly interest is that while the *Metaphysics* (*Kitāb al-ilāhīyāt*) of Avicenna's voluminous *The Healing* (*Kitāb al-shifāʾ*) has six translations in European languages—a medieval Latin translation (ed. S. Van Riet), as well as modern translations into German (Horten), French (Anawati), two Italian (Lizini and Porro, and Bertolacci), and English (Marmura)—the *Physics* (*Kitāb al-samāʿ al-ṭabīʿī*), in contrast, was never completely translated into Latin and has received modern translations only into Persian and Turkish.[2] While Avicenna's *Physics*, for whatever reasons,

---

1. I am happy to say that since the 1980s, there has been a gradually increasing interest in Avicenna's *Physics*. This new trend was spearheaded by Aydın Sayılı ("Îbn Sınâ and Buridan on the Dynamics of Projectile Motion," in *Îbn Sınâ, Doğumunum Bininci Yılı Armağanı*, ed. A. Sayılı, [Ankara: Türk Tarih Kurumu Basımevi, 1984]) and even more so by Ahmad Hasnawi ("La dynamique d'Ibn Sīnā (la notion d'‘inclination’: *mayl*)," in *Études sur Avicenne*, ed. J. Jolivet and R. Rashed, Collection Sciences et philosophies arabes (Paris: Les Belles Lettres, 1984), 103–23; "Le mouvement et les catégories selon Avicenne et Averroès l'arrière-fond grec et les prolongements latins médiévaux," *Oriens-Occidens* 2 (1998): 119–22; "La définition du mouvement dans la 'Physique' du Shifāʾ d'Avicenne," *Arabic Sciences and Philosophy* 11 (2001): 219–55; "La *Physique* du *Shifāʾ*: Aperçus sur sa structure et son contenu," in *Avicenna and His Heritage; Acts of the International Colloquim . . . 1999*, Ancient and Medieval Philosohpy, ser. 1, 28, ed. Jules Janssens and Daniel de Smet (Leuven, Belg.: Leuven University Press, 2002), 678–80; "Le statut catégorial du mouvement chez Avicenne: Contexte grec et postérité médiévale latine," in *De Zénon d'Élée à Poincaré: Recueil d'études en hommage à Roshdi Rashed*, ed. R. Morelon and A. Hasnawi, Les cahiers du MIDEO 1 (Leuven, Belg.: Peeters, 2004), 607–22).

2. The Persian translation is from 1937 by Muhammad Farūghī; and, more recently, there is the Turkish translation by Muhittin Macit and Ferruh Özpilavci (Istanbul: Litera Yayincilik, 2004–2005).

has not fascinated most students of the great Muslim sage, the paucity of work dedicated to it has, in my opinion, impoverished Avicennan studies generally. That is because the *Physics* frequently provides the basis for a full appreciation and proper understanding of many of Avicenna's advancements in other fields. It presents the language, concepts, and presuppositions for the special sciences within natural philosophy, such as the study of the soul. Similarly, it raises the puzzles that were to become the issues and themes central to Avicenna's metaphysics—and in many cases, even offers the first pass at their solutions.

A few examples from his psychology and metaphysics might make this abstract point a bit more concrete. One of the distinguishing characteristics of Avicenna's psychology—at least among medieval theories of the soul—is its substance dualism. For Avicenna, the human intellect is not the form of the body, but an immaterial substance that is the perfection of the body and that uses the body as a tool. In his *Psychology* (*Kitāb al-nafs*) (5.2), Avicenna attempts to demonstrate this claim; and as part of his proof, he shows that what receives intelligible objects—the intellect—cannot be material and indivisible. In the *Psychology* itself, he sketches the argument against this position only loosely and refers his reader back to his *Physics* for the full account. While the argument in the *Psychology* can appear quite baffling, it is in fact merely a truncated version of his fully articulated refutation of atomism from books 3.3–5 of his *Physics.* Thus, while it might be quite difficult on the basis of the *Psychology* passage alone to see that he is arguing against an atomist's account of the soul, a position associated with certain *mutakallimūn,* it is obvious if one has read his *Physics.* In this case, a knowledge of the *Physics* helps one understand Avicenna's argument as well as its place in the history of psychology.

Another example from psychology concerns the role and function of Avicenna's celebrated doctrine of the estimative faculty (*wahm*). Concerning this internal faculty, there has been some scholarly dispute: does it function primarily as animal instinct,[3] or does it have a more robust role in Avicenna's thought?[4] Certainly, in his *Physics,* Avicenna gives it a much greater cognitive role than mere instinct. In fact, *wahm*

3. Dag N. Hasse, *Avicenna's "De anima" in the Latin West: The Formation of a Peripatetic Philosophy of the Soul, 1160–1300,* Warburg Institute Studies and Texts 1 (London: The Warburg Institute; Turin: Nino Aragno Editore, 2000), 127–41.

4. Deborah L. Black, "Estimation in Avicenna: The Logical and Psychological Dimensions," *Dialogue* 32 (1998): 219–58.

and its cognates are Avicenna's preferred vocabulary for setting up so-called thought experiments throughout the *Physics*. Indeed, he makes it the faculty that tracks what initially appears to the human intellect to be possible, even if not always indicating real possibility. Here, then, seeing how Avicenna actually employs the estimative faculty in physical investigation, beyond relying merely on his formulaic examples from his *Psychology*, provides one a deeper insight as to how he envisions the role of this faculty.

As for his metaphysical thought, Avicenna raises literally scores of problems in his *Physics* whose answer he defers to first philosophy. Thus, a proper appreciation of many of the problems treated in his *Metaphysics* requires an understanding of the physical theory or issues that gave rise to them. Certainly, one of the more apparent cases is Avicenna's distinction between "metaphysical causation" as opposed to mere "physical causation." In both the *Physics* and the *Metaphysics*, Avicenna dedicates an entire book to the causes. If I may resort to a bit of "bean counting" to suggest the significance of the *Physics* for understanding Avicenna's theory of causation, I would note that while the relevant book from the *Metaphysics* consists of six chapters, totaling altogether around thirty-five pages, the corresponding book from the *Physics* on the causes and principles of natural things consists of fifteen chapters, coming in at around one hundred pages. Moreover, despite his constant refrain that "these issues are better treated in First Philosophy," the pages of the *Physics* are nonetheless filled with material vital for reading and interpreting Avicenna's conception of metaphysical causation, since he regularly contrasts the physical causation that he is discussing with metaphysical causation. (See especially his account of the efficient and final causes in 1.2 of his *Physics*, but elsewhere as well.) Indeed, even the Necessary Existent in itself makes an appearance when Avicenna provides his own unique interpretation of Parmenides (see 1.4).

Perhaps an even more telling example of metaphysical doctrines that are foreshadowed in the *Physics* is Avicenna's arguments concerning the age of the world and those arguments' reliance on his analysis of possibility (3.11). While today we may think that issues associated with temporal topology (as, for example, "What was there before the Big Bang?") belong to metaphysics or theoretical cosmology, for those working in the ancient and medieval Aristotelian tradition—as Avicenna in a real sense was—this topic fell squarely within the science of physics, or natural philosophy. Thus, while Avicenna does have a strictly "metaphysical" argument for the eternity of the world based upon divine immutability, which he gives only in his *Metaphysics,* he also has two modal proofs for the eternity of the world developed in his *Physics:* one showing that change presupposes the possibility for change and that the possibility for change requires pre-existent matter as a subject; and another that draws upon his unique and provocative analysis of time in terms of possibility. Indeed, his discussion of possibility throughout his *Physics* sheds much light on his corresponding treatments of possibility, power, and potency throughout the *Metaphysics.*

The above is merely intended to give one a taste for the important role that Avicenna's physical theory plays in an overall appreciation of his philosophical synthesis and system. By no means have I exhausted all the instances where an understanding of Avicenna's natural philosophy provides a deeper insight into other areas of his thought. Similarly, I do not make any claim to having uncovered the most important cases. Hopefully, though, I have piqued some interest.

It is also worth noting that Avicenna's treatment of issues physical was arguably the most creative, well conceived, and overarching in all of the medieval Arabic-speaking world. Thus, in addition to aiding our understanding of other facets of Avicenna's thought, a careful study of Avicenna's *Physics* will provide historians of science with a more complete understanding of the history of physics and natural philosophy in general, and particularly its development in the medieval Islamic milieu.[5]

There can be no question that Avicenna's physical theory is deeply indebted to Aristotle's *Physics* and other physical writings by Aristotle such as *On the Heavens, On Generation and Corruption,* and the *Meteorology.* In fact, Avicenna tells us in his preface that he is intentionally going to follow the order of presentation followed by the Peripatetics. There also seems little doubt that he drew on the commentary traditions that grew up around Aristotle's *Physics.* Thus, one sees in Avicenna's *Physics* hints of Alexander of Aphrodisias, Themistius, and other Greek Aristotelian commentators alongside Abū Bishr Mattá and Yaḥyá ibn ᶜAdī, among Peripatetics working in Arabic. Most significantly, however, one sees the influence of the Neoplatonist John Philoponus (ca. 490–570s)—both his *Physics* commentary and his *Contra Aristotelem* (that is, at least based upon what we know from the fragments of that now-lost work). Sometimes

5. For a discussion of Avicenna's significance as a historian of science, see Jules Janssens, "Ibn Sînâ: An Extraordinary Historian of the Sciences," paper presented at the Ibn Sīnā Symposium, Istanbul, Turkey, May 2008, forthcoming.

Philoponus is an object of criticism, as, for example, in Avicenna's refutation of the interval (Grk. *chōra,* Arb. *buʿd*) (2.8), but sometimes he is a source of inspiration, as, for instance, in Avicenna's defense of inclination (Grk. *rhopē,* Arb. *mayl*) or acquired power (4.8 and 12). Additionally, one sees Avicenna addressing issues raised by more thoroughgoing Neoplatonic works, such as the *Enneads* of Plotinus (204–270), a redacted version of which Avicenna knew under the pseudoeponymous title the *Theology of Aristotle.*[6] Avicenna similarly incorporates the medical works of Galen, the astronomical works of Ptolemy, and the works of other Greek intellectual luminaries into his overall physical theory.

The sources upon which Avicenna drew for his *Physics,* however, were by no means limited merely to Greek ones. He likewise knew and responded to works of Arabic composition. Thus, along with al-Kindī, al-Fārābī, and the Baghdad Peripatetics, Avicenna seems to have been intimately familiar with the thought of Abū Bakr Muḥammad al-Rāzī (ca. 864–925 or 932) and particularly with his theory of time as an eternal substance (see 2.10).

Equally of interest to the historian of science is Avicenna's impressive knowledge of *kalām* Atomism, discussions of which permeate the first half of book 3. Indeed, not only does Avicenna rehearse many of the *kalām* arguments found in the notable studies of this topic by Shlomo Pines (1936; 1997), Alnoor Dhanani (1994), and more recently, A. I. Sabra (2006), but he also presents arguments not catalogued by these scholars. In the same vein, Avicenna is conversant with the thought of the anti-Atomist *mutakallim,* Ibrāhīm al-Naẓẓām, and quite likely had read the latter's *Book of the Atom* (*Kitāb al-juzʾ*), now no longer extant. All in all, then, Avicenna's *Physics,* and especially book 3, makes an excellent additional source for the study of Islamic Atomism.

Moreover, Avicenna knew and responded to a growing trend in medieval Islamic courts to mathematize problems in natural philosophy—that is, to consider natural things not only qualitatively, but also quantitatively.[7] One example of Avicenna's knowledge of this newly

6. For a discussion of Plotinus's influence in the Arabic-speaking world, see Peter Adamson, *The Arabic Plotinus: A Philosophical Study of the 'Theology of Aristotle'* (London: Duckworth, 2002).

7. For a general discussion of the application of mathematics to so-called physical problems, see Roshdi Rashed, "The Philosophy of Mathematics," in *The Unity of Science in the Arabic Tradition; Science, Logic, and Their Interactions,* ed. S. Rahman, T. Street, and H. Tahiri Logic, Epistemology, and the Unity of Science 11 (Dordrecht, Germany: Springer, 2008), 155–84.

emerging approach to physical theory is his awareness and use of the method of projection (2.8) as part of his criticism of an infinite void.[8] Another example is his careful and tentative comments on the issue of the *quies media,* the topic of whether a stone, for example, when it is thrown upward must come to rest at the apex of its motion before turning downward (4.8).[9] Finally, while his criticism of certain commonly accepted mathematical formulas for the relations between power exerted and the time and/or distance of the motion produced (4.15) may give the impression that Avicenna is opposed to the application of mathematics to physics—and indeed, in a certain sense he was—his real objection was that these overly simplistic formulas failed to do justice to the actual complexity of the physical phenomena they intended to explain. Consequently, they did not provide the desired necessity that is the hallmark of demonstrative science, at least in Avicenna's mind.

All of these examples, then, hopefully give witness to Avicenna's knowledge of and role in the emergence of this nascent quantified physics: Despite all the sundry influences coming together in Avicenna's *Physics*—or perhaps more precisely because of them—Avicenna's natural philosophy defies being classified as simply either "Aristotelian" or "Neoplatonized Aristotelianism." It is perhaps simply best to say that Avicenna's natural philosophy is "Avicennan." Michael Marmura's elegant comments about Avicenna's *Metaphysics* apply equally to his *Physics:* "The conceptual building blocks, so to speak, of this system are largely Aristotelian and Neoplatonic. The final structure, however, is other than the sum of its parts, and the cosmic vision it portrays has a character all its own."[10]

I have mentioned the role that Avicenna's *Physics* can play in clarifying the Avicennan philosophical system overall as well as its place in

8. Avicenna's argument has certain similarities to an argument of al-Qūhī, though it is much simplified and put to quite a different use. For al-Qūhī's argument, see Roshdi Rashed, "Al-Qūhī vs. Aristotle: On Motion," *Arabic Sciences and Philosophy* 9 (1999): 3–24.

9. See Shlomo Pines, "Études sur Awḥad al-Zamān Abu'l-Barakāt al-Baghdādī," in *The Collected Works of Shlomo Pines,* vol. 1, *Studies in Abu'l-Barakāt al-Baghdādī, Physics and Metaphysics* (Jerusalem: The Magnes Press, 1979), 1–95, esp. 66–71. See also **Y.** Tzvi Langermann, "*Quies media:* A Lively Problem on the Agenda of Post-Avicennian Physics," paper presented at the International Ibn Sīnā Symposium, Istanbul, Turkey, May 2008, forthcoming.

10. Avicenna, *The Metaphysics of* The Healing: *A Parellel English-Arabic Text,* trans. and ed. Michael E. Marmura, Islamic Translation Series (Provo, Utah: Brigham Young University Press, 2005), xxii.

the history of early natural philosophy and physics. Let me now address Avicenna's *Physics* as a work of philosophical and historical interest in its own right. Natural philosophy is one of the three theoretical sciences (Grk. *epistēmē,* Arb. *ᶜilm*) identified among ancient and medieval philosophers. These three theoretical sciences include physics, mathematics, and metaphysics. Each of them is distinguished by its own proper subject matter. In the case of physics, Avicenna identifies that science's proper subject matter with the sensible body insofar as it is subject to change—in effect, those things that have a nature—and specifically the necessary accidents and concomitants of these natural things. He then proceeds to investigate this subject in four books.

The first book concerns the causes and principles of natural things, corresponding roughly with books 1 through 2 of Aristotle's *Physics.* Chapter one provides a nice overview of how a demonstrative science should investigate its proper subject; and as such, this chapter gives a succinct presentation of many of the salient points of Avicenna's *Book of Demonstration* (*Kitab al-burhān,* the work most closely following in the tradition of Aristotle's *Posterior Analytics*). Chapters two through twelve (with a brief hiatus at chapter four, in which Avicenna discusses Melissus and Parmenides) take up the principles of natural things as well as the causes of change, perfection, and generation and corruption. Again, to repeat what I noted earlier, these chapters represent perhaps Avicenna's most extended discussion of the nature of causation. Chapters thirteen through fifteen, which conclude book 1, deal with luck and chance and provide evidence for Avicenna's strong causal determinism.[11]

The second book, which treats motion and its necessary concomitants—namely, place and time—is, in many ways, medieval theoretical physics at its best. It loosely follows the first part of book 3 of Aristotle's *Physics* (up to but not including the discussion of the infinite) and book 3 of that work. In book 2.1 of his *Physics,* Avicenna offers up his double sense of motion with its accompanying theory of motion at an instant.[12] He likewise provides in 2.1 and 3 an in-depth analysis of circular motion and introduces a new genus of motion—that of motion with respect to the category of position—which goes beyond the traditional three mentioned

11. For a fuller discussion of Avicenna's position on this point, see Catarina Belo, *Chance and Determinism in Avicenna and Averroes,* Islamic Philosophy, Theology, and Science 69 (Leiden: E. J. Brill, 2007).

12. See Ahmad Hasnawi, "La définition du mouvement," and my "A Medieval Arabic Analysis of Motion at an Instant: The Avicennan Sources to the *forma fluens/fluxus formae* Debate," *British Journal for the History of Science* 39 (2006): 189–205.

by Aristotle: motion in the categories of quantity, quality, and place. This new element of Avicenna's physics, in its turn, offers him a simple yet elegant answer to one of the great physical questions of the ancient period: "How can the cosmos undergo motion if it has no place?" The problem, which plagued Aristotle and many of his subsequent commentators and was in fact used by Aristotle's detractors to undermine his physics, was this: According to Aristotle, place is the limit of the outermost containing body. Consequently, the cosmos, which has nothing outside of it to contain it, cannot have a place. Yet, according to Aristotle and accepted by virtually every ancient and medieval thinker, the cosmos's outermost celestial sphere was constantly moving, making a complete rotation once approximately every twenty-four hours. It was just this motion that accounted for the rising and setting of the sun, moon, and stars. Clearly, this daily motion is not change in quantity or quality. Given the three canonical types of change identified by Aristotle, the motion of the cosmos must be with respect to place. But Aristotle's analysis of place precluded the cosmos from having a place such that it could undergo change with respect to it. For Avicenna, the solution was simple: The cosmos has no place—thus Aristotle's account of place is preserved—but moves with respect to its position, which, in fact, is just rotation without change of place.[13]

Another point of interest in book 2 is Avicenna's conceptual proof against the existence of a void (2.8), which seems to be a developed version of an argument hinted at by al-Kindī.[14] The difficulty is trying to prove that there *is* something that does *not* exist. Avicenna's argument takes advantage of logical developments he himself made in the *Introduction* (*Kitāb al-madkhal,* 1.13) and *Book of Demonstration* (1.10) of *The Healing.* Using these logical developments, he argues that if some notion is not merely an empty concept in the mind and in fact exists in reality, then one should be able to give a proper Aristotelian definition of it in terms of genus and difference, where both the genus and difference must identify some positive feature and not merely be a negation. Avicenna then shows that every attempt to define a void requires some appeal to its *not being* something else and so fails to yield a proper definition.

13. For the history of this problem and a more developed account of Avicenna's solution see my "Positioning Heaven: The Infidelity of a Faithful Aristotelian," *Phronesis* 51, no. 2 (2006): 140–61.

14. See al-Kindī, *On First Philosophy,* chapter 2.

Consequently, Avicenna concludes that the idea of a void is simply a vain intelligible.[15]

Also of importance in book 2 is Avicenna's proof for the existence of time and his analysis of it in at chapter eleven. Aristotle's temporal theory, upon which Avicenna's draws heavily, begins with a series of puzzles intended to cast doubt on the reality of time. Unfortunately, Aristotle does not follow up his puzzles with solutions. Avicenna, however, not only resolves all the puzzles raised by Aristotle, as well as several others, but he also included an explicit proof for the existence of time. His analysis of time, in turn, shows that it corresponds with the *possibility* for motions of the same speed to vary either in the length of the distance traversed (in the case of motions with respect to place) or the number of rotations made (in the case of motions with respect to position). This conception of time, which is quite intuitive and clever, provides the linchpin for one of Avicenna's proofs for the eternity of the world, presented at 3.11.

The third book of Avicenna's *Physics*, which treats issues of the infinitely large and infinitesimally small, took on a special urgency for Avicenna. This is because John Philoponus, in a series of arguments, had rebutted Aristotle's account of the infinite, especially with respect to the issue of the age of the world and its purported infinite extent into the past, a position that Avicenna himself held. Similarly, Islamic speculative theologians (that is, the practitioners of *kalām*) also denied that anything—whether matter, space, or even time—could be infinitely large. So, like Philoponus, they too denied the eternity of the world. Moreover, they additionally denied that the infinitely small was possible. Consequently, they argued for the existence of atoms that are physically and conceptually indivisible. The first half of book 3 of Avicenna's *Physics* is therefore dominated by Avicenna's rejection of Atomism, whether of the Democritean, Epicurean, or *kalām* variety. Interestingly, however, in chapter twelve of this book, Avicenna does suggest that there are *minima naturalia,* or magnitudes below which an element cannot retain its species-form. In effect, Avicenna is allowing that there are bodies that cannot physically be divided further and so are physical *a-toms* (literally, "something that cannot be cut"), even if they are conceptually divisible *ad infinitum.*

15. See my "Logic and Science: The Role of Genus and Difference in Avicenna's Logic, Science and Natural Philosophy," *Documenti e studi sulla tradizione filosofica medievale* 43 (2007): 165–86.

Most of the second half of book 3 discusses ways that the infinite can and cannot enter reality. In the *Physics* of *The Healing*, Avicenna takes the Aristotelian position that while it is impossible for an actual infinite to enter reality, a potentially infinite not only can but must enter reality in the form of infinite changes and time's stretching into the infinite past. In a second encyclopedic work, the *Salvation* (*Kitāb al-najāt*), Avicenna would, however, argue that an actually infinite number of immaterial souls must exist in reality.[16] Interestingly, a key element in the *Salvation* argument as to why there can be an infinite number of immaterial souls in existence is that such a totality does not involve an essentially ordered series, a point that Avicenna does make even in *The Healing*. The last two chapters of book 3 treats issues from Aristotle's *On the Heavens* concerning natural directions, such as up/down, right/left, and front/back, and how the natural philosophers can establish these directions.

The final book of Avicenna's *Physics*, book 4, is a miscellany. It covers issues such as what makes a motion one, whether with respect to number, species, or genus, as well as accidental motions. It additionally speaks about natural place and what makes a motion natural as opposed to forced. Perhaps two of the more theoretically interesting questions treated in this book are the issues of the *quies media* (4.8), already briefly mentioned, and Avicenna's account of inclination or acquired power (4.12). The issue of *quies media*, or medial rest, again concerns a motion that involves a change from one contrary to another, as, for example, a stone that is thrown upward and then falls downward. In a case of such motion, must the moving thing come to rest at the precise moment of the change from the one contrary to the other, or are the two motions, in fact, continuous without any rest? The issue was hotly debated by Avicenna's time, and Avicenna finds none of the arguments on either side to be wholly demonstrative. In the end, he opts that there must be a rest, but he does so because his own theory of inclination seems to force him in that direction.

The issue of inclination, in its turn, involves projectile motion. More precisely, the question is what keeps a mobile in motion in the cases where it is separated from its initial mover—as, for example, an arrow

16. See Michael E. Marmura, "Avicenna and the Problem of the Infinite Number of Souls," *Mediaeval Studies* 22 (1960): 232–39; reprinted in *Probing in Islamic Philosophy: Studies in the Philosophies of Ibn Sina, al-Ghazali and Other Major Muslim Thinkers* (Binghamton, NY: Global Academic Publishing, 2005), 171–79.

shot from a bow. The problem becomes more acute once it is assumed that the cause of an effect must exist together with the effect. Thus, at every instant that something is moving (the effect), it would need to be conjoined with its mover (the cause). So what is the mover in the case of projectile motion? At least within the Aristotelian tradition, there were two answers: the historical solution proposed by Aristotle, who maintains that the initial mover sets not only the projectile but also the surrounding air into motion, and it is the moving air that then acts as the immediate mover of the projectile's subsequent motion; and that of John Philoponus, who argues that the mover impresses a power, impetus, or inclination (*rhopē*) into the projectile that keeps it in motion. In the end, Avicenna, taking a position closer to that of Philoponus, thinks that there is an acquired power or inclination that keeps the projectile in motion.

This annotated outline of Avicenna's *Physics* by no stretch of the imagination exhausts the subjects treated in it. Hopefully though, it nonetheless provides at least a sense of the importance of this work, whether it be its place in the history of philosophy and science, its value for understanding Avicenna's overall philosophical thought, or its own intellectual merit.

A Note on the Source Texts

There are two editions of *Kitāb al-samāᶜ al-ṭabīᶜī:* that of Saᶜīd Zāyid (1983)[17] and that of Jaᶜfar al-Yāsīn (1996).[18] At least in one respect, al-Yāsīn's edition is preferable to Zāyid's in that al-Yāsīn seems to have a better grasp of the philosophical content of Avicenna's *Physics* and so, as it were, breaks up the text at its natural philosophical junctures. In contrast, in one case, Zāyid divides a single sentence admittedly a long one—into three separate paragraphs; similar instances can be multiplied. Consequently, al-Yāsīn's edition is, at least from a philosophical point of view, an advancement over Zāyid's. Also, Zāyid's edition is replete with typographical errors, some of which can be sorted out easily enough, but many others of which introduce significant misreading. While al-Yāsīn's also has typos, there seems to be fewer of them. Unfortunately,

---

17. *Al-Ṭabīᶜīyāt, al-samāᶜ al-ṭabīᶜī,* ed. Saᶜīd Zāyid (Cairo: The General Egyptian Book Organization, 1983); henceforth **Z**.

18. *Al-Ṭabīᶜīyāt, al-samāᶜ aṭ-ṭabīᶜī,* ed. Jaᶜfar al-Yāsīn (Beirut: Dār al-Manāhil, 1996); henceforth **Y**.

al-Yāsīn's edition suffers from a serious flaw not found in Zāyid's that prevents the former from being used as a basis of translation, at least as it stands: al-Yāsīn's edition is simply rife with instances of dittography and homeoteleuton—inadvertent repetitions and omissions of, at times, lengthy bits of Avicenna's text. Despite this failing of the edition established by al-Yāsīn, I decided to start with it and then emend the text as needs be in light of Zāyid's edition, as well as the Arabic edition of the text found in the Tehran lithograph of *The Healing*[19] and the available medieval Latin translation of Avicenna's *Physics*.[20] It is this collation of these four texts that makes up the edition included here.

As a warning to the reader, however, while I hope that the present edition is an improvement on the available editions of Avicenna's *Physics,* I did not approach the various texts at my disposal in as critical a way as some might hope. In general, I assumed that al-Yāsīn's edition was for the most part acceptable, and in general I only spot checked it, albeit frequently, against Zāyid's edition, the Tehran lithograph, and the Latin. Only when al-Yāsīn's text seemed to me to have obvious philosophical or grammatical problems did I then closely consult all the texts. Thus I have made no attempt to note all of the variations between al-Yāsīn's edition and that of Zāyid, the Tehran lithograph, and the Latin. Still I hope that the end result is at least a serviceable edition of Avicenna's Arabic text, even if more work may still need to be done.

Let me offer one further warning as well. As anyone who has seriously worked on the *Physics* can attest, both Avicenna's Arabic and his argumentation can be extremely difficult at times. Thus, in those cases where translations of certain parts of Avicenna's text were available, I consulted them. These translations include the Latin versions of book 1,[21]

---

19. *Al-Shifāʾ,* 2 vols. (Tehran: n.p., 1886), vol. 1, 2–159; henceforth **T**.

20. The Latin translation includes all of books 1 and 2 and then chapters 1–10 of book 3, after which the Latin translators stopped. For a history of the Latin edition, see Jules Janssens, "The Reception of Avicenna's *Physics* in the Latin Middle Ages," in *O Ye Gentlemen: Arabic Studies on Science and Literary Culture in Honour of Remke Kruk,* ed. Arnoud Vrolijk and Jan P. Hogendijk, Islamic Philosophy, Theology, and Science 74 (Leiden: E. J. Brill, 2007), 55–64. It should be noted that the manuscripts used to establish the Latin edition are approximately as early as some of the earliest Arabic manuscripts used by either Zāyid or al-Yāsīn. Thus, its variants should be considered when trying to establish the text.

21. *Liber Primus Naturalium, tractatus primus: De causis et principiis naturalium,* ed. Simon Van Riet, Avicenna latinus (1.8): (Leuven, Belg.: Peeters; Leiden E. J. Brill, 1992).

2,[22] and 3[23] (which only goes through Chapter 10); Ahmad Hasnawi's partial translation of 2.1[24] and complete translation of 2.2;[25] Yegane Shayegan's translation of 2.10–13;[26] and Paul Lettinck's translation of 3.3–5.[27] While I always greatly benefited from these earlier translations, there are bound to be differences of interpretations among scholars when dealing with a text as difficult as Avicenna's. Any errors or other faults in the translation are wholly my own.

While I believe that, for the most part, I have accurately presented Avicenna's philosophical intention, I must confess that I have not fully grasped every argument and position Avicenna presents. I found this to be particularly the case when he is treating arguments outside of the tradition associated with Aristotle's *Physics* proper. In those cases where I felt uncertain of Avicenna's philosophical intention, I have taken refuge in presenting what I believe to be a very literal translation of the text, hoping that the clear-sighted reader will see more in it than I have.

In contrast, however, where I felt confident in my understanding of the text, I rendered it in what I hope is idiomatic English rather than an overly literal translation of the Arabic. The reason for this liberty is that while Avicenna's prose often has a certain elegance to it, when I translated the text completely literally, it frequently came across as too crabbed or too prolix. Indeed, at times a literal English translation seemed wholly unintelligible, whereas the Arabic made perfect sense. In the end, intelligibility and readability seemed more desirable than being able to reconstruct the Arabic on the basis of the translation. Additionally, Avicenna has a decidedly wry wit about him. After presenting a philosophically rigorous argument against a position, he often draws a humorously absurd image of it. Moreover, he seems to have enjoyed a not-too-infrequent pun. To preserve some of the flavor of Avicenna's

22. *Liber Primus Naturalium, tractatus secundus: De motu et de consimilibus,* ed. S. Van Riet, J. Janssens, and A. Allard, Avicenna Latinus (1.10): (Leuven, Belg.: Peeters, 2006).

23. I am exceptionally grateful to Jules Janssens, who has graciously provided me with his provisional edition of what was available in Latin of book 3.

24. Ahmad Hasnawi, "La définition du mouvement dans la 'Physique' du Shifāʾ d'Avicenne."

25. Ahmad Hasnawi, "Le statut catégorial du mouvement chez Avicenne: Contexte grec et postérité médiévale latine."

26. Yegane Shayegan, "Avicenna on Time" (Ph.D. diss., Harvard University, 1986).

27. Paul Lettinck, "Ibn Sīnā on Atomism," *al-Shajarah* 4 (1999): 1–51.

prose, I was at times slightly loose in rendering a term or image in order to capture a pun or make an example culturally relevant to a modern reader. In the end, I can only hope that the present translation gives one a sense of the thought and the man that was Avicenna.

◆ ◆ ◆

Sigla and Conventions

Z *al-Ṭabīʿīyāt, al-samāʿ al-ṭabīʿī*, ed. Saʿīd Zāyid (Cairo: The General Egyptian Book Organization, 1983)

Y *al-Ṭabīʿīyāt, al-samāʿ aṭ-ṭabīʿī*, ed. Jaʿfar al-Yāsīn (Beirut: Dār al-Manāhil, 1996)

T *al-Shifāʾ*, 2 vols. (Tehran: n.p., 1886), vol. 1, 2–159

In this work, terms of Arabic derivation found in *Webster's Third New International Dictionary* generally follow the first spelling given and are treated as regular English words. Otherwise, Arabic or Persian words and proper names have been transliterated following the romanization tables established by the American Library Association and the Library of Congress (*ALA-LC Romanization Tables: Transliteration Schemes for Non-Roman Scripts.* Compiled and edited by Randall K. Barry. Washington, D.C.: Library of Congress, 1997; available online at www.loc.gov/catdir/cpso/roman.html).

Passages from *The Physics* are referenced by book, chapter, and paragraph number, e.g. (3.9.7).

بسم الله الرحمن الرحيم

In the Name of God, the Merciful and Compassionate.

Praise God, the Lord of the Two Worlds,
and Blessing on Muḥammad and All His Family.

◆

THE FIRST PART OF THE NATURAL SCIENCES

Physics

[in Four Books]

◆

[Preface]

(1) Since, through God's assistance and help, we have completed in this book of ours what was needed by way of preface—namely, in the section on the art of logic—we should begin discussing the science of physics in the manner upon which our opinion has settled and to which our speculation has led. We shall adopt in this regard the order associated with the one that Peripatetic philosophy normally follows, and provide additional support[1] for what is farthest removed from what is immediate and seen on first inspection—namely, that which one is more likely to reject than [simply] to disagree with.

1. Reading *nushaddidu* with **Z** and **T** for **Y**'s *natashaddadu* (we are strengthened).

<تصدير>

بسم الله الرحمن الرحيم

الحمد الله ربّ العالمين والصلاة عاى محمدٍ وآله أجمعين

الفن الأول من الطبيعيات

في السّماع الطبيعي

(١) وإذ قد فرغنا ، بتيسير الله وعونه ، مما وجب تقديمه في كتابنا هذا ؛ وهو تعليم اللُباب من صناعة المنطق ، فحريّ بنا أن نفتتح الكلام في تعليم العلم الطبيعي على النحو الـذي تقرّر عليه رأينا وانتهى إليه نظرنا . وأنْ نجعل الترتيب في ذلك مقارناً للترتيب الذي تجري عليه فلسفة المشائين ، وأنْ نشدّد فيما هو أبعد عن البداية والفِطَر الأُول ، والمخالف فيه أبعد من الجاحد .

(2) We shall indulge whatever the truth itself reveals of its form, giving evidence against the one who disagrees by means of what [the truth] shows and holds back of itself. That our time be not wasted and bound up by repudiating and sufficiently opposing every school of thought—for often we see that when those discussing the sciences treat in their refutation some insignificant treatise or dwell in their exposition upon a question about which the truth is clearly perceived, then they expend every effort, exhibit every subdivision, and list every argument, whereas when they are persistently troubled by some problem or reach some doubtful matter, they tend to ignore it—[in order to avoid all that,] we hope to tread a different path and follow a course opposite theirs.

(3) We shall endeavor, as far as possible, to exhibit the truth arrived at by our predecessors and to excuse what we think they have overlooked unintentionally. This is what has barred us from commenting on their books and interpreting their texts, for we could not guard against coming upon matters that we believe they overlooked and so being compelled to try to find an excuse for them, invent an argument or profess it on their behalf, or simply confront them with a rebuttal. God has spared us all this [trouble] and has, in fact, assigned it to people who have exerted their utmost effort in achieving it and interpreting their books. He who wishes to acquaint himself with their words will find that their commentaries will guide him aright and that their interpretations will suffice him, whereas he who exerts himself in pursuing knowledge and meanings will find them scattered throughout these books. Part of what the measure of our search has yielded, despite the short space of time we spent therein, is given in the books that we have written and called collectively *The Healing.* God is the source of our support and strength, and in Him we place our trust. From here we start our exposition.

(٢) ونتســاهل فيما نفس الحق تكشــف عن صورته ، ونشهد على المخالف ببرائه وجحــده . وأنْ لا يذهــب عمرنا فــي مناقضته كلّ مذهب ، والعــدول عن الاقتصار في مقاومته على البلاغ ، فكثيراً ما نرى المتكلمين في العلوم ؛ إذا تناولوا بنقضهم مقالة واهية ، أو أكبّوا ببيانهم على مسألةٍ يُلحظ الحق فيها عن كثب ، نقضوا كلّ قوة وحقّقوا كلّ قسمة ، وســوّدوا كل حجــة ، وإذا تلججوا في المُشــكل ، وخلصوا إلى جانب المشــتبه ؛ مرّوا عليه صفحاً . ونحن نرجو أنْ نكون وراء ســبيلٍ مقابلةً لسبيلهم ، ونهجْ معارض لنهجهم .

(٣) ونجتهد ما أمكن في أنْ ننشــر عمّن قبلنا الصواب ، ونعرض صفحاً عمّا نظنّهم ســهوا فيه . وهذا هو الذي صدّنا عن شرح كتبهم وتفسير فصوصهم ، إذْ لم نأمس الانتهاء إلــى مواضع نظنّ أنّهم ســهوا فيها ؛ فنضطــر إمّا إلى تكلف اعتــذارٍ عنهم ، أو اختلاق حجّــة وتمحلها لهم ، أو إلى مجاهرتهــم بالنقْض ؛ وقد أغنانا اللّه عن ذلك ونَصَبَ له قوماً بذلوا طوقهم فيه ، وفسّــروا كتبهم . فمَنْ اشتهى الوقوف على ألفاظهم ؛ فشروحهم تهديه ، وتفاسيرهم تكفيه . ومَنْ نشط للعلم وللمعاني فسيجدها في تلك الكتب منثورة . وبعض ما أفادناه مقدار بحثنا مع قصر عمرنا في هذه ، الكتَبَ التي عملناها وسميناها كتاب الشفاء مجموعــاً ، واللّه وليّ تأييدنا وعصمتنا .ومن ها هنا نشــرع فــي عرضنا متوكلين عليه .

FIRST BOOK: ON THE CAUSES AND PRINCIPLES OF NATURAL THINGS

Chapter One

Explaining the means by which to arrive at the science of natural things from their first principles

(1) From the part [of *The Healing*] where we concisely presented the science of demonstration, you have already learned that some sciences are universal and some particular, and that some are related to others.[1] So now what you need to learn is that the science we are engaged in explaining is physics, which is a particular science in relation to what comes later. Since you have learned that each science has a subject matter, the subject matter of [physics] is the sensible body insofar as it is subject to change. What is investigated about it is the necessary accidents belonging to [the body subject to change] as such—that is, the accidents that are termed *essential*[2]— and also the concomitants that

1. See *Kitāb al-burhān* 2.2.

2. Reading *min jihah mā huwa hākadhā wa hiya al-aʿrāḍ allatī tusammī dhātīyah wa hiya* with **Z**, **T**, and the equivalent phrase in the Latin translation (*Avicenna latinus: Liber primus naturalium, tractatus primus de causis et principiis naturalium*, ed. Simon Van Riet [Louvain-la-Neuve: E. Peeters; Leiden: E. J. Brill, 1992]), which is omitted in **Y**.

المقالة الأولى

في الأسباب والمبادىء للطبيعيات

<الفصل الأول>

في تعريف الطريق الذي نتوصل منه إلى العلم بالطبيعيات من مبادئها

(١) قد علمتم من الفن الذي فيه علم البرهان الذي لخّصناه؛ أنَّ العلوم منها كلّية ومنها جزئية؛ وعلمتم مقايسات بعضها إلى بعض. فيجب أن تعلموا الآن أنَّ العلم الذي نحن في تعليمه هو العلم الطبيعي؛ وهو علمٌ جزئي، بالقياس إلى ما نذكره فيما بعد وموضوعه - إذ قد علمتم أنَّ لكل علمٍ موضوعاً - هو الجسم المحسوس من جهة ما هو واقعٌ في التغيّر. والمبحوث عنه فيه هو الأعراض اللازمة له من جهة ما هو هكذا، وهى الأعراض التى تسمى ذاتية وهى واللواحق التي تلحقه بما هو هو؛ كانت صوراً أو

attach to [the body] inasmuch as it is, whether forms, accidents, or derivatives of the two, as you have come to understand. Now, natural things are these bodies considered from this respect and whatever is accidental to them insofar as they are such. All of them are called *natural* in relation to that power called *nature,* which we will define later.[3] Some of them are subjects for [the nature] and some are effects, motions, and dispositions proceeding from it. If, as was explained in the science of demonstration,[4] natural things have principles, reasons, and causes without which the science of physics could not be attained, then the only way to acquire genuine knowledge of those things possessing principles is, first, to know their principles and, from their principles, to know them, for this is the way to teach and learn that gives us access to the genuine knowledge of things that possess principles.

(2) Also, if natural things do possess certain principles, then either each and every one of them has those principles or they do not all share the principles in common. In [the latter] case, it would not be unlikely that the science of physics establishes the existence of those principles and at the same time identifies their essence. If these natural things do share certain first principles in common that are general to all of them—namely, those that are undoubtedly principles of their shared subjects and shared states—then the proof of these principles (if they are in need of proof) will not belong to the discipline of the natural philosophers, as was shown in the part [of *The Healing*] dealing with the science of demonstrations,[5] but to another science, and the natural philosopher must simply accept their existence as a postulate and conceptualize their essence as fully real.[6]

---

3. See 1.5.

4. See *Kitāb al-burhān* 1, passim.

5. The reference appears to be to *Kitāb al-burhān* 1.12.

6. Avicenna's point here is explained more fully at 1.2.8–11, where he observes that principles can be common in two ways. For now, it is enough to note that certain common principles, such as the existence of forms, prime matter, a universal agent such as God, and the like have to be posited by the natural sciences and are not proven within them, whereas other common principles, such the existence of the natural places toward which bodies move naturally or primary qualities such as hot-cold and wet-dry, can be proven within the natural sciences.

أعراضاً أو مشـــتقةً منهما ؛ على ما فهمتم. والأمور الطبيعية؛ هي هذه الأجسام من هذه الجهة، وما يعرض لها ، من حيث هي بهذه الجهة، وتســـمّى كلّها طبيعية بالنســـبة إلى القوة التي تســـمّى طبيعة؛ التي سنعرّفها بعد . فبعضها موضوعات لها ، وبعضها آثار وحركات وهيئات تصدر عنها . فإنْ كان للأمور الطبيعية مبادىء وأســباب وعلل؛ لم يتحقّق العلم الطبيعي إلاّ منها ، فقد شـــرح في تعليم البرهان أنّه لا ســـبيل إلى تحقّق معرفة الأمور ذوات المبادىء إلاّ بعد الوقوف على مبادئها ، والوقوف من مبادئها عليها ، فإنَّ هذا هو النحو من التعليم أو التعلّم الذي نتوصل منه إلى تحقّق المعرفة بالأُمور ذوات المبادىء .

(٢) وأيضـــاً إنْ كانت الأمور الطبيعيـــة ذوات مبادىء ؛ فلا يخلو إمّا أنْ تكون تلك المبادىء لجزئي منها ، و⟨إمّا⟩ لا تشترك كافّتها في المبادىء ، فحينئذ لا يبعد أنْ يفيد العلم الطبيعي إثبات إنِّية هذه المبادىء وتحقيق ماهيّتها معاً . وإنْ كانت الأُمور الطبيعية تشـــترك في مبـــادىء أُول تعمّ جميعها ، وهي التي تكون مبادىء لموضوعها المشـــترك ولأحوالها المشـــتركة لا محالة، فلا يكون إثبات هذه المبادىء – إنْ كانت محتاجة إلى الإثبات – إلى صناعة الطبيعيين كما عُلِم في الفن المكتوب في علم البرهان، بل إلى صناعةٍ أُخرى، وأمّا قبول وجودها وضعاً ، وتصوّر ماهيتها تحقيقاً ، فيكون على الطبيعي.

(3) Moreover, if natural things possess certain principles common to all of them as well as possessing principles that are more specific than those (for instance, belonging to one of their genera, such as the principles of growth), and some [principles] are [even] more specific than [those] more specific [ones] (for instance, belonging to one of their species, such as the principles of humanity), and if, in addition, they possess essential accidents common to them all and others that are common to the genus, and still others that are common to the species, then the right course of intellectual teaching and learning consists in starting with what is more common and then proceeding to what is more specific. That is because you know that the genus is part of the definition of the species, and so the knowledge of the genus must be prior to the knowledge of the species, because the knowledge of the part of the definition precedes the knowledge of the definition, and conceptualizing it precedes the knowledge of what is defined, since we mean by *definition* that which identifies the essence of what is defined. Consequently, the principles of common things must first be known in order that common things be known, and the common things must first be known in order to know the specific things.

(4) We must, then, begin with the explanation of the principles belonging to common things, since common things are better known to our intellects even if they are not better known by nature.[7] In other words, [common things] were not in themselves the things intended in the natures for the completion of existence, for what is intended in the nature is not the existence of an animal absolutely or a body absolutely, but rather that the natures of the specific things exist, and when the specific nature exists in the concrete particulars, there is some individual. So, then, what is intended is that the natures of the specific things exist as certain individuals in the concrete particulars. Now, the

7. For discussions of *better known* and *prior to us* and *by nature* or *in themselves*, see Aristotle, *Posterior Analytics* 1.2.71b33–72a5 and *Kitāb al-burhān* 1.11.

(٣) وأيضاً إنْ كانت الأمور الطبيعية ذوات مبادىء عامة لجميعها، وذوات مبادىء أخص منها؛ تكون مثلاً لجنس من أجناسـها؛ مثل مباديء النامية منها. وذوات مباديء أخص من الأخص تكون مثلاً لنوع من أنواعها؛ مثل مباديء النوع الإنساني منها. وكانت أيضـاً ذوات عوارض ذاتية عامّة لجميعها، وأُخرى عامة لجنس، وأخرى عامة لنوع؛ فإنَّ وجه التعلّم والتعليم العقلي فيها أنْ تبتدىء ممّا هو أعم وتسلك إلى ما هو أخصّ. لأنك تعلم أنَّ الجنس جزء حدّ النوع؛ فتعرّف الجنس يجب أنْ يكون أقدم من تعرّف النوع؛ لأنَّ المعرفة بجزء الحدّ قبل المعرفة بالحدّ وتصوّره قبل الوقوف على المحدود، إذ كنّا نعني بالحدّ ما يحقق ماهيّـة المحدود. فإذا كان كذلك، فالمبـاديء التي للأمور العامّة يجب أنْ تعرف هي أولاً حتى تعرف الأمور العامة، والأمور العامّة يجب أنْ تعرف أولاً حتى تعرف الأمور الخاصّة.

(٤) فيجب أنْ نبتدىء في التعليم من المباديء التي للأمور العامّة، إذ الأمور العامّة أعـرف عند عقولنا، وإنْ لم تكن أعرف عنـد الطبيعة. أي لم تكن الأمور المقصودة في الطبـاع لتتمّة الوجود بذاتها. فإنَّ المقصـود في الطبيعة ليس أنْ يُوجد حيوانٌ مطلقاً ولا جسـمٌ مطلقاً، بل أنْ توجد طبائع النّوعيـات، والطبيعة النوعية إذا وجدت في الأعيان كان شـخصاً ما. فالمقصـود إذن أنْ توجد طبائع النوعيات أشـخاصاً ما في الأعيان؛ وليس المقصود هو الشّـخص المعيّن إلا في الطبيعة الجزئية الخاصّة بذلك الشـخص، ولو

concrete individual is not what is intended except with respect to the particular nature proper to that individual; if the concrete individual [itself] were what was intended [by nature], then through its corruption and nonexistence the order of existence would be diminished. Likewise, if the common and generic nature were what was intended, then existence and order would be completed through its [singular] existence, whether it is, for example, the existence of some body or some animal, however it might be. So it is nearly self-evident that what is intended is the nature of the species, in order that it cause the existence of some individual (even if not some particular individual). In other words, [what is intended] is the perfection and the universal end. It is this that is better known by nature, while not being prior by nature (if, by *prior,* we mean what is stated in the *Categories*[8] and we do not mean the end).

(5) Now, all men are as good as alike in knowing the common and generic natures, whereas they are distinguished only insofar as some men know and reach the specific things and apply themselves to making differentiations, while others stop at the generic things. So, for example, some might know [only] animality, whereas others might additionally know humanity and equinity. When knowledge reaches the specific natures and what is accidental to them, inquiry stops and is not followed by the fleeting knowledge of individuals to which our souls[9] are not at all inclined.

---

8. For the different meanings of *prior,* cf. Aristotle's *Categories* 12.

9. Reading with **Z** and added to **T**'s *nufūsnā,* which does not appear in **Y** or the Latin.

كان المقصود هو الشّـــخص المعيّن، لكان الوجود ينتقص نظامه بفســـاده وعدمه، كما لو كان المقصود هو الطبيعة العامّة والجنسية؛ لكان الوجود والنظام يتمّ بوجوده؛ مثل وجود جســـم كيف كان، أو حيوانٍ كيف كان. فما أقرب من البيان أنَّ المقصود هو طبيعة النوع لتوجد شخصاً - وإنْ لم يعيّن - وهو الكامل، وهو الغاية الكلّية، فالأعرف عند الطبيعة هـــو هذا، وليس هو أقدم بالطبـــع؛ إنْ عنينا بالأقدم ما قيل فـــي قاطيغورياس، ولم نعنِ بالأقدم الغاية.

(٥) والناس كلّهم كالمشـــتركين في معرفة الطبائع العامة والجنسية، وإنّما يتميّزون بأنَّ بعضهم يعرف النوعيات وينتهي إليها ويمعن في التفصيل، وبعضهم يقف عند الجنســـيات؛ فبعضهم مثلاً يعرف الحيوانية، وبعضهم يعرف الإنسانية أيضاً والفرسية. وإذا انتهت المعرفة إلى الطبائع النوعية وما يعرض لها، وقف البحث ولم يُبَلْ بما يفوتها من معرفة الشخصيات، ولا مالتْ إليها نفوسنا البتّة.

(6) It is clear, then, that when we compare common and specific things and then compare them together with [what is better known to] the intellect, we find that common things are better known to the intellect. When, on the other hand, we compare them together with the order of existence and what is intended in the universal nature, we find that specific things are better known by nature. When we compare the concrete individuals with the specific things and relate both to the intellect, we find that the concrete individuals have some place of priority or posteriority in the intellect only if we include the internal sensitive faculty.[10] In that case, then, the individuals are better known to us than universals, for individuals are impressed on the internal sense faculty from which the intellect subsequently learns what things are shared in common and what things are not, and so extracts the natures of things common in species.[11] When we relate them both to the nature, we find [that] the thing common in species[12] is better known, even if its actuality begins with determinate individuals. So nature's intention concerning the existence of body is precisely that it arrives at the existence of man and what is generically similar. [Similarly,] its intention concerning the existence of the generable and corruptible particular individual is that the nature of the species exists; and when it is possible to achieve that end through a single individual whose matter is not subject to change and corruption, as, for example, the Sun, the Moon, and the like, then there is no need for another individual to belong to the species.

---

10. Avicenna probably has the imagery faculty (*qūwah khayālīyah*) specifically in mind, for it is this faculty that provides the material intellect with the potential intelligible, which, when "illuminated" by the active intellect, becomes a universal corresponding with some specific thing, such as humanity.

11. For discussions of the roles of sensation and abstraction in Avicenna's noetic, see *Kitāb al-burhān* 3.5, and *Kitāb al-nafs* 2.2.

12. For example, the humanity or equinity common to all the individuals within the species.

(٦) فبيِّنٌ أنّا إذا قايسنا ما بين الأُمور العامّة والخاصّة، ثمّ قايسنا بينهما معاً وبين العقل، وجدنا الأُمور العامّة أعرف عند العقل. وإذا قايسنا بينهما معاً وبين نظام الوجود والأمر المقصود في الطبيعة الكلّية، وجدنا الأُمور النوعية أعرف عند الطبيعة. وإذا قايسنا بين الشخصيات المعيّنة، وبين الأُمور النوعية، ونسبناهما إلى العقل؛ لم نجد للشخصيات المعيّنة عند العقل مكان تقدم وتأخر إلاّ أنْ نستشرك القوة الحاسّة في الباطن، فحينئذ تكون الشخصيات أعرف عندنا من الكلّيات، فإنَّ الشخصيات ترتسم في القوة الحاسّة التي في الباطن، ثم يقتبس منه العقل المشاركات والمباينات، فينتزع طبائع العامّيات النوعية. وإذا نسبناهما إلى الطبيعة وجدنا العامّة النوعية أعرف وإنْ كان إبتداء فعلها من الشخصيات المعيّنة. فإنَّ الطبيعة إنّما تقصد من وجود الجسم أنْ تتوصل به إلى وجود الإنسان وما يجانسه، وتقصد من وجود الشخص المعينّ الكائن الفاسد أنْ تكون طبيعة النوع موجودة. وإذا أمكنها حصول هذا الغرض في شخص واحد، وهو الذي تكون مادّته غير مُذْعنة للتغيّر والفساد، لم يحتج إلى أنْ يوجد للنوع شخص آخر كالشمس والقمر وغيرهما.

(7) Although in perceiving particulars, sensation and imagination initiate the most important part of conceptualizing an individual, it is more like the common notion until they reach the conceptualization of the individual that is absolute in every respect. An illustration of how this is would be that *body* is a common notion to which it belongs, *qua* body, to be individualized and thus become this or that body. Similarly, *animal* is a common notion, but more particular than *body,* and it belongs to it, *qua* animal, to be individualized and thus become this or that animal. *Man* is also a common notion that is more particular than *animal,* and it belongs to it, *qua* man, to be individualized and thus become this or that man. Now, if we relate these orderings to the power of perception and observe therein two kinds of order, we find that what is closer to and more like the common thing is better known. Indeed, it is impossible that one should sensibly or imaginatively perceive that this is *this* man unless one perceives that he is *this* animal and *this* body. [Similarly,] one would not perceive that this is *this* animal, unless one perceives that it is *this* body, whereas if one perceives him from afar, one might perceive that he is *this* body without perceiving that he is *this* man. It is clearly obvious, therefore, that the case of sensation in this respect is similar to the case of the intellect and that what corresponds with the general is better known in itself even for sensation as well.

(8) With respect to time, however, sensation provides imagination with only an individual member of the species that is not uniquely delimited. So, from among those sensible forms impressed on the imagery faculty, the first one impressed on the child's imagery faculty is the form of an individual man or woman, without his being able to distinguish a man who is his father from a man who is not and a woman who is his mother from a woman who is not. Eventually, he is able to distinguish a man who is his father from one who is not, and a woman who is his mother from one who is not, and then by degrees the individuals remain differentiated for him.

(٧) على أنَّ الحسّ والتخيّل، في إدراكهما للجزئيات، يبتدئان أول شيءٍ من تصور شـخصٍ هو أكثر مناسـبة للمعنى العامّي، حتى يبلغا تصوّر الشّخص الذي هو شخص صـرف من كل وجـه. وأمَّا بيان كيفية هذا؛ فهو أنَّ الجسـم معنى عـام وله - بما هو جسم - أنْ يتشخص فيكون هذا الجسم، والحيوان أيضاً معنى عام وأخصّ من الجسم، وله بما هو حيوان أنْ يتشـخص فيكون هذا الحيوان، والإنسـان أيضاً معنى عام وأخصّ من الحيوان، وله بما هو إنسان أنْ يتشخّص فيكون هذا الإنسان. فإذا نسبنا هذه المراتب إلى القوة المدركة، وراعينا في ذلك نوعين من الترتيب؛ وجدنا ما هو أشبه بالعام وأقرب مناسبة له هو أعرف. فإنّه ليس يمكن أنْ يدرك بالحسّ والتخيّل أنَّ هذا هو هذا الإنسان، إلاّ وأدرك أنّه هذا الحيوان وهذا الجسـم، وقد يدرك أنّه هذا الجسـم إذا لمحه من بعيد ولا يدرك أنّه هذا الإنسـان فقد بانَ وَوَضح أنَّ حـال الحسّ أيضاً من هذه الجهة كحال العقل، وأنَّ ما يناسب العام أعرف في ذاته أيضاً عند الحسّ.

(٨) وأمـا في الزمان؛ فإنَّ التخيّل إنّما يسـتفيد من الحسّ شـخصاً من النوع غير محدودٍ بخاصّيته. فأول ما يرتسـم في خيال الطفل من الصورة التي يحسّـها على سبيل تأثّرٍ من تلك الصور في الخيال؛ هو صورة شـخص رجل، أو صورة شـخص إمرأة، من غير أن يتميّز له رجلٌ هو أبوه عن رجلٍ ليس هو أباه، وامرأةٍ هي أُمّه عن امرأة ليست هي أُمّه. ثم يتميّز عنده رجل هو أبوه، ورجلٌ ليس هو أباه، وامرأةٌ هي أُمّه وامرأة ليست هي أُمّه، ثم لا يزال تنفصل الأشخاص عنده يسيراً يسيراً.

(9) Now, this image, in which a wholly indistinct likeness of the individual human is imprinted, is the image of something that is termed *vague*. When *vague individual* is said of [(1)] this [indistinct likeness] and of [(2)] an individual imprinted upon sensation from a distance (assuming the impression is that it is a body without perceiving whether it is animal or human), then the expression *vague individual* is applied equivocally to them. The reason is that what is understood by the expression *vague individual* in [the first] case is one of the individuals of the species to which it belongs, without determining how or which individual; and the same holds for a certain man and woman. It is as though the sense of *individual*, while not being divided into the multitude of those who share in its definition, has been combined with the account of nature applied relative to the species or the kind. From them both, there is derived a single account termed *a vague indeterminate individual*—just as is indicated by our saying, "Rational, mortal animal is one," which does not apply to many when it is defined in this way, since the definition of individuality is attributed to the definition of the specific nature. In short, this is an indeterminate individual. In [the second] case, however, it is this determinate corporeal individual. It cannot be other than it is, save that, owing to the mind's uncertainty, either the account of being animate or inanimate can be attributed to it in thought, not because the thing in itself can be such—that is, such that any one of the accounts could be attributed indiscriminately to that corporeality.[13]

(10) So the vague individual in [the first] case can be thought to be any existing individual of that genus or the one species. In [the second] case, however, it cannot be thought to be just any individual of that species, but can only be this single, determinate one. Be that as it may, the mind can still be susceptible to uncertainty, making it possible that, relative to [the mind, the individual] is designated, for example, either by determinately being animate to the exclusion of being inanimate or determinately being inanimate to the exclusion of being animate, even after it is judged that in itself it cannot be both things but is determinately one or the other of them.

---

13. The first case of a vague image of *human* is that which appears before the mind's eye when one is asked to imagine *human*, but not any particular human, whereas the second case of a vague image is of some particular human seen from afar, even though one might not be able to make out which particular human it is.

(٩) وهذا الخيال الذي يرتسم فيه مثلاً من الشخص الإنساني مطلقاً غير مخصّص، هو خيال المعنى الذي يسمى منتشراً، وإذا قيل شخص منتشر لهذا؛ وقيل شخص منتشر لما ينطبع في الحسّ من شخصٍ لائح من بعيد إذا ارتسم أنّه جسم، من غير إدراك حيوانيةٍ أو إنسـانيةٍ؛ فإنّما يقع عليها اسم الشخص المنتشـر باشتراك الاسم. وذلك أنَّ المفهوم من لفظة الشّـخص المنتشر بالمعنى الأول هو أنّه شـخصٌ مّا من أشخاص النوع الذي يُنسب إليـه، غير معيّنٍ كيف كان وأي شـخص كان، وكذلك رجلٌ مّـا وامرأةٌ مّا. فيكون كأن معنى الشـخص - وهو كونه غير منقسم إلى عدة مَنْ يشاركه في الحدّ - قد انضم إلى معنى الطبيعة الموضوعة للنوعية أو للصِنْفية، وحصل منهما معنى واحد يسـمى شخصاً منتشـراً غير معيّن، كأنّه ما يدل عليه قولنا حيـوانٌ ناطقٌ مائتٌ هو واحدٌ؛ لا يقال على كثرةٍ، ويحد بهذا الحدّ، فيكون حدّ الشخصية مضافاً إلى حدّ الطبيعة النوعية. وبالجملة هذا هو شخص غير معيّن. وأمّا الآخر؛ فهو هذا الشخص الجسماني المعيّن، ولا يصلح أنْ يكون غيره، إلاّ أنّه يصلح عند الذّهن أنْ يضاف إليه معنى الحيوانية أو معنى الجمادية لشّك الذهن، لالأنَّ الأمر في نفســه صالـح أنْ يكون كذلك؛ أي يكون بحيث يصلح أن يضاف إلى تلك الجسمية أي المعنيين منهما كان.

(١٠) فالشّـخص المنتشر بالمعنى الأول، يصلح عند الذّهن أنْ يكون في الوجود أي شـخص كان من ذلك الجنس أو النوع الواحد. وبالمعنى الثاني، ليس يصلح في الذّهن أنْ يكون أي شخصٍ كان من ذلك النوع، بل لا يكون غير هذا الواحد المعيّن. لكنه يصلح عند الذّهن صلوح الشـك والتجويز - أنْ يتعيّن بحيوانيةٍ معيّنةٍ مثلاً دون جمادية، أو جمادية معيّنة دون حيوانية، تُعيّنا بالقياس إليه، بعد حكمه أنّه في نفسه لا يجوز أنْ يكون صالحاً للأمرين، بل هو أحدهما متعيّناً.

(11) There is also a correlation here between causes and effects and a correlation between simple parts and composites. So when the causes enter into the constitution of the effects as parts of them—as, for example, the case of wood and shape relative to the bed—then their relation to the effects is that of simple parts to composites. As for when the causes are separate from the effects—as, for example, the carpenter who makes the bed—then it is a different issue.

(12) Now, both correlations have a relation to sensation, intellect, and nature. As for the correlation between sensation and causes and effects where the causes are separate, then, if the causes and effects are sensible, neither one has more priority or posteriority over the other as a sensation. If they are insensible, then neither one of them has a relation to sensation. The same holds for the status of the image.

(13) Vis-à-vis the intellect, however, the cause might reach it before the effect, whereupon [the intellect] moves from the cause to the effect. Examples are when someone sees the Moon in conjunction with a planet whose degree is near the lunar nodes, while the Sun is at the opposite extreme of the [celestial] arc, and so the intellect judges that there is an eclipse. Again, [another example would be that] when [a person] knows that matter [within his body] has undergone putrefaction, he knows that fever has set in. Often the effect reaches [the intellect] before the cause—sometimes through deduction, sometimes through sensation—in which case [the intellect] moves from the effect to the cause. Also, [the intellect] often recognizes an effect first, and then moves from it to the cause, and then thereafter moves from the cause to another effect. We have already explained these notions clearly in our study of demonstration.[14]

---

14. See *Kitāb al-burhān* 1.7, where Avicenna distinguished between the *burhān lima* (demonstration *propter quid*), which goes from cause to effect, and the *burhān inna* (demonstration *quia*), which goes from effect to cause and is itself divided into the "absolute *burhān inna*" and the "indication," which correspond with the accounts given here.

(١١) هذا، وها هنا مقايسة أيضاً بين العلل والمعلولات، ومقايسة بين الأجزاء البسيطة والمركّبات. فإذا كانت العلل داخلة في قوام المعلولات، وكالأجزاء لها؛ مثل حال الخشب والشكل بالقياس إلى السرير، فإن نسبتها إلى المعلولات نسبة البسائط إلى المركّبات، وأما إذا كانت العِلل مباينة للمعلولات؛ مثل النّجار للسرير، فهناك نظرٌ آخر.

(١٢) ولكلتا المقايستين نسبة إلى الحسّ والعقل وإلى الطبيعة؛ فأمّا مقايسة ما بين الحسّ وبين العِلل والمعلولات - على أنَّ العِلل مباينة - فإنْ كانت العِلل والمعلولات محسوسة؛ فلا كثير تقدّم وتأخّر لأحدهما على الآخر حسّاً، وإنْ كانت غير محسوسة فلا نسبة لأحدهما إلى الحسّ، وكذلك حكم الخيال.

(١٣) وأما عند العقل فإنَّ العقل ربما وصلتْ إليه العِلّة قبل المعلول؛ فسلك من العلّة إلى المعلول؛ كما إذا رأى الإنسانُ القمرَ مقارناً لكوكب درجته عند الجوزهر، وكانت الشمس في الطرف الآخر من القطر، فيحكم العقل بالكسوف، وكما إذا علم أنَّ المادة متحركة إلى عَفَنٍ فيعلم أنَّ الحمى كائنة. وربّما وصل إليه المعلول قبل العلّة، فسلك من المعلول إلى العِلّة، وقد يعرف المعلول قبل العِلّة؛ تارة من طريق الاستدلال، وتارة من طريق الحسّ. وربّما عرف أولاً معلولاً فسلك منه إلى العِلّة، ثم سلك من العلّة إلى معلولٍ آخر. وكأنا قد أوضحنا هذه المعاني في تعليمنا لصناعة البرهان.

(14) As for the correlation of those separate causes analogous to nature with the effects, those that are causes in the sense of an end are better known by nature. Also better known by nature than the effect are those causes that are an agent—that is, the one that acts for the sake of what it makes, not [merely] given that it exists. [As for] that [cause] whose existence in nature does not [act] for the sake of [what is made] itself, but, rather, whatever comes from it is made such that not only does it have [that agent] as an end with respect to its [own] activity but also with respect to its very existence (assuming that there is such a thing in nature), it would not be better known than the effect; and in fact, the effect would be better known by nature than it.

(15) As for the relation of the parts of the composites to what is composed from them, the composite is better known according to sensation, since sensation first grasps and perceives the whole and then differentiates. When it grasps the whole, it grasps it in the most general sense (namely, that it is a body or an animal), and thereafter it differentiates it. In the intellect, however, the simple is prior to the composite, since it knows the nature of the composite only after it knows its simple components. If [the intellect] does not know [the composite's] simple components, then it really knows it through one of the accidents or genera [of the composite] without having reached the thing itself—as, for instance, if it knew it as a round or a heavy body and the like but did not know the essence of its substance. As for by nature,[15] the composite is what is intended in most things and parts in such a way that from them, the composite comes to subsist.

15. Reading *fa* with **Z** and **T**, which is omitted in **Y**.

(١٤) واما مناسبة هذه العلل المفارقة للمعلولات، بحسب القياس إلى الطبيعة، فإنَّ ما كان منها عِلّة على أنَّه غاية فهو أعرف عند الطبيعة. وما كان منها علّة؛ على أنّه فاعل وكان فاعلاً لا على أنَّ وجوده ليكون فاعلاً لما يفعله، فإنَّه أعرف عند الطبيعة من المعلول. وما كان وجوده في الطبيعة ليس لذاته، بل ليفعل ما يكون عنه، حتى يكون المفعول غاية لا له في فعله فقط، بل له في وجود ذاته - إنْ كان في الطبيعة شيء هذا صفته - فليس هو أعرف من المعلول، بل المعلول أعرف عند الطبيعة منه.

(١٥) وأمّا نسبة أجزاء المركبات إلى المركبات منها، فإنَّ المركّب أعرف بحسب الحسّ؛ إذ الحسّ يتناول أولاً الجملة ويدركها ثم يفصّل، وإذا تناول الجملة تناولها بالمعنى الأعم؛ أي أنّه جسمٌ أو حيوانٌ، ثم يفصّلها. وأمّا عند العقل، فإنَّ البسيط أقدم من المركّب؛ فإنّه لا يعرف طبيعة المركّب إلاّ بعد أنْ يعرف بسائطه، فإنْ لم يعرف بسائطه فقد عرفه بعرض من أعراضه أو جنسٍ من أجناسه ولم يصل إلى ذاته؛ كأنه عرفه مثلاً جسماً مستديراً أو ثقيلاً، أو ما أشبه ذلك، ولم يعرف ماهيّة جوهره. وأمّا عند الطبيعة، فإنَّ المركّب هو المقصود فيها في أكثر الأشياء والأجزاء؛ تُقصد ليحصل منها قوام المركّب. فالأعرف عند العقل

(16) So, from among the general and specific things and the simple and compound things, the general and simple are better known to the intellect, whereas the specific property and composite are better known by nature. Now, just as nature begins in the way of discovery with the general and simple and from them discovers the things that are themselves differentiated according to species and themselves composite, so likewise instruction begins with the general and simple and from them comes to know specific things and composites. The primary aim of both, then, is reached upon acquiring specific and compound things.

(١٦) مــن الأمور العامّــة والخاصّة ومن الأمور البســيطة والمركّبــة - هو العامّة والبســيطة، وعند الطبيعة هو الخاصّة والنوعيّة والمركّبة. لكنه، كما أنّ الطبيعة تبتدىء في الإيجاد بالعوام والبســائط، ومنها توجــد ذوات المفصّلات النوعية وذوات المركّبات، فكذلــك التعلّم يبتدىء من العوام والبســائط، ومنها يوجد العلــم بالنوعيات والمركبات، وكلاهما يقف قصده الأول عند حصول النوعيات والمركبات.

Chapter Two

Enumerating the principles of natural things by assertion and supposition

(1) Natural things have certain principles that we shall enumerate, setting forth what is necessary about them and providing their essences.

(2) We say, then, that the natural body is a substance in which one can posit one dimension, and another crossing it perpendicularly, and a third dimension crossing both of them perpendicularly, where its having this description is the form by which it becomes a body. The body is not a body by virtue of having a given [set of] three posited dimensions, since a body can exist and remain as a body even if the dimensions belonging to it are actually changed. So, [for example], a piece of wax or a drop of water may be such that there exist in it the actual dimensions of length, breadth, and depth determined by its extremities; but then, if it changes in shape, each of these definite dimensions ceases, and other dimensions or extensions exist. Yet the body continues as body, without corruption or change, and the form that we predicated of it as necessary—namely, that those dimensions can be posited in it—continues unchanged. This has been referred to in another place,[1] where you learned that those definite extensions are the quantity of its sides, which are concomitant with it and change, while its form and substance do not change—although this quantity may follow[2] a change in certain accidents or forms in it, just as water, when heated, increases in volume.

---

1. See *Kitāb al-burhān* 1.10.
2. Reading *tabiᶜat* with **Z** for **Y**'s *tabᶜathu* (emit).

<الفصل الثاني>

في تعديد المبادىء للطبيعيات على سبيل المصادرة والوضع

(١) ثم أنَّ للأمور الطباعية مبادىء وسنعدّها ونضعها وضْعاً على ما هو الواجب فيها، ونعطي ماهيّاتها

(٢) فنقول: إنَّ الجسم الطبيعي هو الجوهر الذي يمكن أنْ يُفرض فيه إمتدادٌ، وامتدادٌ آخر مقاطعٌ له على قوائم، وامتدادٌ ثالثٌ مقاطعٌ لهما جميعاً على قوائم. وكونه بهذه الصفة هو الصورة التي بها صار الجسم جسماً. وليس الجسم جسماً بأنّه ذو امتدادات ثلاثة مفروضة، فإنَّ الجسم يكون موجوداً جسماً وثابتاً وإنْ غُيّرت الامتدادات الموجودة فيه بالفعل. فإنَّ الشمعة، أو قطعة من الماء، قد تحصل فيها أبعاد بالفعل؛ طولاً وعرضاً وعمقاً محدودة بأطرافها. ثم إذا استبدل شكلاً، بَطل كلّ واحدٍ من أعيان تلك الأبعاد المحدودة، وحصلت أبعادٌ وامتدادات أخرى، والجسم باقٍ بجسميته لم يفسد ولم يتبدل. والصورة التي أوجبناها له - وهي أنّه بحيث يمكن أنْ تُفرض فيه تلك الأبعاد - ثابتة لا تبطل، وقد أُشير لك إلى هذا في غير هذا الموضع. وعلمت أنَّ هذه الامتدادات المعيّنة هي كميّة أقطاره، وهي تلحقه وتتبدل، وصورته وجوهره لا تتبدل. وهذه الكميّة ربّما تبعت تبدل أعراض فيه أو صور، كالماء يسخن فيزداد حجماً.

(3) This natural body has certain principles *qua* natural body, as well as additional principles *qua* generable and corruptible or in general alterable. The principles by which it acquires its corporeality include whatever are parts of its existence as actually present in [the natural body] itself, and these are more appropriately called *principles,* according to [the natural philosopher]. They are two: one of them is like the wood of the bed, while the other is like the form or shape of the bed. What is like the wood of the bed is called *material, subject, matter, component,* and *element,* according to various considerations, whereas what is like the form of the bed is called *form.*

(4) Since the form of corporality is either prior to all the other forms that belong to natural things and their genera and species or is something inseparably joined with them, what belongs to the body as the wood belongs to the bed also belongs to all those other things that possess the forms in this way, since all of them exist in fact together with corporality; and so that [namely, the material] is a substance. When [the material] is considered in itself, without reference to anything, it exists devoid in itself of these forms. Still, it is susceptible to receiving these forms or being joined with them in either of two ways. On the one hand, it may be from the susceptibility of [the material's] universal absolute nature, as if it were a genus for two species, one prior and one joined, each one of which is specified by a receptivity to some forms to the exclusion of others, after the [form] of corporality. On the other hand, from the susceptibility of the nature, [the material] itself may be something common to all [the forms]; and so, by means of its universality, it is susceptible to receiving all of these forms, some of them collectively and successively and others just successively. In this case, there would be a certain correspondence with the forms in its nature—namely, that [the material] is receptive to them, where this receptivity is like an impression in it and a shadow and specter of the form, while it is the form that actually perfects this substance.

(٣) لكن هذا الجسم الطبيعي - من حيث هو جسمٌ طبيعي - له مبادىء، ومن حيث هو كائن وفاسد، بل متغيّرٌ بالجملة، له زيادة في المبادىء. فالمبادىء التي بها تحصل جسميته؛ منها ما هي أجزاء من وجوده وحاصلة في ذاته، وهذه أولى عندهم بأنْ تسمى مبادىء وهي إثنان: أحدهما قائمٌ منه مقام الخشب من السرير، والآخر قائمٌ منه مقام صورة السريرية وشكلها من السرير. فالقائم منه مقام الخشب من السرير يسمى هيولى، وموضوعاً، ومادّة، وعنصراً، وأُسْطُقسّاً؛ بحسب اعتباراتٍ مختلفة، والقائم منه مقام صورة السريرية يسمّى صورة.

(٤) وإذ صورة الجسمية إمّا متقدمة لسائر الصور التي للطبيعيات وأجناسها وأنواعها، وإمّا مقارنة لا تنفك هي عنه؛ فيكون هذا الذي هو للجسم كالخشب للسرير، هو أيضاً لسائر ذوات تلك الصور بهذه المنزلة، إذ كلّها متقرّره الوجود مع الجسمية فيه. فيكون ذلك جوهراً إذا نُظر إلى ذاته غير مضافٍ إلى شيءٍ وُجد خالياً في نفسه عن هذه الصور بالفعل، ويكون من شأنه أنْ يقبل هذه الصور أو يقترن بها. أمّا من شأن طبيعته المطلقة الكلّية كأنها جنسٌ لنوعين: للمتقدمة وللمقارنة، وكلّ واحدٍ منهما يختص بقبول بعض الصور دون بعض بعد الجسمية. أمّا من شأن طبيعةٍ هي بعينها مشتركة للجميع، فتكون بكلّيتها من شأنها أنْ تقبل كلّ هذه الصور، بعضها جمعة ومتعاقبة، وبعضها بتعاقب فقط. فتكون في طبيعتها مناسبةٌ مّا مع الصور؛ على أنّه قابلٌ لها، وتكون هذه المناسبة كأنّها رسمٌ فيها وظلٌ وخيالٌ من الصورة، وتكون الصورة هي التي تكمل هذا الجوهر بالفعل.

(5) Let it be posited for the science of physics, then, that body *qua* body has a principle that is material and a principle that is form, whether you intend an absolute corporeal form, or a species form from among the forms of bodies, or an accidental form ([as] whenever you regard body, insofar as it is white, strong, or healthy). Let it also be posited for [this science] that what is material is never separated from form so as to subsist in itself. In other words, [the material] does not actually exist unless form is present and so actually exists through [the form]. If it were not the case that the form departs from it only with the arrival of another form that takes over and replaces it, then the material would actually cease to be.

(6) Now, from the perspective that this material is potentially receptive to a form or forms, it is called their *material.* From the perspective that it is actually bearing some form, it is called in this context its *subject.* (The sense of *subject* here is not the same as the meaning of *subject* that we gave in logic as part of the description of substance, since matter is never subject in that sense.)[3] From the perspective that it is common to all forms, it is called *matter* and *stuff.* It is called *element* because, through a process of analysis, it is resolved into [constitutive elements], in which case [the material] is the simple part of the whole composite receptive to form; and the same is true of everything of that sort. Finally, because composition in this precise sense starts with [the material], it is called *component,* and the same is true of everything of that sort. It is as though, when the composition in this precise sense starts from it, it is called a *component;* whereas when it starts with the composite and ends with [the material], it is called *element,* since the element is the simplest part of the composite. These, then, are the internal principles that constitute the body.

3. The reference is to *Kitāb al-burhān* 1.10.

(٥) فليوضع للطبيعي أنَّ للجسم - بما هو جسم - مبدأ هو هيولى، ومبدأ هو صورة إن شئت صورة جسمية مطلقة، وإنْ شئت صورة نوعية من صور الأجسام، وإن شئت صورة عرضية، إذا أخذت الجسم من حيث هو؛ كالأبيض أو القوي أو الصحيح. وليُوضَع له أنَّ هذا الذي هو هيولى لا يتجرد عن الصورة قائمة بنفسها البتّة، ولا تكون موجودة بالفعل إلاّ بأنْ تحصل الصورة؛ فيوجد بها بالفعل وتكون الصورة التي تزول عنها، لولا أنَّ زوالها إنَّما هو مع حصول صورة أُخرى تنوب عنها وتقوم مقامها، لفسدت معها الهيولى بالفعل. وهذه الهيولي.

(٦) من جهةِ أنّها بالقوة قابلة لصورةٍ أو لصور فتسمّى هيولى لها، ومن جهة أنّها بالفعل حاملة لصورةٍ فتسمّى - في هذا الموضع - موضوعاً لها، وليس معني الموضوع هاهنا، معني الموضوع الذي أخذناه في المنطق جزء رسم للجوهر، فإنَّ الهيولى لا تكون موضوعاً بذلك المعنى البتة. هذا؛ ومن جهة أنّها مشتركة للصور كلّها تسمّى مادّة وطينة، ولأنّها تنحّل إليها بالتحليل؛ فتكون هي الجزء البسيط القابل للصورة من جملة المركّب؛ تسمّى أُسْطُقُسّاً، كذلك كلّ ما يجري في ذلك مجراها. ولأنّها يبتدىء منها التركيب في هذا المعنى بعينه؛ تسمّى عنصراً، وكذلك كل ما يجري في ذلك مجراها. وكأنّها إذا ابتدأ منها التركيب في هذا المعنى بعينه تسمّى عنصراً، وإذا ابتديء من المركّب وانتهى إليها تسمّى أُسْطُقُسّاً؛ إذ الأُسْطُقُس هو أبسط أجزاء المركب. فهذه هي المبادىء الداخلة في قوام الجسم.

(7) The body also has additional principles: an agent and an end. The agent is that which impresses the form belonging to bodies into their matter, thereby making the matter subsist through the form, and from [the matter and form] making the composite subsist, where [the composite] acts by virtue of its form and is acted upon by virtue of its matter. The end is that for the sake of which these forms are impressed into the matters.

(8) Now, since our present discussion concerns the common principles, the agent and end considered here are common to them. Now, what is common may be understood in two ways. One is the way in which the agent is common as producing the first actuality from which all other actualities follow, such as that actuality that provides Prime Matter with the initial corporeal form. If there is such a thing (as you will learn in its proper place),[4] it would provide the initial foundation subsequent to which what comes next reaches completion. The end would be common [in this sense], if there is such an end (as you will learn in its proper place),[5] in that it is the end toward which all natural things tend. This is one way. The other way that something is common is by way of generality, as the universal [predicate] *agent* is said of each of the particular agents of particular things, and the universal [predicate] *end* is said of each one of the particular ends of particular things.

---

4. See the *Ilāhīyāt* of the *Najāt* 2.12, for what is perhaps Avicenna's most succinct version of his celebrated proof for a common, efficient cause in this first sense, which can safely be identified with the Necessary Existent in Itself, or God. The version of the proof found in the *Ilāhīyāt* of the *Shifā᾽* its spread throughout that work, although in general see book 8. See also Michael E. Marmura, "Avicenna's Proof from Contingency for God's Existence in the *Metaphysics* of the *Shifā᾽*," in his *Probing in Islamic Philosophy: Studies in the Philosophies of Ibn Sina, al-Ghazali and Other Major Muslim Thinkers* (Binghamton, NY: Global Academic Publishing, 2005), 131–48.

5. Cf. *Ilāhīyāt* 8.6.

(٧) وللجسم مبادىء أيضاً، فاعلة وغائية. والفاعلة هي التي طبعت الصورة التي للأجسام في مادّتها فقوّمت المادة بالصورة، وقوّمت منهما المركّب يفعل بصورته، وينفعل بمادّته، والغائية هي التي لأجلها ما طُبعت هذه الصور في المواد.

(٨) ولما كان كلامنا ها هنا في المبادىء المشتركة، فيكون الفاعل المأخوذ ها هنا هو المشترك، والغاية المعتبرة ها هنا هي المشترك فيها، والمشترك فيه ها هنا يُعقل على نحوين: أحدهما أنْ يكون الفاعل مشتركاً فيه على أنّه يفعل الفعل الأول الذي تترتب عليه سائر الأفاعيل؛ كالذي يفيد المادّة الأولى الصورة الجسمية الأولى - إنْ كان شيء كذلك؛ على ما تعلمه في موضعه - فيكون يفيد الأصل الأول، ثم من بعد ذلك يتمّ كون ما بعده. وتكون الغاية مشتركاً فيها بأنّها الغاية التي تؤمها جميع الأمور الطبيعية، إنْ كانت غاية كذلك، على ما تعلمه في موضعه؛ فهذا نحوٌ. والنحو الآخر؛ أنْ يكون المشترك فيه مشتركاً فيه بنحو العموم؛ كالفاعل الكلّي المقول على كلّ واحد من الفاعلات الجزئية للأمور الجزئية، والغاية الكليّة المقولة على كل وحدةٍ من الغايات الجزئية للأمور الجزئية.

(9) The difference between the two is that in the first sense, *common* denotes a determinately existing entity that is numerically one [and] which the intellect indicates that it cannot be said of many, whereas in the second sense, *common* does not denote a single determinately existing entity in reality, but an object of the intellect that applies to many that are common in the intellect in that they are agents or ends, and so this common thing is predicated of many.

(10) The efficient principle common to all in the first sense (if natural things have an efficient principle in this sense) would not be part of the natural order, since everything that is part of the natural order is subsequent to this principle, and it is related to all of them as their principle [precisely] because they are part of the natural order. So, if that principle were part of the natural order, then either it would be a principle of itself, which is absurd, or something else would be the first efficient principle, which is a contradiction. Consequently, the natural philosopher has no business discussing [such an efficient principle], since it has nothing to do with the science of physics.[6] Also, if there is such a thing, it may be a principle of things that are part of the natural order as well as things that are not part of the natural order, in which case its causality will be of a more general existence than [both] the causality of what specifically causes natural things and the things that are specifically related to natural things.

6. Here Avicenna is anticipating his position put forth in book 1 of his *Ilāhīyāt* (1.1–2), that discussions of the First Efficient and/or Final Cause—God—properly belong to the subject matter of metaphysics, and that Aristotle and the tradition following him erred when they discussed the deity in the science of physics.

(٩) والفرق بين الأمرين ؛ أنَّ المشترك بحسب المعني الأول - يكون في الوجود ذاتاً واحدةً بالعدد ؛ يشير العقل إليها بأنَّها هي من غير أنْ يجوز فيها قولاً على كثيرين. والمشترك، بحسب المعنى الثاني، لا يكون في الوجود ذاتاً واحدةً، بل أمراً معقولاً يتناول ذواتاً كثيرة تشترك عند العقل في أنَّها فاعلة أو غاية ؛ فيكون هذا المشترك مقولاً على كثيرين.

(١٠) فالمبدأ الفاعلي المشترك للجميع بالنحو الأوّل - إنْ كان للطبيعيات مبدأ فاعلي من هذا النحو - فلا يكون طبيعياً، إذْ كان كل طبيعي فهو بعد هذا المبدأ ؛ وهو منسوبٌ إلى جميعها بأنه مبدؤه ؛ لأنّه طبيعي. فلو كان ذلك المبدأ طبيعياً لكان حينئذ مبدأ لنفسه، وهذا محال، أو يكون المبدأ الأول الفاعلي غيره، وهذا خُلْفٌ. فإذا كان كذلك، لم يكن للطبيعي بحثٌ عنه بوجهٍ ؛ إذ كان لا يخالط الطبيعيات بوجه. وعساه يكون مبدأ للطبيعيات وللموجودات غير الطبيعيات، فتكون عليته أعمّ وجوداً من علّية ما هو علّة للأمور الطبيعية خاصّة، ومن الأمور التي لها نسبة خاصّة إلى الطبيعيات، إنْ كان شيء كذلك.

(11) Certainly, it might be possible that, with respect to the totality of natural things, what is an efficient principle of everything within the natural order other than itself is not such absolutely but is the common efficient principle in the latter sense, in which case it would not be at all out of place if the natural philosopher were to investigate [this efficient principle or agent]. The method of that investigation would be [(1)] to discover the state of whatever is an efficient cause of some given natural thing, the manner of its power, its relation to its effect in point of proximity, remoteness, when it is in direct contact and not in direct contact, and the like; and [(2)] to demonstrate it. When he does this, he will have learned the nature of the general [term] *agent* that is common to natural things in the latter sense, since he will know the state that is particular to whatever is an agent among natural things. So also, in an analogous fashion, let him discover the state of the final principle. That the principles are these four[7] (and we shall discuss them in detail later)[8] is a matter postulated in physics but demonstrated in first philosophy.[9]

(12) The body has an additional principle insofar as it is changeable or perfectible or comes to be or is generable, where its being changeable is different from its being perfectible, and both are again different from what is understood by its being something that comes to be and is generable. Now, what is understood by *changeable* is that it had a specific attribute that ceased to exist, and it came to have another attribute. In this case, then, there are [three factors]: [(1)] something that remains—namely, what undergoes the change; [(2)] a state that existed and then ceased to exist; and [(3)] a non-existent state that came to exist. Clearly, then, insofar as [a body] undergoes change, there must be [(1)] something susceptible to [both] that from which and that into which it changed; [(2)] a presently existing form; and [(3)] its privation, which occurred together with the form that departed. An example [of these three factors] would be the robe that became black, the whiteness, and the blackness, where there was a privation of blackness when the whiteness existed.

7. That is, the four causes of Aristotelian physics: the material, formal, efficient, and final; cf. Aristotle, *Physics* 2.3.

8. See 1.10.

9. See *Ilāhīyāt* 6.1.

(١١) نعم، قد يجوز أنْ تكون في جملة الأُمور الطبيعبة ما هو مبدأ فاعلي لجميع الطبيعيات غير نفسه، لا مبدأ فاعلي لجميع الطبيعات مطلقاً. والمبدأ الفاعلي المشترك بالنحو الآخر فلا عجب أنْ لو بحث الطبيعي عن حاله، ووجه ذلك البحث أنْ يتعرف حال كلّ ما هو مبدأ فاعلي لأمرٍ من الأمور الطبيعية؛ أنَّه كيف قوته وكيف تكون نسبته إلى معلوله، في القرب والبعد، والموازاة والملاقاة، وغير ذلك، وأنْ يبرهن عليه. فإذا فعل ذلك فقد عرف طبيعة الفاعل العام المشترك للطبيعيات بهذا النحو، إذْ عرف الحال التي تخصّ ما هو فاعل في الطبيعيات من الطبيعيات، وعلى هذا القياس. فاعرف حال مبدأ الغائي. وأمّا أنَّ المبادىء هي هذه الأربعة - وسنفصل الكلام فيها بعد - فهو موضوع للطبيعي مبرهنٌ عليه في الفلسفة الأُولى.

(١٢) هذا؛ وأمّا الجسم من جهة ما هو متغيّرٌ أو مستكملٌ أو حادثٌ كائنٌ؛ فإنَّ له زيادة مبدأ. وكونه متغيّراً غير كونه مستكملاً، والمفهوم من كونه حادثاً وكائناً هو غير المفهوم من كليهما جميعاً. فإنَّ المفهوم من كونه متغيّراً أنّه كان بصفةٍ حاصلةٍ بطلت وحدثت له صفةٌ أخرى. فيكون هناك شيءٌ ثابتٌ هو المتغيّر، وحالة كانت موجودة فعُدمت، وحالة كانت معدومة وجدت. فبيّنٌ أنَّه لا بدّ له - من حيث هو متغيّر - من أنْ يكون له أمرٌ قابل لما تغيّر عنه ولما تغيّر إليه، وصورة حاصلة، وعدم لها كان مع الصورة الزائلة؛ كالثوب الذي اسودّ، والبياض والسواد، وقد كان السواد معدوماً إذْ كان البياض موجوداً.

(13) What is understood by [a body's] being *perfectible* is that it comes to have something that did not exist before, without itself losing anything. An example would be the object at rest that is moved, for so long as it rested, there was only a privation of the motion that belonged to it possibly or potentially, whereas when it is moved, nothing is lost of it except the privation only. [Another] example is the empty slate once one has written on it. That which undergoes perfection must also include [three factors]: [(1)] a determinate being that was imperfect and then was perfected, [(2)] something presently existing in it, and [(3)] a privation that preceded [what is presently existing in it].

(14) Privation, in fact, is a precondition for something's being subject to change and perfection, since, were there no privation, it would be impossible for it to be perfected or changed, but rather, there would always be the presently existing perfection and form. Therefore, what is changed and what is perfected require that a certain privation precede them to the extent that they really are something changeable or perfectible, whereas the privation in that it is a privation does not require that a change or perfection occur. So, the elimination of privation requires the elimination of the changeable and perfectible, insofar as they are changeable and perfectible, whereas the elimination of the changeable and perfectible does not requires the elimination of privation. So privation in this respect is prior, and so is a principle, if *principle* is whatever must exist, however it might exist, in order that something else exist, but not conversely. If that is not sufficient for being a principle, and a principle is not whatever must exist, however it might exist, but rather is whatever must exist simultaneously with the thing whose principle it is without being prior or posterior, then privation is not a principle. We achieve nothing by quibbling over terminology, so in lieu of *principle,* let us use *whatever must... but not conversely*. So we find that, in order for the body to be subject to change and perfection, there needs to be [(1)] that which is susceptible to change or perfection, [(2)] privation, and [(3)] form.[10] This is clear to us on the slightest reflection.

10. Cf. Aristotle, who regarded privation, *sterēsis,* as principle in an accidental sense; see his *Physics* 1.7.190b27, and 1.7.191a13–15.

(١٣) والمفهوم من كونه مستكملاً هو أنْ يحدث له أمرٌ لم يكن فيه، من غير زوال شيءٍ عنه، مثل الساكن يتحرك؛ فإنَّه حين ما كان ساكناً لم يكن إلاّ عادماً للحركة التي هي موجودة له بالإمكان والقوة، فلمّا تحرّك لم يزل منه شيءٌ إلاّ العدم فقط، ومثل اللوح الساذج كُتِبَ فيه. والمستكمل لا بُدَّ أنْ يكون له ذاتٌ وجدت ناقصة ثم كَمُلت، وأمرٌ حصل فيه وعدمٌ تقدمه،

(١٤) إذ العدم شرطٌ في أنْ يكون الشيء متغيّراً أو مستكملاً. فإنَّه لو لم يكن هناك عدمٌ لاستحال أنْ يكون مستكملٌ أو متغيّرٌ، بل كان يكون الكمال والصورة حاصلة له دائماً. فإذن المتغيّر والمستكمل يحتاج إلى أنْ يكون قبله عدمٌ حتى يتحقّق كونه متغيّراً أو مستكملاً، والعدم ليس يحتاج في أنْ يكون عدماً إلى أنْ يحصل تغيّرٌ أو استكمالٌ، فرفْع العدم يوجب رَفْع المتغيّر والمستكمل، من حيث هو متغيّر ومستكمل، ورفْع المتغيّر والمستكمل لا يوجب رَفْع العدم. فالعدم من هذا الوجه أقدم؛ فهو مبدأ - إنْ كان كلّ ما لا بُدّ من وجوده، أي وجود كان، ليوجد شيءٌ آخر، من غير إنعكاسٍ - مبدأ، وإنْ كان ذلك لا يفي في كون الشيء مبدأ. ولا يكون المبدأ كلّ ما لا بُدَّ من وجوده للأمر أيّ وجود كان، بل لا بُدّ من وجوده مع الأمر الذي هو له مبدأ، من غير تقدّم ولا تأخّر فليس العدم مبدأ، ولا فائدة لنا في أنْ نناقش في التسمية، فلنستعمل بدل المبدأ المحتاج إليه من غير انعكاس، فنجد القابل للتغيّر والاستكمال، ونجد العدم، ونجد الصورة؛ كلّها محتاجاً إليه في أنْ بكون الجسم متغيّراً أو مستكملاً، وهذا يتّضح لنا بأدنى تأمل.

(15) What is understood by the body's being something subject to generation and coming to be compels us to affirm something that has come to be as well as a preceding privation. As for whether the generation and coming to be of what is subject to such requires a preceding substance that [initially] was associated with the privation of the generable form and then ceased to be associated with it once the privation of [the form] ceased, that is not something that is obvious to us on immediate inspection. In fact, for physics we must simply posit it and content ourselves with inductive proof, but we will demonstrate it in first philosophy.[11] Dialectic may sometimes provide a useful bit of information to quiet the soul of the student, but [take care] not to confuse the demonstrative with the dialectical.

(16) Among the principles, the body has those that are inseparable from it and by which it subsists: it is these that we specifically term *principles*. Insofar as [the body] is a body absolutely, these are the aforementioned material and corporeal form, which necessarily entails the accidental quantities, or the specific form that perfects it. Insofar as [the body] is the subject of change, perfection, and generation, it is additionally related to the privation associated with its material, which is a principle in the sense previously mentioned. Now, if we consider what is common to the changeable, the perfectible, and the generable, the principles are a certain material, a disposition, and a privation. Now, if we confine ourselves to the changeable, then the principles are a certain material and some contrary, for the thing that changes out of and into the intermediate does so only inasmuch as it contains a certain contrariness. The difference between contrariety, disposition, and privation is apparent from what you have learned and can be acquired by you from what you have been taught.

---

11. Avicenna has argued that, in the case of what is changeable and what is perfected, there must be three factors: (1) an underlying thing, (2) a form, and (3) a privation. In the case of what is generated and comes to be in time it is likewise obvious that there is something that comes to be—the form—and a privation. What is not obvious is whether there must always be a pre-existing underlying thing. That is because if there must be, then it is quite easy to show that the world is eternal, which, in fact, is the issue to which Avicenna is alluding here. See *Ilāhīyāt* 4.2, where he argues that matter must precede all generation and temporal coming to be and thus provides the key premise in his argument for the eternity of the world; see also 3.11 of the present work where, despite his claim that this issue should not be treated in the science of physics, he provides arguments much like those found in the *Ilāhīyāt*.

(١٥) والمفهوم من كون الجسم كائناً وحادثاً يضطرنا إلى إثبات أمر حَدَث وإلى عدم سبق. وأمّا أنَّ هذا الحادث وهذا الكائن هل يحتاج إلى أنْ يتقدم كونَه وحدوثَه وجودُ جوهرٍ كان مقارناً لعدم الصورة الكائنة ثم فارقه وبَطل عنها العدم، فهو أمرٌ ليس يبين لنا عن قريب بيان ذلك، بل يجب أنْ نضعه للطبيعي وضْعاً ونقنعه بالاستقراء، ونبرهن عليه في الفلسفة الأولى. وربّما قامت صناعة الجدل في إفادة نفس المتعلم طرفاً صالحاً من السكون إليه، إلاّ أنَّ الصنائع البرهانية لا تُخلط بالجدل.

(١٦) فالجسم له من المبادىء التي ليست مفارقة له ولما فيه بالقوام؛ وإياها نخصّ باسم المبادىء. أمّا من حيث هو جسم مطلقاً؛ فالهيولى والصورة الجسمية المذكورة التي تلزمها الكميّات العرضية، أو الصورة النوعية التي تكمله. وأمّا من من حيث هو متغيّرٌ أو مستكملٌ أو كائنٌ، فقد تزيد له نسبة العدم المقارن لهيولاه قبل كونه، ويكون مبدأ على ما قيل. فإنْ أخذنا ما يعمّ المتغيّر والمستكمل والكائن؛ كانت المبادىء هيولى وهيئة وعدما، وإنْ خصّصنا المتغيّر كانت المبادىء هيولى ومضادّة، فإنَّ المتوسط إنّما يتغيّر عنه وإليه من حيث فيه ضدّية مّا. ويشبه أنْ يكون الفرق بين المضادّة والهيئة والعدم ممّا قد عرفته، ويحصل لك فيما قد عُلّمتَه.

(17) Now, the disposition of the substance, insofar as it is substance, is a form, whereas the disposition of that which is undergoing nonsubstantial change and perfection is an accident, and we have already explained to you the difference between form and accident. In this context, however, it is standard to call every disposition *form;* and so let us do so, where by *form* we mean anything that comes to be in a recipient such that [the recipient] comes to have a certain specific description. The material is distinct from both [that is, form and privation] in that it has its own existence together with each of them. Form is distinct from privation in that the form is, in itself, a certain essence that adds to the existence belonging to matter, whereas privation does not add to the existence that belongs to matter, but rather is a certain accompanying state that corresponds with this form when [that form] does not exist but the potential to receive it does exist. This privation, however, is not absolute privation, but one having a certain mode of being, since it is a privation of some thing, bringing along with itself a certain predisposition and preparedness in some determinate matter. So, [for example,] human does not come to be from whatever is nonhuman, but only from nonhuman in what is receptive to [the form of] humanity. So generation [comes about] by the form, not the privation, whereas it is through the privation, not form, that there is corruption.

(١٧) والجوهـــر - من حيث هو جوهـــر - فهيئته صورة؛ وقد عرّفناك الفرق بين الصورة والعرض، وأمّا المتغيّرات والمستكملات لا في الجوهرية فهيئتها عرض. وقد جرت العادة أنْ تسمّى كل هيئة في هذا الموضع صورة، فلنسم كلّ هيئة صورة؛ ونعني به كلّ أمرٍ يحدث في قابل يصير له موصوفاً بصفةٍ مخصوصة، والهيولى تفارق كلّ واحدٍ منهما بأنّها توجد مع كلّ واحدٍ منهما بحالها، والصورة تفارق العدم؛ بأنَّ الصورة ماهيّة مّا بنفســـها، زائـــدة الوجـــود على الوجود الذي للهيولـــى، والعدم لا يزيد وجـــوداً على الوجود الذي للهيولى، بل يصحبه حال مقايسته إلى هذه الصورة إذا لم تكن موجودة، وكانت القوة على قبولها موجودة. وهذا العدم ليس هو العدم المطلق؛ بل عدم له نحوٌ من الوجود؛ فانّه عدم شيءٍ مع تهيؤٍ واستعدادٍ له في مادة معيّنة. فإنّه ليس الإنسان يكون عن كلّ لا إنسانية، بل عن لا إنسانية في قابلٍ للإنسانية، فالكَوْن بالصورة لا بالعدم، والفساد بالعدم لا بالصورة.

(18) It is often said that something was from the material and from the privation, whereas it is not said that it was from the form. So it is both said that the bed was from the material—that is, from the wood—and from the non-bed. Now, in many cases it is all right to say that [the thing] came to be from the material, but in many others it is not, whereas it is always said that it was from the privation. A case in point is that we do not say that a writer was *from* the man, but that the man *was* a writer, whereas we say that a man was from the semen, and a bed was from the wood. The reason for this, in the case of the semen, is that the seminal form is cast off. In this instance, *from* is equivalent to *after,* just as the claim *it was from the privation* signifies the same thing as *a man was from the not-man*—that is, *after* the not-man. As for the case of wood, and so again where a bed is said to be from the wood, it was because the wood is devoid of a certain form, even if it is not devoid of the form of wood, since unless the wood changes with respect to some description and shape, through carving and woodworking, neither will there be the bed from it nor will it take on the shape of [the bed]. So, in some sense, [the wood] resembles the semen, since both of them changed from their current state, and so we use *from* also in the case of [the wood].

(١٨) وقد يقال إنَ الشـــيء كان عن الهيولى وعن العدم، ولا يقال كان عن الصورة، فيقال إنَّ السرير كان عن الهيولى أي عن الخشب، ويقال كان عن اللاسرير، وفي كثير من المواضـع يصح أنْ يقال إنَّه كان عـن الهيولى، وفي كثير منها لا يصح، ودائماً يقال إنَّه كان عن العدم. فإنَّه لا يقال كان عن الإنسان كاتب، بل يقال إنَّ الإنسان كان كاتباً، ويقال عن النُطْفة كان إنســان، ويقال عن الخشب كان سرير. والسبب في ذلك؛ أمّا في النُطْفة فلأنَّهـا خلعتْ صورة النُطْفية، فتكون ها هنـا لفظة «عن» تدل على «بعد»؛ كما يدلّ في قولهم: كان عن العدم، كما يقال إنَّه كان عن اللاإنســان إنسانٌ، أي بعد اللاإنسانية. وأمّا في الخشـب فحيث يقال أيضاً عن الخشب كان سريرٌ، فكان؛ لأنَّ الخشبَ، وإنْ لم يخلُ عن صورة الخشب، فقد خلا عن صورةٍ مّا، إذ الخشب ما لم يتغيّر في صفةٍ من الصفات وشـكلٍ من الأشكال بالنحت والنّجر، لا يكون عنه السرير، ولا يتشكل بشكله، فيشبه النُطْفة من وجه، إذ كل منهما قد تغيّر عن حاله، فنستعمل فيه أيضا لفظة عن.

(19) About these two kinds of subjects and material things, we may say *from* in the sense of *after,* but there is another kind of subjects about which we use *from* and *out of* in another sense. To illustrate that, when one of the forms has certain subjects that are produced for it only by means of mixture or composition, then what is generated is said to be *from* them, and so *from* and *out of* signify that what is generated is constituted by [those subjects], just as we say that the ink was *from* vitriol and gall. It also would seem that *from,* [both] in the sense of something composed of after-ness and in this latter sense, is said about the first kind [of subjects and material things]. [That] is because what is meant by something's having been from the semen or wood is that it was after they were in a certain state as well as that something was drawn from them, where the generated thing, which was said to be from them, was made to subsist. Now, it is not said about what is like the semen or vitriol that it was the generated thing, such that the semen would be said to be a man or vitriol to be ink, as it is said[12] that man was a writer, save in some figurative sense meaning that [the semen or vitriol] *became* (that is, changed), whereas both ways are said about what is like the wood. So [in the case of wood], it is said that a bed was *from* the wood and the wood *was* a bed, because the wood, as wood, does not undergo corruption in the way the semen does, and so [the wood] is like the man insofar as he is susceptible to being something that writes. Still, if it is not devoid of a certain shape, it cannot receive the shape of a bed, and so it is like the semen insofar as [the semen] is altered into being something that is a man.

12. Following **Z**, **T**, and the Latin, which do not have the negation *lā*.

(١٩) فهذان الصنفان من الموضوعات والهيوليّات يقال فيهما «عن» بمعنى «بعد»، وصنفٌ من الموضوعات يستعمل فيه لفظة «عن» ولفظة «من» على معنى آخر. وبيان ذلك أنَّه إذا كانت موضوعات مّا لصورةٍ من الصور؛ إنّما توضع لها بالمزاج والتركيب، فقد يقال إنَّ الكائن يكون عنها، فيدل بلفظة «عن» وبلفظة «من» على أنَّ الكائن متقوّم منها كقولنا عن الزّاج والعفص كان المداد. ويشبه أيضاً أن يكون الصنف الأوّل يقال فيه لفظة «عن» بمعنى مركّب من البَعْدية وهذا المعنى. إنَّ النُطْفة أو الخشب كان عنهما ما كان؛ بمعنى أنَّه كان بعد أنْ كانت على حال، ثم أُستل منهما شيءٌ وقوّم به الكائن الذي قيل إنَّه كان عنهما. فما كان مثل النُطْفة أو الزاج فلا يقال فيه أنَّه كان الشيء الكائن؛ فلا يقال إنَّ النُطْفة كانت إنساناً، أو الزّاج كان حبراً. كما يقال إنَّ الإنسان كان كاتباً إلاّ بنوعٍ من المجاز وبمعنى صار أي تغيّر. وما كان مثل الخشب فقد يقال فيه كلا الوجهين: فيقال عن الخشب كان سريرٌ، وأنَّ الخشب كان سريراً؛ وذلك لأنَّ الخشب - من حيث هو خشب - لا يفسد فساد النُطْفة، فيشبه الإنسان من حيث يقبل الكتابة. ولكنه ما لَمْ يُخلِ شكلاً لم يقبل شكل السرير، فيشبه النُطْفة من حيث تَستحيل إلى الإنسانية.

(20) Now, in those cases where it is not acceptable to say *from,* it becomes so once the privation is added to it—as, for example, saying, *a writer was from a* nonwriting *man.* It is never acceptable, however, to say it about the privation itself, except together with *from,* for we do not say that the nonwriter *was* a writer; otherwise, a nonwriter would be a writer. Certainly, if one does not mean by *nonwriter* the nonwriter himself, but simply the subject who is described as a nonwriter, then we can say that, and [of course] it is always acceptable to use *from* in this case. Still, I do not insist on this and similar cases, since languages may differ in the license and proscription of these uses. I only say that when we mean by *from* the two aforementioned senses, they are permissible where we allowed and not permissible where we did not allow.

(21) In the place corresponding with the present one,[13] there is sometimes mentioned the material's desire for the form and its imitating the female, while the form imitates the male, but this is something I just do not understand [for the following reasons]. As for the desire associated with having a soul, there is no dispute about denying it of the material. Equally improbable is the natural compulsive desire whose incitement is in the way of a drive, as, for example, belongs to the stone to move downward in order that it be perfected after being displaced from its natural place.

---

13. That is, in Aristotle's *Physics* 1.9.192a22–25.

(٢٠) وحيث لا يصح من ذلك أنْ يقال فيه «عن»، فإذا أُضيف إليه العدم صحّ، كما يقال عن الإنسان غير الكاتب كان كاتب. والعدم نفسه لا يصح فيه البتّة أنْ يقال إلاّ مع لفظة «عن»، فإنّه لا يقال إنَّ غير الكاتب كان كاتباً، وإلاّ فيكون كاتباً غير كاتب، نعم إنْ لم نعن بغير الكاتب نفس غير الكاتب، بل الموضوع الموصوف بأنّه غير كاتبٍ، فربما قيل دلك. وأما لفطه «عن» فيصح استعمالها فيه دائما، على أني لا اتشدد في هذا وما أشبهه: فعسى اللغات تختلف في إباحة هذه الاستعمالات وحظرها. بل أقول إذا عُني بلفظة «عن» المعنيان اللّذان ذكرناهما جازا حيث أجزنا، ولم يجوزا حيث لم نُجز.

(٢١) وقد يذكر في مثل هذا الموضع حال شوق الهيولى إلى الصورة وتشبيهها بالأُنثى وتشبيه الصورة بالذكر؛ وهذا شيء لستُ أفهمه. أمّا الشوق النفساني فلا يُختلف في سلبه عن الهيولى، وأمّا الشوق التسخيري الطبيعي الذي يكون إنبعاثه على سبيل الإنْسياق، كما للحجر إلى التسفّل ليستكمل به بعد نقْصٍ له في أيْنه الطبيعي.

(22) Again, then, it would have been possible for the material to desire forms, were it free of all forms, or [if] it grew weary of a given form joined to it or lost the sense of contentment with the presently existing forms that perfect it as a species, or [if] it could move itself toward the acquisition of form, in the way the stone acquires [its natural] place (assuming that it possessed a motive power). Now, it is not the case that [the material] is devoid of all forms. Also, [the material] is not the sort of thing that grows weary of the presently existing form so as to work for its dismissal and destruction. [That] is because if the weariness is the necessary result of the very presence of this form, then [the form] is necessarily undesirable, whereas if [the weariness] results from the length of time, the desire would not be something in the substance of [the material] but something that accidentally happens to it after a period of time, in which case there is a cause necessitating it. Equally impossible is that [the material] grew discontent with the presence [of the form] and rather desired to gather contraries into itself, which is absurd. The [real] absurdity, however, seems to be having supposed that it desires [in] the way [that] the soul desires. The [natural] compulsive desire, on the other hand, is only for some end in the perfecting nature. Now, natural ends are inevitable, and so, notwithstanding this, how can the material be moved toward the form when its being disposed to the form arises only from some cause that nullifies the existing form, not its acquiring [that form] through its own motion? Had [the Peripatetics] not made this desire a desire for the forms that make [the material] subsist, which are first perfections, but rather, [made it] a desire for the secondary concomitant perfections, it would have been difficult enough understanding the sense of this desire; but how [is it possible at all] when they have made this desire a desire for the forms that cause [the material] to subsist?

(٢٢) فهذا أيضاً بعيدٌ عنه فلقد كان يجوز أنْ تكون الهيولى مشتاقة إلى الصور؛ وكان هناك خلوٌ عن الصور كلّها، أو ملال صورة مقارنة، أو فقدان القناعة بما يحصل من الصور المكملة إيّاها نوعاً، وكان لها أنْ تتحرك بنفسها إلى اكتساب الصورة كما للحجر في اكتساب الأين - إنْ كان فيها قوة محرّكة - وليست خالية عن الصور كلّها. ولا يليق بها الملال للصورة الحاصلة، فتعمل في نقضها ورفضها. فإنَّ حصول هذه الصورة - إنْ كان موجباً للملال لنفس حصولها - وَجَبَ أنْ لا يشتاق إليها، وإنْ كان لمدةٍ طالت، فيكون الشوق عارضاً لها بعد حين لا أمراً في جوهرها؛ ويكون هناك سببٌ يوجبه. ولا يجوز أيضاً أن تكون غير قانعةٍ بما يحصل، بل مشتاقة إلى اجتماع الأضداد فيها؛ فإنَّ هذا محال، والمحال ربّما ظُنَّ أنّه يُشتاق إليه الاشتياق النفساني. وأما الاشتياق التسخيري فإنّما يكون إلى غايةٍ في الطبيعة المكمّلة، والغايات الطبيعية غير محالة. ومع هذا؛ فكيف يجوز أنْ تكون الهيولى تتحرّك إلى الصورة؟ وإنّما تأتيها الصورة الطارئة من سببٍ يُبطل صورتها الموجودة لا أنّها تكسبها بحركتها. ولو لم يجعلوا هذا الشوق إلى الصورة المقوُمة التي هي كمالات أُولى، بل إلى الكمالات الثانية اللاحقة، لكان تصوّر معنى هذا الشوق من المتعذر، فكيف وقد جعلوا ذلك شوقاً لها إلى الصورة المقوّمة؟

(23) For these reasons, it is difficult for me to understand this talk, which is closer to the talk of mystics[14] than that of philosophers. Perhaps someone else will understand it as it should be, so that one might refer to him in this matter. If the material [understood] absolutely were replaced with a certain material that is [already] perfected by the natural form so that, from the natural form that belongs in it, it comes to have an incitement toward the perfections of that form—like, for example, earth's moving downward and fire's moving upward—there would be some sense to this talk, even if it attributed that desire to the active form. In an absolute sense [of the material], however, I cannot understand it.

14. Literally, Sufis.

(٢٣) فمن هذه الأشـياء يعْسُـر عليَّ فهم هذا الكلام الذي هو أشبه بكلام الصوفية منه بكلام الفلاسـفة! وعسـى أنْ يكون غيري يفهم هذا الكلام حقَّ الفهم؛ فليُرجع إليه فيه. ولو كان بدل الهيولى بالإطلاق؛ هيولى ما تسـتكمل بالصورة الطبيعية حتى يحدث من الصورة الطبيعية التي فيها لها إنبعاثٌ نحو إسـتكمالات تلك الصورة، مثل الأرض في التسـفّل والنار في التصعّد، لكان لهذا الكلام وجهٌ - وإنْ كان مرجع ذلك الشـوق إلى الصورة الفاعلة - وأمّا على هذا الإطلاق فممّا لستُ أفهمه.

Chapter Three

How these principles are common

(1) Since our inquiry is about common principles only, we should inquire into which of the two aforementioned ways[1] these three common principles [that is, matter, form, and privation] are common.

(2) It will become apparent to us later[2] that some bodies are susceptible to generation and corruption (namely, those whose material acquires a new form and loses another), while others are not susceptible to generation and corruption and instead exist as a result of an atemporal creation.[3] If that is the case, then there is no common material in the first of the two senses, since there is no single material that is sometimes susceptible to the form of what undergoes generation and corruption and at other times is susceptible to the form of what is naturally incorruptible and has no material generation. So that is impossible. (In fact, however, it might be possible that the class of bodies subject to generation and corruption has material that is common to those that are generated out of and corrupted into one another, as we shall show in the case of the four properly called *elements* [namely, earth, water, air, and fire].)[4] Or at best, [if a common material for both what is and what is not subject to generation and corruption is not impossible], we would have to concede that the nature of the subject that belongs to the form of what is incorruptible and the subject that belongs to the form of what is corruptible is a single nature that, in itself, is able to receive every form, except that what is incorruptible was accidentally joined with a form that has no contrary.

1. See 1.2.8–11, where again Avicenna notes two distinct ways in which natural things can share something in common. So, on the one hand, *common* might be understood as a numerically singular thing common to all natural things, or, on the other hand, *common* might be understood as some specifically or generically similar notion applying to all natural things equally. So, for example, *agent* understood as something common to natural things in the first instance would signify God, whereas in the second instance it would signify the universal predicate *agent*.

2. This point is discussed in detail throughout *Kitāb fī al-kawn wa-l-fasād*.

3. For a detailed discussion of Avicenna's use of *ibdā*ᶜ (atemporal creation), see Jules Janssens, "Creation and Emanation in Ibn Sīnā," *Documenti e studi sulla tradizione filosofica medievale* 8 (1997): 455–77.

4. See for instance *Kitāb fī al-kawn wa-l-fasād,* chs. 9 and 14.

<الفصل الثالث>

في كيفية كوْن المبادىء مشتركة

(١) لمّا كان نظرنا هذا إنّما هو في المباديء المشتركة فيحق علينا أن ننظر في هذه المبادىء الثلاثة المشتركة ؛ إنّها على أي نحوٍ من النحوين المذكورين تكون مشتركة.

(٢) لكنه سيظهر لنا أنَّ الأجسام منها ما هي قابلة للكون والفساد ، أي منها ما هيولاها تستجدّ صورة وتُخلي صورة ، ومنها ما ليس قابلاً للكون والفساد ، بل وجودها بالإبداع. فإذا كان كذلك ؛ لم تكن هيولى مشتركة على النحو الأوّل من النحوين ؛ فإنّه لا تكون هيولى واحدة تارة تقبل صورة الكائنات الفاسدة وتارة تقبل صورة ما لا يفسد في طباعه ولا له كونٌ هيولاني ، فإنَّ ذلك مستحيل. بل ربّما جاز أنْ تكون الهيولى المشتركة لمثل الأجسام الكائنة الفاسدة التي يتكون بعضها من بعض ، ويفسد بعضها إلى بعض ، كما سنبيّن من حال الأربعة التي تسمّى الأُسْطُقسّات. اللّهم إلاَّ أنّ نجعل طبيعة الموضوع التي لصورة ما لا يفسد والموضوع لصورة ما يفسد ، طبيعةً واحدةً في نفسها صالحة لقبول كل صورة إلاَّ أنَّ ما لا يفسد قد عرض أنْ قارنته الصورة التي لا ضدّ لها ، فيكون السبب في

In this case, the reason that things not subject to generation and corruption are such would be owing to their form, which, as a result of what is in their natures, hinders the matter, not because the matter is passive. Assuming that—which is unlikely, in light of what will become clear later[5]—a common material, would exist in this way. The common material in this way—whether common to all natural things or just those subject to generation and corruption—would be something resulting from an atemporal creation, neither being generated out of nor corrupted into anything; otherwise, [this common material] would need another material, in which case that [other material] would be prior to it and common.

(3) As for whether natural things have a formal principle common in the first of the two ways, only the corporeal form among the forms that we imagine belongs to them as such. So if you turn to the bodies undergoing generation and corruption that are only in what immediately follows the corporeal form (so that the corporeal form that is in water, for example, when air undergoes alteration [into water], is something that itself remains in the water), then the bodies so described would have formal principles that are numerically common to them, whereas [the bodies] thereafter have individual formal principles specific to each of them. If this is not the case, and instead when the form of water is corrupted, there is in the corruption of the form of water the corruption of the corporeality that belonged to [the water's] material and some other corporeality different in number but similar in kind comes to be, then bodies would not have this kind of common, formal principle. The truth concerning the two cases will become apparent to you in its proper place.[6] Should bodies, or some subset of bodies, or even a single body, have a formal principle of this kind as an inseparable form, then that formal principle would eternally be joined with matter and would not undergo generation and corruption, but instead it would again result from an atemporal creation.

---

5. It is not clear to what Avicenna is referring. Perhaps he means the necessary role of matter with respect to those things that at are *ḥādith*—that is, what comes to be in time and, as such, would be subject to generation and corruption. See 3.11 below and *Ilāhīyāt* 4.2.

6. The reference would appear to be to *Kitāb fī al-kawn wa-l-fasād* 14 and *Ilāhīyāt* 9.5.

أنّها لا تكون ولا تفسد من جهة صورتها المانعة لمادّتها عمّا في طباعها ، لا من جهة المادّة المطاوعة . فإنْ كان كذلك - وبعيدٌ أنْ يكون كذلك على ما سيتضح بعد - فسيكون حينئذ هيولى مشتركة بهذا الوجه ، والهيولى المشتركة بهذا الوجه ، سواء كانت مشتركة للطبيعيات كلّها أو للكائنات الفاسدة منها ، فإنّها متعلقة بالإبداع ، وليست تكون من شيء وتفسد إلى شيء ، وإلاّ كانت تحتاج إلى هيولى أخرى ، فتكون تلك متقدمة عليها ومشتركة .

(٣) وأمّا هل للطبيعيات مبدأ صوري مشترك بالنحو الأول من النحوين ، فليس يوجد لها من الصور ما نتوهمه أنّه ذاك إلاّ الصورة الجسمية . فإنْ كان تصرف الأجسام في الكون والفساد إنّما يكون فيما وراء الصورة الجسمية ؛ حتى تكون مثلاً الصورة الجسمية التي في الماء - إذا استحال هواء - باقية بعينها في الماء . فيكون للأجسام مبادىء صورية على هذه الصفة مشترك لها بالعدد ، ووُجد لها بعده مبادىء صورية يخص كلّ واحدٍ منها واحدة منها ، وإنْ كان الأمر ليس كذلك ، بل إذا فسدتْ المائية فسدت الجسمية التي كانت لهيولاه في فساد المائية ، وحدثت جسمية أُخرى مخالفة بالعدد موافقة في النوع ، فلا يكون للأجسام مثل هذا المبدأ الصوري المشترك ، وسيظهر لك الحقّ من الأمرين في موضعه . ولو كان للأجسام مبدأ صوري بهذه الصفة ، أو لطائفةٍ من الأجسام ، أو لجسم واحدٍ صورة لا تفارق ، لكان ذلك المبدأ الصوري مداوم الاقتران بالهيولى ، ولم يكن ممّا يكون ويفسد ، بل يتعلق أيضاً بالإبداع .

(4) As for privation, it is clearly altogether impossible that there be a common privation in the first sense. [That is] because this privation is the privation of something, *x*, that regularly comes to be through a process of generation; and if it is such, then it is likely that *x* will be generated, and so at that time this privation will no longer remain. In that case, however, it is not something common.

(5) As for that which is common in the second of the two senses, the three principles are common to what is subject to generation and change, since it is common to all [of those sorts of things] that they all have matter, form, and privation.

(6) *Being neither generable nor corruptible* is predicated of what is common in the same way that it is predicated of universals—namely, in two ways. One way by which we mean that the universal is neither generable nor corruptible is that with respect to the world, there is no moment that is the first moment at which some first individual or number of first individuals of whom the universal is predicated existed and before which there was a moment at which none of [those individuals] existed. The case would be the parallel opposite to this with respect to corruption. In this way, some people (namely, those who require that as long as the world exists there always be generation, corruption, and motion in it) say that these common principles are neither generable nor corruptible. The second way is to inquire into their essence—as, for example, the essence of man—and then consider whether [man] *qua* man is subject to generation and corruption. In this case, accounts of generation and corruption are found that are not the account of man *qua* man, and so both are denied of the essence of man *qua* man, because something that is necessarily joined to him is not intrinsic to him. The same is said about these principles that are common in the second of the two ways that [the predicate] *being common* was used; and in the present context, our inquiry and discussion are about the principles from this perspective and not the first one.

(٤) وأمّــا العدم فواضحٌ مــن حاله أنَّه لا يجوز أنْ يكون من جملته عدمٌ مشــتركٌ بهذا النحو الأول؛ لأنَّ هذا العدم هو عدم شــيءٍ من شــأنه أنْ يكون، وإذا كان من شأنه أنْ يكــون، لم يبعد أنْ يكــون، فحينئذ لا يبقى هذا العدم، فحينئذ لا يكون مشــتركاً.

(٥) وأمّا المشترك على النحو الآخر من المعنيين فإنَّ المباديء الثلاثة توجد مشتركة للكائنات والمتغيّرات، إذ تشترك كلّها في أنَّ لكل منها هيولى وصورة وعدماً.

(٦) وهذا المشــترك يقال إنَّه لا يكون ولا يفســد، على نحو ما يقال للكلّيات أنَّها لا تكون ولا تفسد ويقال للكلّيات أنَّها لا تكون ولا تفسد على وجهين؛ فنعني بأحد الوجهين أنَّ الكلّي لا يكون ولا يفسد، أي أنَّه لا يكون وقتٌ في العالم هو أول وقتٍ وجد فيه أول شــخصٍ أو عدّة أوائل أشــخاص يُحمل عليها ذلك الكلّي، وكان قبله وقت، وليس ولا واحد منها موجوداً فيه، وفي الفساد ما يقابل هذا. فبهذا الوجه، من الناس من يقول إنَّ هذه المباديء المشتركة لا تكون ولا تفسد، وهم القوم الذين يوجبون في العالم دائماً؛ كوناً وفســاداً وحركة، ما دام العالم موجوداً. والوجه الثانــي؛ أنْ يُنظر إلى ماهيّتها، كماهيّة الإنســان، فينظر هل هو من حيث هو إنســان يكون أو يفســد، فيوجد معنى إنّه يكون ومعنى إنَّه يفسد ليس معنى الإنسان من حيث هو إنسان، فيسلبان عن ماهيّة الإنسان من حيث هو إنســان لأنَّه أمرٌ يلزمه ليس داخلاً فيه. وكذلك يقال في هذه المباديء المشتركة بالنحو الثاني من نحويْ الاشتراكِ المذكور. ونظرنا ها هنا في المباديء هو من هذه الجهة، وليس كلامنا هاهنا في الجهة الأُولى.

(7) In the case where we intend the existence of concrete particulars, then, there will be materials that are subject to generation and corruption, such as the wood of the bed and gall for ink; whereas Prime Matter, to which we have gestured,[7] is not subject to generation and corruption, but exists only as a result of an atemporal creation. As for forms, some are generated and corrupted (namely, the ones subject to generation and corruption), whereas others are not (namely, those that are atemporally created). In another sense, however, it might be said of [forms] that they are neither generable nor corruptible, for it might be said of the forms that involve generation and corruption that they are not generated and corrupted in the sense that they are not a composite of form and material so as to undergo generation and corruption, since in this case, one means by *generation* (and the parallel opposite for *corruption*) that a subject comes to have a form, where it is through the [form and matter] together that something is generable.

(8) As for privation, its generation (if it has one) is its being present after it was not. Also, its being present and existing cannot be as some presently existing determinate entity in itself; but rather, it exists accidentally, because it is a privation of some determinate thing, *F*, in some determinate thing, *x*, in which there is the potential of [*F*]. Therefore, [privation] likewise has some accidental mode of generation and corruption. Its generation, then, is that the form is removed from the matter through corruption, in which case a privation [conversely related] to this attribute becomes present, whereas its corruption is that the form becomes present, at which time the privation that [is conversely related] to this attribute no longer exists. Now, this privation has an accidental privation, just as it has an accidental existence, where its privation is the form. Be that as it may, the form's subsistence and existence are not relative to it; but rather, that belongs to it accidentally through a certain

7. See 1.2.8 and par. 2 of the present chapter.

(٧) وأمّا إذا قصدنا إلى الأعيان الموجودة؛ فها هنا هيوليات تكون وتفسد؛ كالخشب للسرير والعفص للحبر. والهيولى الأُولى - التي أشرنا إليها - لا تكون ولا تفسد، إنّما هي متعلقة الوجود بالإبداع. وأمّا الصور فبعضها يكون ويفسد، وهي التي في الكائنة الفاسدة، وبعضها لا يكون ولا يفسد وهي التي في المبدعات. وقد يقال لها إنّها لا تكون ولا تفسد بمعنى آخر؛ فإنَّه ربّما قيل للصور التي في الكائنة الفاسدة إنّها لا تكون ولا تفسد؛ بمعنى أنَّها غير مركبة من هيولى وصورة حتى تكون وتفسد؛ إذ يراد بالكون حينئذ حصول صورة لموضوع ويكون الكائن مجموعهما، وبالفساد ما يقابله.

(٨) وأمّا العدم، فإذا كان كونه - إنْ كان له كوْن - هو حصوله بعد ما لَمْ يكن، وكان حصوله ووجوده ليس وجود ما له ذات حاصلة بنفسه بل كان وجوده بالعرض، لأنَّه عدم شيءٍ معيّن في شيءٍ معيّن هو الذي فيه قوّته، فنكون له نحوٌ من الكون أيضاً بالعرض ومن الفساد بالعرض. فكونه هو أنْ تفسد الصورة عن المادّة فيحصل عدمٌ بهذه الصفة، وفساده أنْ تحصل الصورة فلا يكون حينئذٍ العدم الذي بهذه الصفة موجوداً. ولهذا العدم عدمٌ بالعرَض، كما أنَّ له وجوداً بالعرض، وعدمه هو الصورة، لكن ليس قوام الصورة ووجودها هو بالقياس إليه، بل ذلك يعرض له باعتبار مّا، وقوام هذا العدم

consideration—that is, the subsistence and existence of this privation is a result of the relation itself to this form. So it is as if the privation of privation is a certain consideration accidentally belonging to the form from among the infinitely many relational considerations that might accidentally belong to something. The potentiality for privation is of the same type, since real potential is relative to actuality and perfection, and there is no perfection by privation, nor does it really have an actuality.

(9) Concerning these three common principles, we should also know in what way they are common in relation to whatever falls under any one of the things in which they are common. The claim of those[8] who say that each one of them is an equivocal term we find distressing, since, if that is the case, then the efforts of this group would be limited to finding three terms for the many principles, each one of which would include a subset of the principles, while the three terms [together] would encompass all. Had it been possible that this were enough, the important issue would have been that, among ourselves, certain terms are adopted as a matter of convention and there is agreement upon them. Whether we should have been the ones to do that or not and instead we accepted what others did, we would have nothing available to us but three terms and would not be one step closer [to understanding] what the principles signify. What an awful thing to inflict upon whoever would content himself with this! Equally, we cannot say that each one of them indicated what is included in it by way of sheer univocity. How could that be, when different kinds of various categories fall under each one of them, differing with respect to the meaning of *principles* by way of priority and posteriority? In fact, they must signify by way of analogy,[9] just as *being, principle,* and *unity* signify. We have already explained in the section on logic the difference between what is analogous as compared with what is agreed upon and what is univocal.[10]

---

8. I have not been able to identify the referent here. The claim that *matter, form,* and *privation* are equivocal terms does not appear in Aristotle's explicit discussion of the principles of nature in book 1 of the *Physics,* nor have I found it in the earlier extant Arabic commentaries on the *Physics.*

9. *Tashkīk* literally has the sense of "being ambiguous" or even "equivocal." In the present context, it would seem that Avicenna is using it in the sense of the Aristotelian *pros hen* equivocation. For a discussion of *pros hen* equivocation see G. E. L. Owens, "Logic and Metaphysics in Some Earlier Works of Aristotle," in *Logic, Science and Dialectic: Collected Papers in Greek Philosophy* (Ithaca, NY,: Cornell University Press, 1986), 180–99.

10. The reference seems to be to *Kitāb al-jadal* 2.2.

ووجـــوده هو بنفس القياس إلى هذه الصورة. فكأنَّ عـــدم العدم اعتبار ما يعرض للصورة من الاعتبارات الإضافية التي ربّما عرضت للشـــيء إلى غير نهاية. والقوة على العدم هي بهذه المنزلة؛ لأنَّ القوة الحقيقية هي بالقياس إلى الفعل والاســـتكمال، ولا استكمال بالعدم ولا فعلٌ حقيقاً له.

(٩) ويجـــب أن نعلم أيضاً: أنَّ هذه المبادىء الثلاثة المشـــتركة على أي نحو يكون مشـــتركاً فيها بالقياس إلى ما تحت كلّ واحدٍ ممّا فيه تكون الشـــركة. فإنَّه يعظم علينا ما يقولونه من أنَّ إســـم كلّ واحدٍ منها مشـــترك؛ فإنه إن كان كذلك، فيكون سعي الجماعة مقصـــوراً علـــى أنْ يوجدوا للمبادىء الكثيرة ثلاثة أســـماء يعمّ كلّ إســـم منها طائفة من المبـــادىء، وتحتوي الأســـماء الثلاثة على الجميع. فإنَّ هذا قـــد كان يمكن أنْ يكفي المهمّ فيه بأنْ نصطلح فيما بيننا على أســـماء ويتواطأ عليها - ولو فعلنا ذلك أو لم نفعله - بل قبلنا ما فعلوه؛ لم يكن في أيدينا إلاّ أسماء ثلاثة، وما كان يحصل لنا من معاني المبادىء شيءٌ البتّة، وبئس ما فعل مَنْ رضي بهذا لنفسه. وليس يمكننا أيضاً أنْ نقول إنَّ كلّ واحدٍ منها يدلّ على ما يشمله بالتواطؤ الصرف، فكيف وقد وقع تحت كلّ واحدٍ منها أصنافٌ شـــتى من مقولاتٍ شـــتى، تختلف في معنى المبدئية بالتقديم والتأخير، وبالأحرى - بل يجب - أنْ تكون دلالتها دلالة التشـــكيك، كدلالة الوجود والمبدأ والوحدة، وقد عرفنا الفرق بين المشكّك وبين المتّفق والمتواطىء في المنطق.

(10) Everything of which *material* is predicated has a nature that is common in that [the material] [(1)] is a certain factor that is capable of acquiring some other factor in itself that it previously did not have, [(2)] is that from which something is generated, and [(3)] is in [that thing] nonaccidentally. Sometimes it is simple,[11] and at other times it is something composite, such as the wood that belongs to the bed, that follows after the simple. It is also something that might acquire a substantial form or an accidental disposition. Everything of which *form* is predicated is a disposition that has been acquired by an instance of this previous factor [namely, the material] and from which, together with it, a given thing actually exists as a result of this type of composition. Everything of which *privation* is predicated is the nonexistence of some instance of what we have called *form* in that which is capable of acquiring it [that is, in the material].

(11) Now, in the present context [namely, with respect to the science of physics], our entire inquiry into and approach to form and its being a principle is strictly limited to its being a principle in the sense that it is one of the two parts of something that undergoes generation, not that it is an agent, even if it is possible that a form be an agent. Also, we have already shown that the natural philosopher does not deal with the efficient and final principles that are common to all natural things in the first way [mentioned in the previous chapter], and so we should concentrate our efforts on the second [way] that the efficient principle is common to all natural things.

(12) Having finished [the discussion] of those principles that most properly are called *principles*—namely, those that are constitutive of what is subject to generation or of the natural body—we should next focus on those principles that most deserve the title *causes*. Of these, let us define the efficient principle common to natural things—namely, the nature.

11. As, for example, the elements fire, air, water, and earth.

(١٠) فلجميع ما يقال إنَّه هيولّى طبيعة تشترك في أنّها أمرٌ من شأنه أنْ يحصل له أمرٌ آخر في ذاته بعد أنْ لا يكون له، وهو الذي يكون منه الشيء وهو فيه لا بالعرض. فربّما كان هو بسيطاً، وربّما كان مركّباً بعد البسيط كالخشب للسرير، وربّما كان الحاصل له صورة جوهرية، أو هيئة عرضية. وجميع ما يقال له إنَّه صورة فهو الهيئة الحاصلة لمثل هذا الأمر المذكور؛ الذي يحصل منهما أمرٌ من الأُمور بهذا النحو من التركيب، وجميع ما يقال له عدمٌ فهو لا وجود مثل هذا الشيء الذي سميناه صورة فيما من شأنه أنْ تحصل له.

(١١) وجميع نظرنا في الصورة ها هنا، واعتبار مبدئيتها، مصروفٌ إلى كونه مبدأ؛ بأنَّه أحد جزئي الكائن، لا أنَّه فاعل - وإنْ جاز أنْ تكون صورة فاعلاً. وقد كنا بيّنا أنَّ الطبيعي لا يشتغل بالمبدأ الفاعلي والغائي المشتركين بالنحو الأول للأمور الطبيعية كلّها. فحريٌ بنا أنْ نشتغل بالمبدأ الفاعلي المشترك للطبيعيات التي بعده،

(١٢) إذ قد فُرغ من المباديء التي هي أحرى أنّ تسمّى مبادىء، أي المقوّمة للكائن أو للجسم الطبيعي، فيجب أنْ بشتغل بالمباديء التي هي أولى بأنْ تسمى عللاً، ولنعرف منها المبدأ الفاعلي المشترك للطبيعيات؛ وهو الطبيعة.

Chapter Four

Examination of what Parmenides and Melissus said regarding the principles of being

(1) Once we reached this point, some of our colleagues asked us to talk about the more troublesome schools of the Ancients concerning the principles of natural things, given that the custom has been to mention them at the opening of the science of physics before discussing nature. Those schools of thought are, for example, the one associated with Melissus and Parmenides—namely, that what exists is one and unmovable, of which Melissus further says that it is infinite, while Parmenides says that it is finite.[1] Other examples are the school of those who said that [the principle] is one finite thing, whether water, air, or the like,[2] that is able to move. Again, there are also those who maintained an infinite number of principles, either atoms dispersed in the void[3] or small bodies, whether water, flesh, air, or the like, that are homogeneric with what results from them and all of which are mixed together in the whole.[4] There are also the rest of the schools of thought mentioned in the books of the Peripatetics. [Finally, we were asked] to talk about how [the Peripatetics] refuted these views.

(2) As for the view of Melissus and Parmenides, I do not get it. I can neither state what their aim is nor believe that they reached the level of foolish nonsense that their words, taken at face value, might indicate, since they also spoke about natural things and about [those natural things'] having more than one principle—as, for example, Parmenides, who held that there is earth and fire and that from them, there is the composition of things subject to generation. It is almost to the point that what they mean by *what exists* is the Necessary Existent, the Existence that truly is what exists, as you will learn in its proper place,[5] and

1. For the views of Melissus and Parmenides, cf. Aristotle, *Physics* 1.2–3.

2. For instance, Thales, who believed that everything came from water (Aristotle, *Metaphysics* 1.3.983b20–21), or Anaximenes, who said that everything is some manifestation of air (Aristotle, *Metaphysics* 1.3.984a5).

3. This is the view of the Atomists such as Democritus and Leucippus; cf. Aristotle, *Physics* 1.5.188a22ff. and Aristotle, *On Generation and Corruption* 1.8.325a5ff.

4. The view of Anaxagoras; cf. Aristotle, *Physics* 1.4.187a24ff.

5. See the *Ilāhiyat* 8.

<الفصل الرابع>

في تعقّب ما قاله برمنيدس ومليسوس في أمر مبادىء الوجود

(١) وإذْ قد بلغنا هذا المبلغ، فقد سألنا بعض أصحابنا أن نتكلم على المذاهب المستفسدة التي للقدماء في مبادىء الطبيعيات، وقد جرت العادة بذكرها في فاتحة العلم الطبيعي، قبل الكلام في الطبيعية. وتلك المذاهب مثل المذهب المنسوب إلى مليسوس وبَرمِنيدس أنَّ الموجود واحد غير متناه قابل للحركة، إمّا ماء أو هواء، أو غير ذلك. ومذهب مَنْ جعل المباديء غير متناهية العدد، إمّا أجزاء لا تتجزأ مبثوثة في الخلاء، وإمّا أجساماً صغاراً مشابهة لما يكون عنها؛ مائية ولحميّة وهوائية وغير ذلك، مخالطٌ كلّها للكل. وسائر المذاهب المذكورة في كتب المشائين، وأنْ نتكلم على النحو الذي نقضوا به مذاهبهم؛ فنقول:

(٢) أمَّا مذهب مليسوس وبَرمِنيدس؛ فإنّا غير محصّلين له، ولا يمكننا أن نصّ على غرضهما فيه، ولا نظنهما يبلغان من السَفَه والغباوة المبلغ الذي يدل عليه ظاهر كلامهما فلهما كلامٌ أيضاً في الطبيعيات وعلى كثرة المباديء لها؛ مثل قول بَرمِنيدس بالأرض والنار وعلى تركيب الكائنات منهما. فيوشك أنْ تكون إشارتهما إلى الموجود هي إلى الموجود الواجب الوجود الذي هو بالحقيقة موجودٌ - كما تعلمه في موضعه - وأنّه غير متناهٍ

that it is what is infinite, immobile, and infinitely powerful, or that it is "finite" in the sense that it is an end at which everything terminates, where that at which [something] terminates is imagined to be finite insofar as [something] terminates at it. Or their aim [could] be something different—namely, that the nature of existence *qua* the nature of existence is a single account in definition and description, and that the other essences are different from the nature of existence itself because they are things, such as humanity, to which [existence] just happens to belong while being inseparable from them (for humanity is an essence that is not itself what exists, but neither does existence belong to it as a part). Instead, existence is something outside of the definition of [humanity], while concomitant with its essence that it happens to have, as we have explained in other places.[6] So it seems that whoever says that [what exists] is finite means that what is defined in itself is not the natures that pass into the many, whereas whoever says that it is infinite means that it happens to belong to infinitely many things.

(3) Now you know very well from other places that the man *qua* man is not what exists *qua* what exists, which, in fact, is something extrinsic to [man *qua* man]. The same holds for any one of the states that fall within the categories; and in fact, anything involving them is a subject for existence, [albeit] the existence is inseparable from it. If, however, this is not their opinion and they obstinately hold the view [that is foolish nonsense], then I actually cannot refute them. That is because the syllogism by which I would refute their view is inevitably composed of premises. Now, those premises either [(1)] must be better known in themselves than the conclusion or [(2)] must be granted by the opponent. As for the first, I do not find anything that is more evident than this conclusion [namely, that existence *qua* existence is different, for example, from what humanity is *qua* humanity]. As for the second, it is not up to me to suggest which of the premises these two should concede, since if they can live with this absurdity, then who is to assure me that they would not unabashedly deny any premise used in the syllogism against them?

6. Cf., for instance, *Kitāb al-madkhal* 1.2, and then later at *Ilāhīyāt* 1.5.

ولا متحرك ، وأنَّه غير متناهي القوة . أو أنَّه متناهٍ على معنى إنَّه غاية ينتهي إليها كلّ شيء ؛ والذي ينتهَى إليه يتخيل أنّه متناهٍ من حيث أنّه ينتهَى إليه . أو يشبه أنْ يكون غرضهما شيئاً آخر ، وهو أنَّ طبيعة الموجود ، بما هي طبيعة الوجود ، معنى واحد بالحدّ أو بالرسم ، وإنَّ سائر الماهيات هي غير نفس طبيعة الوجود ، لأنَّها أشياء يعرض لها الوجود ويلزمها كالإنسانية ؛ فإنَّ الإنسانية ماهيّة وليست نفس الموجود ، ولا الوجود جزء لها ، بل الوجود خارجٌ عن حدّها لا حقٌ لماهيّتها ، كما بيّنا في مواضع أُخرى ، عارض لها . فيشبه أنْ يكون مَنْ قال إنَّه متناهٍ عنى أنَّه محدود في نفسه ليس طبائع ذاهبة في الكثرة ، ومَنْ قال إنَّه غير متناهٍ عنى أنّه يعرض لأشياء غير متناهية .

(٣) وليس يخفى عليك ممّا تعلّمته في مواضع أُخرى إنَّ الإنسان - بما هو إنسان - ليس هو الموجود بما هو موجود ، بل معناه خارجٌ عنه ، وكذلك حال واحدٍ من الأُمور الداخلة في المقولات ، بل كلّ شيء منها موضوع للوجود يلزمه الوجود . فإنْ لم يذهبا إلى هذا وكابرا فليس يمكنني أنْ أناقضهما ؛ وذلك لأنَّ القياس الذي نناقض به مذهبهما يكون لا محالة مؤلَّفاً من مقدمات ، ويجب أنْ تكون تلك المقدمات إمّا في أنفسها أظهر من النتيجة - ولا أجد شيئاً يكون أظهر من هذه النتيجة - أو تكون مسلّمة عند الخصم ، وليس يمكنني أنْ أعرض أي المقدمات يسلّمانها هذان . فإنَّهما إنْ جوّزا إرتكاب هذا المحال ، فمَنْ يؤمنني إقدامهما على إنكار كلّ مقدمة من المقدمات المستعملة في القياس عليهما .

(4) In fact, I find many of the premises by which they are "refuted" less known than the intended conclusion. An example is the statement that if what exists is substance only, it would be neither finite nor infinite (since this is an accident of quantity, and quantity is an accident of substance), and so, in that case, there would be an existent quantity and an existent substance, and so *being existent* would stand above both.[7] Now, if you think about it, you find that the existence of the finite and infinite, is in fact, sufficient for there to be a continuous quantity—namely, the observable magnitude. What we really need to do is to show that the observable magnitude subsists in matter or a subject, not that it is something existing, save in a subject (for this is not known in itself); but then we would need to undertake the difficult task of proving [this premise], which is preparatory to [the conclusion]. So how can this be taken as a premise [used] in drawing a conclusion that is self-evident? The same is true of their claim that what is defined is divisible into the parts of its definition and the like.[8]

(5) As for the remaining groups, let us just gesture at where the problems with their views are, and then, in our subsequent discussions, we will treat the details of their errors more thoroughly. For now, then, the refutation of those who claimed that there is a single principle comes from two sides: one is their statement that the principle is one, and the other is their statement that that principle is water or air. The refutation with respect to that principle's being water or air more naturally comes in the place where we discuss the principles of things subject to generation and corruption[9] rather than [where we discuss] the general principles, since [this group] also assumed that that principle is a principle

7. Cf. Aristotle, *Physics* 1.2.185a20ff.

8. Cf. Aristotle, *Physics* 1.3.186b14ff. and John Philoponus, *In Aristotelis Physicorum,* ed. Hieronymus Vitelli (Berlin: George Reimer, 1887), *ad* 186b14ff. (henceforth, Philoponus, *In Phys.*).

9. Cf. *Kitāb fī al-kawn wa-l-fasād* 3.

(٤) على أني أجد كثيراً من المقدمات التي يناقضان بها أخفى من النتيجة التي تُراد منهـا ، مثل ما يقال إنَّه إنْ كان الموجود جوهراً فقط ؛ فلا يكون متناهياً ولا غير متناهٍ لأنَّ هذا عارض للكمّ، والكـمّ عارضٌ للجوهر ، فيكون حينئذ كمٌّ موجودٌ وجوهرٌ موجود ، فيكـون الموجـود فوق اثنين . وأنت إذا تأملتَ وجدتَ التناهـي وغير التناهي يكفي في تحقّق وجوده أن يكون كمّاً متصلاً ، وهو المقدار المشاهد . وبنا حاجة شديدة إلى أن نبيّن أنَّ المقـدار المشـاهد قائمٌ في مادةٍ وموضوع، وليس موجـوداً إلاّ في موضوع؛ فإنَّ هذا ليس ممّا يبين بنفسه ، بل يحتاج في إبانته إلى تكلّفٍ يُعتد به ، فكيف يؤخذ هذا مقدمة في إنتاج ما هو بيّنٌ بنفسه؟ وكذلك ما قالوا من أنَّ المحدود يتجزأ بأجزاء حدّه، وغير ذلك .

(٥) وأمّا سائر القوم؛ فلننشر إشارة خفيفة في هذا الموضع إلى فساد مذاهبهم ثم في مسـتقبل ما نكتبه كلامٌ يُوقف منه على جليّة الحال في زيغهم وقوفاً شـافياً ، ونقول الآن : أمّا القائلون منهم بأنَّ المبدأ واحد ، فيتوجه إليهم النقْض من وجهين : أحدهما من جهة أنَّهم قالـوا إنَّ المبـدأ واحد ، والثاني من جهة أنَّهم قالوا إنَّ ذلـك المبدأ هو ماء أو هواء . فأمّا النقض عليهم من جهة أنَّ ذلك المبدأ هو ماء أو هواء ، فالأخلق به الموضع الذي نتكلم فيه على مبادىء الكائنات الفاسدات ، لا على المباديء العامة ، فإنّهم وضعوا ذلك المبدأ مبدأ

of things subject to generation and corruption. What suggests that they are in error concerning the principle's being one is that their view would make all things the same in substance, varying [only] in accidents. That is to say that [this group] would eliminate the species-making differences among various bodies, whereas it will become clear to us that bodies do vary through species-making differences. As for those who hold that there is an infinite number of principles from which these generable things are generated, they [themselves] conceded that they have no scientific understanding of the things subject to generation, since their principles are infinite, in which case there is no way to comprehend them scientifically and so grasp what is generated from them. Now, since [purportedly] there is no way to know the things subject to generation, then how could they also know that the principles of [these things] are infinite? As for refuting them with regard to their specific assertion that those infinite things are either atoms scattered in the void[10] or embedded in the mixture,[11] it is again more fitting that we deal with it when we inquire into the principles of things subject to generation and corruption.[12]

(6) Having reached this point, let us conclude this chapter, which was included [almost] by chance; and so whoever wants to retain it, do so, and whoever does not, then do not.

---

10. The view of Democritus and the Atomists generally.

11. The view of Anaxagoras, who posited an infinite number of germs or seeds (*spermata, chrimata*) jumbled together in the mixture. Cf. Aristotle, *On Generation and Corruption* 1.1.314a20ff.

12. Cf. *Kitāb fī al-kawn wa-l-fasād* 4, but also see the more general argument against Atomism presented below at 3.3–5.

للكائنات الفاسدات أيضاً . وأمّا الدلالة على فساد قولهم إنَّ المبدأ واحدٌ ، فهو أنَّ مذهبهم يجعل الأمور كلّها متفقة في الجوهر مختلفة في الأعراض ، ويبطلون مخالفة الأجسام بالفصول المنوّعة ، وسيتّضح لنا أنَّ الأجسام تختلف بالفصول المنوّعة . وأمّا القائلون بأنَّ المبادىء التي تتكون عنها هذه الكائنات غير متناهية ؛ فقد اعترفوا أنَّهم لا علم لهم بالكائنات ، إذ مبادئها غير متناهية فلا يحاط بها علماً ، فلا يحاط بما يتكون عنها . وإذ لا سبيل إلى معرفة الكائنات ، فكيف علموا أيضاً أنَّ مبادئها غير متناهية؟ وأمّا مناقضتهم من جهة تخصيصهم تلك غير المتناهية بأنّها أجزاء لا تتجزأ مبثوثة في الخلاء أو مودعة بالخليط ؛ فالأحرى أنْ نشتغل به ، حيث ننظر في مبادىء الكائنات الفاسدة أيضاً .

(٦) وإذ بلغنا هذا المبلغ ، فلنختم هذا الفصل ، وهذا الفصل داخلٌ في كتابنا بالعرض ، فمن شاء أنْ يثبته أثبته ، ومَنْ شاء أنْ لا يثبته ، فلا يثبته .

Chapter Five

On defining nature

(1) Certain actions or movements occur in the bodies that are immediately present to us. Now, on the one hand, we find that some of those actions and movements proceed from certain external causes that make their occurrence in [the bodies] necessary, as, for example, water's being heated and a stone's rising. On the other hand, we find that other actions and movements proceed from [the bodies] owing to [the bodies] themselves in such a way that they are not traced back to some foreign cause—as, for example, when we heat water and then leave it alone, it cools through its own nature; and when we raise the stone and then leave it alone, it falls through its nature. This belief is also fairly close to our belief that there are plants because of the alteration of seeds and animals because of the generation of semen. Similarly, we find that animals, through their own volition, have a freedom of action in their [various] kinds of movements, [since] we do not see some external agent forcibly directing them to those actions. So, there is impressed upon our souls an image that those [movements], and, on the whole, the actions and passivities that proceed from bodies, are sometime caused by a foreign, external agent and sometimes are a result of the things themselves without an external agent.

<الفصل الخامس>

في تعريف الطبيعة

(١) نقول: إنّه قد يقع عن الأجسام التي قِبَلنا أفعالٌ وحركاتٌ، فنجد بعضها صادرة عن أسباب خارجة عنها توجب فيها تلك الأفعال والحركات، مثل تسخّن الماء وصعود الحجر. ونجد بعضها تصدر عنها لأنفسها من غير أنْ يستند صدورها عنها، إلى سببٍ غريبٍ كالماء؛ فإنّا إذا سخّنّاه ثم خلّينا عنه يبرد بطباعه. والحجر إذا أصعدناه ثم خلّينا عنه يهبط بطباعه. وعسى أنْ يكون ظنّنا بالبذور في استحالتها نباتاً، والنُطف في تكوّنها حيوانات، قريباً من هذا الظّن. ونجد أيضاً الحيوانات تتصرف في أنواع حركاتها بإرادتها، ولا نرى أنَّ قاسراً لها من خارج يصرفها تلك التصاريف. فيرتسم في أنفسنا تخيل أنَّ تلك الحركات - وبالجملة الأفعال والانفعالات الصادرة عن الأجسام - قد تكون بسببٍ خارجٍ غريب، وقد تكون عن ذاتها لا من خارج.

(2) Moreover, we initially deem it possible that [(1)] some of [the motions and actions] that result from the things themselves without an external agent hold to a single course from which they do not deviate, while [(2)] others change[1] and vary their courses. We additionally deem it possible that both cases might be through volition as well as not through volition (and, rather, are like the bruising that arises from a falling stone or the burning from a blazing fire), and so these two are also impressed on our soul. Furthermore, what becomes known to us after diligent [inquiry] is that there are those bodies that we come across [seemingly] without external movers that are [in fact] moved and acted upon only by an external mover that we neither perceive nor recognize. Instead, [that mover] might be some imperceptible separate thing, or perhaps something perceptible in itself but having an imperceptible influence. In other words, there is an imperceptible relation between it and what is acted on by it indicating that it necessitates [the effect]. An example would be anyone who has never sensibly observed a magnet's attracting iron or who does not intellectually recognize that it attracts iron (since the intellect's inquiry [alone] cannot grasp that). In this case, when he sees the iron being moved toward [the magnet], he will most likely suppose that [the iron] is undergoing motion as a result of itself. Whatever the case, it should be obvious that what is producing the motion is not in fact the body *qua* body but, rather, the result of a power in [the body].

---

1. Following **Z**'s *mufannin* (**T**'s *mutafannin,* "to be changed or vary"; Latin *instituta ad multa*), which is omitted in **Y**.

(٢) ثـم الذي يكون عن ذاتها لا من خارج؛ فنحن، في أول الأمر؛ نجوّز أنْ يكون بعضه لازماً طريقة واحدة لا ينحرف عنها، وبعضه يكون مفنن الطرائق، مختلفة الوجوه. ومـع ذلك فنجوّز أنْ يكـون كلّ واحدٍ من الوجهين صادراً بإرادة، وصادراً لا عن إرادة، بل كصدور الرّض عن الحجر الهابط، والإحراق عن النار المشـتعلة؛ فهذا ما يرتسـم في أنفسـنا . ثم ما يدرينا أنْ تكون هذه الأجسـام، التي لا نجد لها محركات من خارج، إنّما تتحرّك وتنفعل عن محرّكٍ من خارج لا ندركه ولا نصل إليه، بل عساه أن يكون مفارقاً غير محسوس، أو عساه أنْ يكون محسوس الذات غير محسوس التأثير، أي غير محسرس النسـبة التي بينه وبين المنفعل عنه الدالّة على أنّه موجبٌ له؛ كمَنْ لم يرَ مَغناطيس يجذب الحديد حسّـاً، أو لم يعرف عقلاً أنَّه جاذبٌ للحديـد - إذ ذلك كالمتعذر إدراكه بطلب العقل - فإذا رأى الحديد يتحرك إليه لم يبعد أنْ يظّن أنه يتحرّك إليه عن ذاته. أي أنّه من الظاهر أنَّ المحرّك لا يصح أنْ يكون جسماً بما هو جسم، وإنَّما يحرّك بقوةٍ فيه.

(3) We set it down as a posit, which the natural philosopher accepts and the metaphysician demonstrates, that the bodies undergoing these motions are moved only as a result of powers in them that are principles of their motions and actions.[2] They include [(1)] a power that brings about motion and change and from which the action proceeds according to a single course, without volition; [(2)] a power like that, with volition; [(3)] a power that, without volition, varies in the motion and action it produces; and [(4)] a power that, with volition, varies in the motion and action it produces. (The same divisions also hold with regard to rest.) The first division is like what belongs to the stone in falling and coming to rest at the center and is called a *nature*. The second is like what belongs to the Sun in its rotations, [at least] according to the view of accomplished philosophers, and is called a *celestial soul*. The third is like what belongs to plants in their generation, growth, and ceasing to grow further (since they involuntarily move in various directions in the form of branching and the spreading of trunks in both breadth and height) and is called a *vegetative soul*. The fourth is like what belongs to animals and is called an *animal soul*. Sometimes the term *nature* is applied to every power from which its action proceeds without volition, in which case the vegetative soul is called a *nature*. Sometimes *nature* is applied to everything from which its action proceeds without deliberation or choice, so that the spider [may be said] to weave by nature; and the same holds for similar animals. The nature by which natural bodies are natural and that we intend to examine here, however, is nature in the first sense.

2. See *Ilāhīyāt* 9.2.

(٣) لكنا نضع وضْعاً يتسلّمه الطبيعي ويبرهن عليه الإلهي؛ أنَّ الأجسام المتحرّكة هذه الحركات إنّما تتحرّك عن قُوى فيها؛ هي مبادىء حركاتها وأفعالها. فمنها قوة تحرّك وتغيّر ويصدر عنها الفعل على نهجٍ واحدٍ من غير إرادة، وقوة كذلك مع إرادة، وقوة متفنّنة التحريك والفعل من غير إرادة، وقوة متفنّنة التحريك والفعل مع إرادة، وكذلك القسمة في جانب السكون. فالأول من الأقسام كما للحجر في هبوطه ووقوفه في الوسط ويسمّى طبيعة، والثاني كما للشمس في دورانها عند محصّلي الفلاسفة، ويسمّى نفساً فلكية، والثالث كما للنباتات في تكوّنها ونشوئها ووقوفها، إذ تتحرك لا بالإرادة حركات إلى جهات شتى، تفريعاً وتشعيباً للأُصول، وتعريضاً وتطويلاً؛ وتسمّى نفساً نباتية، والرابع كما للحيوان وتسمّى نفساً حيوانية. ربّما قيل إسم الطبيعة على كلّ قوةٍ يصدر عنها فعلها بلا إرادة، فتسمى النفس النباتية طبيعة. وربّما قيل طبيعة لكل ما يصدر عنه فعله من غير رويةٍ واختيار، حتى يكون العنكبوت إنّما يشبك بالطباع، وكذلك ما يشبهه من الحيوانات. لكن الطبيعة التي بها الأجسام الطبيعية طبيعية، والتي نريد أنْ نفحص عنها ها هنا هي الطبيعة بالمعنى الأول.

(4) The statement, "The one who seeks to prove [that nature exists] deserves to be ridiculed,"[3] is odd. I suppose [that] what is meant is that the one who seeks to prove it while engaged in investigating the science of physics should be ridiculed, since he is trying to demonstrate the principles of a discipline from within that discipline itself. If this or some other related interpretation is not what was meant, and [if] instead the intention was that this power's existence is self-evident, then it is not something that I am willing to listen to and support. How could it be, when we frequently find ourselves forced to undertake a great deal of preparatory work to prove that every [body] undergoing motion has a mover? How, then, could we ridicule the one who sees a motion and looks for the argument proving that it has a mover, let alone [one who] clearly shows that there is a mover and makes it external? Still, the claim that nature exists is, as a matter of fact, a principle of the science of physics, [and so] it is not up to the natural philosopher to address anyone who denies it. Proving [that natures exists] belongs only to the metaphysician, whereas it belongs to the natural philosopher to study its essence.[4]

(5) Nature has been defined as *the first principle of motion and rest in that to which it belongs essentially rather than accidentally*[5]—not in the sense that in everything there must be a principle of motion and rest together, but in the sense that it is an essential principle of anything having a certain motion, if there is motion, or rest, if there is rest. One who came afterwards found this description inadequate and decided to add to it, claiming that [the initial account] indicated only the nature's action, not its substance, since it indicates only its relation to what proceeds from it.[6] So, to its definition, [he thought,] one must also add the words *nature is a power permeating bodies that provides the forms and temperament, which is a principle of* . . . and so forth. We begin by explaining the

3. Cf. Aristotle, *Physics* 2.1.193a3.

4. Cf. *Ilāhīyāt* 9.5.

5. Aristotle, *Physics* 2.1.192b21–23.

6. Cf. Philoponus, *In Phys. ad* 192b8ff. For a discussion of Philoponus's reinterpretation of Aristotle's definition of nature, see E. M. Macierowski and R. F. Hassing, "John Philoponus on Aristotle's Definition of Nature: A translation from the Greek with Introduction and Notes," *Ancient Philosophy* 8 (1988): 73–100.

(٤) وما أعجب ما قيل إن الباحث عن إثباتها من حقه أن يهزأ به، وأظنّ أن المراد بذلك أن الباحث عن إثباتها – وهو فاحصٌ عن العلم الطبيعي – يجب أنْ يستهزأ به؛ إذ يريد أنْ يبرهنٍ من الصناعة نفسها على مبادئها. وأمّا إنْ لم يرد هذا، أو تأويل آخر مناسب لهذا، بل أُريد أنْ وجود هذه القوة بيّنٌ بنفسه؛ فهو مما لا أُصغي إليه ولا أقول به. وكيف وقد تلزمنا كُلْفة شاقّة في أنْ نثبت أنَّ لكلّ متحرّكٍ محرّكاً، وقد تجشّم ذلك مفيدنا هذه الآراء تجشّماً يُعند به، فكيف يُستهزأ بِمَنْ يرى حركة ويلتمس الحجّة على إثبات محرّكٍ لها فضلاً عن أنْ يسلّم محرّكاً ويجعله خارجاً؟ إلاّ أنَّ الحقَّ هو أنَّ القول بوجود الطبيعة مبدأ العلم الطبيعي، ليس على الطبيعي أنْ يكلّم مَنْ ينكرها، وإنّما إثباتها على صاحب الفلسفة الأولى، وعلى الطبيعي تحقيق ماهيّتها.

(٥) وقد حُدّت الطبيعة بأنّها مبدأ أول لحركة ما يكون فيه وسكونه بالذات لا بالعرض. ليس على أنّها يجب في كلّ شيء أنْ تكون مبدأ للحركة والسكون معاً، بل على أنَّها مبدأ لكلّ أمرٍ ذاتي يكون للشيء من الحركة إنْ كانت، والسكون إنْ كان. ثمَّ بدا لبعض مَنْ ورد من بَعْد أنْ استنقص هذا الرسم، وتوخى أنْ يزيد عليه زيادة فقال: إنَّ هذا إنّما يدلّ على فعل الطبيعة لا على جوهرها، فإنَّه إنّما يدل على نسبتها إلى ما يصدر عنها، ويجب أنْ يزاد في حدّها فيقال إنَّ الطبيعة قوةٌ ساريةٌ في الأجسام تفيد الصور والخلق،

sense of the description taken from the First Master[7] and thereafter turn to whether the addition is worth all this effort, making clear that what [this later philosopher] did was disastrously flawed and that neither it nor his emendation is required.

(6) So we say: The meaning of *principle of motion* is, for instance, an efficient cause from which proceeds the production of motion in another (namely, the moved body), and the meaning of *first* is that it is proximate, with no intermediary between it and the production of the motion. So perhaps the soul is the principle of certain motions of the bodies in which it is, albeit mediately. One group, however, supposed that the soul produces local motion through the intermediacy of the nature. I, however, do not think that the nature is altered so as to become the limbs' mover, obeying the soul contrary to what it itself requires. If the nature were so altered, then it would be able to perform [any] action that the soul imposes upon it, [even if that action] is different from what is proper to [the nature], and what is proper to the soul would never be at odds with what is proper to the nature. If it is meant that the soul brings about a certain inclination and through the inclination produces motion, then the nature does that as well, as we shall make clear to you.[8] It is as if, for instance, this inclination is not a mover, but something through which the mover produces motion. So if the soul has some intermediary in producing motion, then that will not involve the production of local motion, but generation and growth. Now, if this is meant to be a general definition applying to the production of any motion, then *first* is added to it. [That] is because, although soul may be in that which is moved and may produce motion in that in which it is, so as to bring about growth and change, it does not do so as *first,* but rather does so by

7. That is, Aristotle.

8. This is the general thesis of 4.12 below.

هي مبدأ لكذا وكذا. ونحن مبتدئون بإبانة معنى الرسم المأخوذ عن الإمام الأول، ثم نقبل على كفاية هذا المتكلّف للزيادة كلفته موضحين أنَّ ما فعله رديء فاسد غير محتاجٍ إليه ولا إلى بدله.

(٦) فنقول: إنَّ معنى قولنا مبدأ للحركة؛ أي مبدأ فاعلي يصدر عنه التحريكُ في غيره، وهو الجسم المتحرّك. ومعنى قولنا أوّل إنّه قريب لا واسطة بينه وبين التحريك. فعسى أنْ تكون النفس مبدأ لبعض حركات الأجسام التي هي فيها ولكن بوساطة. وقد ظنَّ قومٌ أنَّ النفس تفعل حركة الانتقال بتوسط الطبيعة، ولا أرى الطبيعة تستحيل محرّكة للأعضاء خلاف ما توجبه ذاتها طاعة للنفس. ولو استحالت الطبيعة كذلك، لما حدث الإعياء عند تكليف النفس إياها غير مقتضاها، ولما تجاذب مقتضى النفس ومقتضى الطبيعة. وإنْ عنى بذلك أنَّ النفس تُحدث ميلاً وبالميل تحرّك، فالطبيعة تفعل ذلك أيضاً، على ما سيتضح لك، وكأنَّ مثل هذا الميل ليس هو المحرّك بل أمرٌ به تحرّك المحرّك. فإنْ كان للنفس متوسطٌ في التحريك، فذلك في غير التحريكات المكانية، بل في تحريك الكون والإنماء، وإذا أُريد أنْ يكون هذا الحدّ عاماً لكل تحريك زيد فيه الأول. فإنَّ النفس قد تكون في المتحرّك وتحرّك ما هي فيه، تحريكها الإنماء والإحالة؛ ولكن لا أولاً، بل

using the natures and qualities, which we shall explain to you later.[9] The phrase *in that to which it belongs* is [needed] to distinguish nature, art, and agents that act by force. His use of *essentially* was predicated in two ways, one of which is in relation to the mover [and] the other in relation to what is moved. The first way of predicating it is that the nature produces motion, whenever it is immediately producing motion, owing to itself and not as the result of some agent compelling it to do so by force. So it is impossible that [the nature] not produce a given motion, apart from forced motion, if nothing is hindering it. The second way that [*essentially*] is predicated is that the nature produces motion owing to what is moved of itself and not as a result of some external agent. His use of *accidentally* was also predicated in two ways, one of which is in relation to the nature and the other in relation to what is moved. Now, the way that it is predicated in relation to the nature is that the nature is a principle of that whose motion is real and not accidental, where *accidental motion* is like the motion of one who is standing still on a boat while the boat is moving him. The other way is when the nature moves the statue, and so moves it accidentally, because it essentially produces motion in the copper, not the statue; so the statue *qua* statue is not something moved naturally, like stone is. That is why the knowledge of the physician is not a nature when [the physician] cures himself. It is the medical knowledge in him that produces the change, because [that knowledge] is in him not *qua* patient, but *qua* physician; for when the physician cures himself and so is healed, his being healed is not as a physician, but rather because he is the one who underwent a cure. So he is one thing *qua* one who applies the cure and [another] *qua* one who undergoes the cure, for *qua* one who applies the cure he produces the cure that he knows, whereas *qua* one who undergoes the cure he is a patient who receives the cure.

9. See, for instance, *Kitāb al-nafs* 1.1, where Avicenna provides a general account of the soul's role when it is a principle of natural body.

باســتخدام الطبائع والكيفيات ، ويتبيّن هذا لك بعد . وقوله ما يكون فيه ليفرق بين الطبيعة والصناعة والقاســرات . وأمّا قوله بالذات فقــد حُمل على وجهين : أحدهما بالقياس إلى المحرّك ، والآخــر بالقياس إلى المتحرّك . ووجه حمله على الوجه الأول ؛ أنَّ الطبيعة تحرّك لذاتها حين ما تكون بحال تحريك لا عن تســخير قاســر فيستحيل أنْ لا تحرّك إنْ لم يكن مانع حركة مباينة للحركة القاســرة . وحمله على الوجه الثاني أنَّ الطبيعة تحرّك لما يتحرّك عــن ذاته لا عن خارج. وقوله بالعرض قد حُمل أيضاً على وجهين : أحدهما بالقياس إلى الطبيعة والآخر بالقياس إلى المتحرّك ، ووجه حمله الذي يحمل عليه بالقياس إلى الطبيعة ، أن الطبيعة مبدأ لما كانت حركته بالحقيقة لا بالعرض ، والحركة بالعرض مثل حركة الساكن في الســفينة تحرّكه الســفينة . والوجه الآخر ، أنّه إذا حركت الطبيعة صنماً فهي تحرّكه بالعرض ؛ لأنَّ تحريكها بالذات للنحاس لا للصنم ، فليس الصنم من حيث هو صنمٌ متحركاً بالطبيعــة كالحجر . فلذلك لا يكون الطبيب طبيعة إذا عالج نفســه ، وحرّك الطب ما هو فيــه ، لأنّه فيه لا من حيث هو مريض ، بل من حيث هــو طبيب . فإنَّ الطبيب إذا عالج نفســه فبرىء لم يكن برؤه لانّه طبيب ، بل لأنّه متعالج . فإنّه من حيث هو معالج شــيء ، ومــن حيث هو متعالج ، فإنّه من حيث هو معالــج صانع العلاج عالم به ، ومن حيث هو متعالج قابل للعلاج مريض .

(7) The addition that one of the successors of the Ancients thought to add was done in vain, for the power he took to be like a genus in the description of nature is the active power, which is defined as the principle of motion from another in another as other. Now, the sense of *power* is nothing but a principle of producing motion that is in something, and the sense of *permeating* is nothing but being in something. Also, the sense of *providing the temperament and form* is already included in *producing motion,* and the sense of *preserving the temperament and form*[10] is already included in *producing rest.* If this man had said that nature is a principle existing in bodies so as to move them to their proper perfections and make them rest therein, which is a first principle of motion and rest of what it is in essentially, not accidentally, it would be only a repetition of a lot of unnecessary things. Similarly, when he replaced this phrase with a single term that has the same meaning as that phrase, he had unwittingly repeated a lot of things. Additionally, since this man wanted to correct an alleged defect in this description [namely, that the initial account of *nature* describes it only relative to its actions rather than what it is in itself], he reckoned that when he used *power,* he had indicated a certain entity that is not related to anything. He did not, for nothing more is meant by *power* than a principle of producing rest and motion. Also, *power* is described only with respect to relative association. So his belief that he had escaped that by introducing *power* is not at all true, and so what this man thought is just idle chatter.

---

10. Avicenna has shifted from the standard, philosophical Arabic term for form, *ṣūra* (used in par. 5 when he first introduced Philoponus's position) to the nontechnical term *shakl,* which frequently means just "shape."

(٧) فأمّا الزيادة التي رأى بعض اللاحقين بالأوائل أنْ يزيدها ؛ فقد فعل باطلاً . فإنَّ القوة التي جعلها كالجنس في رسم الطبيعة هي القوة الفاعلة ، وإذا حُدّتْ حُدّتْ بأنَّها مبدأ الحركة من آخر في آخر بأنُّه آخر . وليس معنى القوة إلاّ مبدأ تحريك يكون في الشيء ، وليس معنى السريان إلاّ الكون في الشيء ، وليس معنى التخليق والتشكيل إلاّ داخلاً في معنى التحريك ، وليس معنى حفظ الخَلق والأشكال إلاّ داخلاً في 〈معنى〉 التسكين . ولو كان هذا الرجل قال إنّ الطبيعة هي مبدأ موجودٌ في الأجسام لتحريكها إلى كمالاتها وتسكينها عليها . هو مبدأ أول لحركة ما هو فيه وسكونه بالذات لا بالعرض ، لم يكن إلاّ مكرّراً لأشياء كثيرةٍ من غير حاجةٍ إليها . فكذلك إذا أورد بدل طائفةٍ من كلامه لفظاً مفرداً مواطئاً لتلك الطائفة ، فيكون قد كرّر أشياء كثيرة وهو لا يشعر . ومع ذلك فإنَّ هذا المتدارك لخلل هذا الرسم ، بزعمه قد حسب أنّه إذا قال قوة فقد دلَّ على ذاتٍ غير مضافةٍ إلى شيء ؛ وما فعل . فإنَّ المفهوم من القوة هو مبدأ التحريك والتسكين لا غير ، والقوة لا ترتسم إلاّ من جهة النسبة الإضافية ، فلا يكون ما ظنّه حقاً من أنّه هَرَب من ذلك بإيراد القوة ، فما عمله هذا الرجل باطلٌ فاسد .

(8) Finally, the one who first proposed the definition—namely, that it is a principle of motion and rest—did not mean the principle that belongs to local motion to the exclusion of the principle that belongs to qualitative motion. On the contrary, he meant that every principle of any essential motion whatsoever is a nature, such as motion in [the categories of] quantity, quality, place, and any other, if there is such. (The kinds of motion will be explained to you later.)[11] So a principle of motion with respect to quality is the nature's state that determines either an increase of rarefaction and extension in the volume or a condensation and contraction in the volume, since this produces a motion from one quantity to another. If you wish to make augmentation natural and apply the term *nature* to it, taking nature in one of the aforementioned senses, then do so. A principle of motion with respect to quality is like the state of water's nature when the water accidentally acquires some foreign quality that is not proper to its nature (coolness being proper to its nature), and then, when the impediment is removed, its nature returns and transforms it into its proper quality and preserves it therein. Similarly, when the humoral mixtures of bodies deteriorate, once their nature becomes strong, it returns them to the proper humoral balance. The case with respect to place is obvious—namely, like the state of the stone's nature when it moves it downward, and the state of fire's nature when it moves it upward. A principle of motion with respect to substance is like nature's state that brings about motion toward the form, being prepared by the modification of quality and quantity, as you will learn.[12] It might be the case, however, that the nature does not actually bring about the form but is only disposed to it, acquiring it from elsewhere. This, however, is more fittingly learned in another discipline.[13]

(9) This, then, is the definition of *nature,* which is like the generic [sense] and provides each of the natures beneath it with its meaning.

---

11. See 2.3.

12. Cf. 2.3, where he discusses substantial change, and *Kitāb al-ḥayawān* 9.5, where much of the same material is treated again in more detail, albeit specifically in relation to substantial changes during prenatal development.

13. "Another discipline" certainly refers to metaphysics, and the reference seems be to *Ilāhīyāt* 9.5, where the "Giver of Forms" (*wāhib al-ṣuwar*) is discussed.

(٨) ثم معنى قول الحادّ الأول إنّه مبدأ للحركة والسكون؛ ليس يعني المبدأ الذي للحركة المكانية دون المبدأ الذي للحركة في الكيف، بل يعني أن كلّ مبدأ لأي حركةٍ كانت بالذات فهو طبيعة. كالمبدأ للحركة التي في الكم والتي في الكيف والتي في المكان وفي غير ذلك - إن كان حركة - وسيتضح لك بعد أصناف الحركات. فأمّا كونها مبدأ للحركة في الكمّ؛ فهو حال الطبيعة الموجبة لزيادة تخلخلٍ وانبساطٍ في الحجم، أو تكاثفٍ وانقباضٍ في الحجم؛ فإنَّ هذا تحريكٌ عن كميّة إلى كميّة. وإنْ شئت أنْ تجعل النمو بالطبيعة وتطلق إسم الطبيعة على ذلك، وتأخذ الطبيعة على أحد المعاني المذكورة، فافعل. وأمّا كونه مبدأ للحركة في الكيف؛ فمثل حال طبيعة الماء إذا عرض للماء أنْ استفاد كيفية غريبة لم تكن مقتضى طبيعة - لكون البرودة مقتضى طبيعته - فإنَّ العائق إذا زال ردّته طبيعته إلى كيفيته وأحالته إليها وحفظته عليها. وكذلك الأبدان إذا ساءتْ أمزجتها وقويت طبيعتها؛ ردّتها إلى المزاج الموافق. وأمّا في المكان فظاهرٌ، وهو مثل حال طبيعة الحجر إذا حركته إلى أسفل، وحال طبيعة النار إذا حركتها إلى فوق. وأمّا كونه مبدأ للحركة في الجواهر؛ فمثل حال الطبيعة التي تحرّك إلى الصورة مُعدّة بإصلاح الكيف والكم، على ما تعلم. وأمّا حصول الصورة فعسى أنْ لا تكون الطبيعة مفيدتها، بل تكون مهيئة لها، وتُستفاد من مواضع أُخر، والأَوْلى أنْ يُعلم هذا من صناعةٍ أُخرى.

(٩) فهذا هو حدّ الطبيعة التي كالجنسية، ويُعطى كلّ واحدٍ من الطبائع التي تحتها معناها.

Chapter Six

On nature's relation to matter, form, and motion

(1) Every body has a nature, form, matter, and accidents. Its nature, again, is the power that gives rise to its producing motion and change, which are from [the body] itself, as well as its being at rest and stable. Its form is its essence by which it is what it is, while its matter is the thing bearing its essence. Accidents are those things that, when [the body's] form shapes its matter and completes its specific nature, either necessarily belong to it as concomitants or accidentally belong to it from some external agent.

(2) In some cases, the nature of the thing is just its form, whereas in others it is not. In the case of the simples [that is, the elements], the nature is the very form itself, for water's nature is [for example] the very essence by which it is water. Be that as it may, it is a nature only when considered in one way, whereas it is a form when considered in another. So when it is related to the motions and actions that proceed from it, it is called *nature;* whereas, when it is related to its bringing about the subsistence of the species water, and if the effects and motions that proceed from it are not taken into account, it is then called *form.* So the form of water, for instance, is a power that makes the water's matter to subsist as a species—namely, water. The former [namely, the nature] is imperceptible, but the effects that proceed from it are perceptible—namely, perceptible coolness and weight (which is actually the inclination and does not belong to the body while it is in its natural location). So the nature's action in, for example, the substance of water is either relative to its passive influence and so is coolness; or is relative to its active influence, giving it its shape, and so is wetness; or is relative to its proximate place and so brings about motion; or is relative to its proper place and so brings about rest. Now, this coolness and wetness are necessary accidents of this nature, given that there is no impediment. Not all accidents in

<الفصل السادس>

في نسبة الطبيعة إلى المادة والصورة والحركة

(١) إنَّ لكل جسم طبيعة وصورة ومادة وأعراضاً، فطبيعته هي القوة التي يصدر عنها تحرّكه أو تغيّره الذي يكون عن ذاته، وكذلك سكونه وثباته، وصورته هي ماهيّته التي بها هو ما هو، ومادته هي المعنى الحامل لماهيّته، والأعراض هي الأمور التي إذا تصورت مادته بصورته وتمّت نوعيته، لزمته أو عرضت له من خارج.

(٢) وربّما كانت طبيعة الشيء هي بعينها صورته، وربّما لم تكن. أمّا في البسائط؛ فإنَّ الطبيعة هي الصورة بعينها، فإنَّ طبيعة الماء هي بعينها الماهيّة التي بها الماء هو ماء، لكنها إنّما تكون طبيعة باعتبار، وصورة باعتبار: فإذا قيست إلى الحركات والأفعال الصادرة عنها سميت طبيعة، وإذا قيست إلى تقويمها لنوع الماء - وإنْ لم يُلتفت إلى ما يصدر عنها من الآثار والحركات - سُميت صورة. فصورة الماء مثلاً قوة أقامت هيولى الماء نوعاً هو الماء، وتلك غير محسوسة وعنها تصدر الآثار المحسوسة من البرودة المحسوسة والثقل الذي هو الميل بالفعل الذي لا يكون للجسم وهو في حيّزه الطبيعي، فيكون فعل الطبيعة مثلاً في جوهر الماء. أما بالقياس إلى المتأثّر عنه فالبرودة، وأمّا بالقياس إلى المؤثر فيه المشكل له فالرطوبة، وأمّا بالقياس إلى مكانه القريب فالتحريك، وبالقياس إلى مكانه المناسب فالتسكين. وهذه البرودة والرطوبة أعراضٌ تلزم هذه الطبيعة إذا لم يكن هناك عائق. وليس كلّ الأعراض تتبع الصورة في الجسم، بل ربّما كانت الصورة

the body follow upon the form; and in fact, frequently the form is something that prepares the matter in order that it be acted upon by some cause from an external agent that is accidental, just as it is prepared to receive artificial as well as numerous natural accidents.

(3) In the case of composite bodies, the nature is something like the form but not the true being of the form. [That] is because composite bodies do not become what they are by a power belonging to them that essentially produces motion in a single direction, even if they inevitably have those powers inasmuch as they are what they are. So it is as if those powers are part of their form and as if their form is a combination of a number of factors, which then become a single thing. An example would be humanness, since it includes the powers of nature as well as the powers of the vegetative, animal, and rational soul; and when all of these are in some way "combined," they yield the essence of humanness. (The particulars of this manner of combining are more fittingly explained in first philosophy.)[1] If, however, we do not intend *nature* in the sense that we defined it but instead mean anything from which something's activities proceed in whichever way it by chance may be, whether according to the previously mentioned condition of nature or not, then perhaps the nature of each thing is its form. Our present intention in using the term *nature,* however, is the definition that we previously gave.[2]

(4) Some of those accidents happen to be from an external agent. Others accidentally occur from the thing's substance, some of which might follow upon the matter—such as, for example, the blackness of the Negro, the scars left by wounds, and standing upright. [Still] others frequently follow upon the form—such as wit, mirth, risibility, and the like in humans (for even if the existence of risibility inevitably requires

1. See *Ilāhīyāt* 2.2–4.

2. See 1.5.3: "a power that brings about motion and change and from which the action proceeds according to a single course, without volition."

معدّة للمادة لأنْ تنفعل عن سببٍ من خارجٍ يعرض، كما تعدّ لقبول الأعراض الصناعية ولكثيرٍ من الأعراض الطبيعية.

(٣) وأمّا في الأجسام المركّبة. فالطبيعة كشيءٍ من الصورة ولا تكون كُنْه الصورة؛ فإنَّ الأجسام المركّبة لا تصير هي ما هي بالقوة المحركة لها بالذات إلى جهةٍ وحدها، وإنْ كان لا بدّ لها في أن تكون هي ما هي من تلك القوّة. فكأن تلك القوى جزء من صورتها، وكأن صورتها تجتمع من عِدّة معانٍ فتتحد كالإنسانية؛ فإنّها تتضمن قوى الطبيعة وقوى النفس النباتية والحيوانية <والناطقة>، وإذا اجتمعت هذه كلّها نوعاً من الاجتماع أعطت الماهيّة الإنسانية. وأمّا كيفية نحو هذا الاجتماع فالأولى أنْ يبيّن في الفلسفة الأولى اللّهم إلاّ أنْ نعني بالطبيعة لا هذا الذي حدّدناه، بل كل ما يصدر عنه أفاعيل الشيء، على أي نحو اتفق وكان، على الشرط المذكور في الطبيعة، أو لم يكن. فعسى أن تكون طبيعة كلّ شيء صورته، ولكن غرضنا ها هنا في إطلاق اسم الطبيعة هو ما حدّدناه.

(٤) ومن هذه الأعراض ما يعرض من خارج، ومنها ما يعرض من جوهر الشيء. وقد يتبع بعضها المادّة كالسواد في الزنجي، وآثار القروح، وانتصاب القامة، وقد يتبع بعضها الصورة كالذكاء والفرح وقوة الضحك، وغير ذلك، في الإنسان. وقوة الضحك - فإنَّ هذه وإنْ لم يكن بُدّ في وجودها عن أنْ تكون مادة موجودة - فإنَّ انبعاثها من الصورة

matter's existence, it originates and begins with the form). You will also discover that some accidents that necessarily follow upon the form (whether originating from it or accidentally occurring owing to it) do not require the participation of matter, which the science of psychology will verify for you.[3] Some accidents jointly begin owing to both [matter and form], like being asleep and being awake, although some of them are closer to the form, like being awake,[4] whereas others are closer to the matter, like being asleep. The accidents that follow on the part of the matter might remain after the form, such as the scars caused by wounds or the Ethiopian's blackness when he dies.

(5) So the true nature is that to which we have gestured, where the difference between it and form is what we have indicated, and the difference between it and motion is even all that much more obvious. Still, the term *nature* might be used in many ways, three of which we shall mention [as] deserving that title most. So [(1)] *nature* is said of the principle, which we mentioned;[5] [(2)] *nature* is said of that by which the substance of anything subsists; and [(3)] *nature* is said of the very being of anything. Now, when by *nature* one means *that by which the substance of anything subsists,* there will inevitably be differences of opinion about it according to the various schools of thought and beliefs. So whoever thinks that the part is more entitled than the whole substance to be [considered as] that which makes it to subsist—namely, its [elemental] component or material—will say that anything's nature is its [elemental] component. Whoever thinks that the form is worthier of that will make [form] the thing's nature. Among the speculative thinkers,[6] there might even be a group who supposed that motion is the first principle providing substances with their subsistence. Now, whoever thinks that anything's nature is its form will, in the case of simple substances, make it their simple essence and, in the case of composites, make it the [elemental or

3. The reference may be to *Kitāb al-nafs* 2.2, where Avicenna discusses the role of material accidents in making things particular and then the degrees of abstraction from matter that are involved in the various kinds of perception (*idrāk*).

4. Reading with **Z**, **T**, and the Latin equivalent *wa-in kāna qad yakūnu baʿḍuhā aqrab ilá al-ṣūrah mithl al-yaqaẓah,* corresponding with "although some . . . being awake," which is (inadvertently) omitted in **Y**.

5. See chapter 1.5.5–6.

6. The locution *ahl al-baḥth* may be a synonym for *ahl al-naẓar,* in which case the reference would be to the Islamic speculative theologians.

وابتداءها منها . وستجد أعراضاً تلزم الصورة وتنبعث منها أو تعرض لها بوجهٍ آخر لا تحتاج إلى مشاركة المادة؛ وذلك إذا حُقّق لك علم النفس . وقد تكون أعراضٌ مشتركةٌ تبتدىء من الجهتين جميعاً كالنوم واليَقَظَة ، وإن كان قد يكون بعضها أقرب إلى الصورة مثل اليقظة وبعضها أقرب إلى المادة مثل النوم . والأعراض اللاحقة من جهة المادة ، قد تبقى بعد الصورة كأنداب القروح وسواد الحبشي إذا مات .

(٥) فالطبيعة الحقيقية هي التي أومأنا إليها ، والفرق بين الصورة وبينها ما أشرنا إليه ، والفرق بين الحركة وبينها أظهر بكثير . لكن لفظ الطبيعة قد يستعمل على معانٍ كثيرة؛ أحقّ ما نذكر منها هو ثلاثة: فيقال طبيعة للمبدأ الذي ذكرناه ، ويقال طبيعة لما يتقوّم به جوهر كلّ شيء ، ويقال طبيعة لذات كلّ شيء . وإذا أُريد بالطبيعة ما يتقوّم به جوهر كلّ شيء حُقّ أنْ يُختلف فيها بحسب اختلاف المذاهب والآراء ؛ فَمَنْ رأى أنْ يجعل الجزء الأحق من كلّ جوهر بأنَّ تقوّمه هو عنصره وهيولاه ، قال إنَّ طبيعة كلّ شيء عنصره ، ومَنْ رأى أنْ يجعل الصورة أحرى بذلك جعلها طبيعة للشيء . وعسى أنْ يكون في أهل البحث قومٌ ظنّوا أنَّ الحركة هي المبدأ الأول لإفادة الجواهر قواماتها ، فجعلوها طبيعة كل شيء . ومَنْ جعل طبيعة كل شيء صورته جعلها في البسائط ماهيّتها البسيطة ، وفي المركبات المزاج.

humoral] mixture. (Although you will later learn what *mixture* is, for now we will just quickly point you in the right direction.[7] So we say that that *mixture* is the quality resulting from the interaction of contrary qualities in neighboring bodies.)

(6) The earliest of the Ancients were quite ardent in giving preference and support to matter and making it nature. Among them was Antiphon, whom the First Teacher [that is, Aristotle] mentioned, relating that he insisted that matter is the nature and that it is what makes substances subsist.[8] [He defended this by] saying that if the form were the nature in the thing, then when a bed decomposes and reaches the point where it would sprout forth branches and grow, it would sprout forth a bed. That, however, is not the case, and instead it reverts to the nature of the wood, and wood grows. It is as if this man thought that nature is the matter—but not just any matter, but, rather, whatever is itself preserved through every change, as if he had not distinguished between the artificial form and nature. In fact, he did not even distinguish between what is accidental and the form, not recognizing that what makes something subsist must inevitably be present while the thing exists, not that it is what must inevitably be present when the thing ceases to exist, which does not separate but remains even when the thing ceases to exist. What need have we for something that remains during changes but whose existence is not enough actually to result in something? This is like the material, which does not provide the actual existence of anything but, rather, provides only its potential existence, whereas it is in fact the form that actually makes it [exist]. Don't you see that when the wood and bricks exist, then the house has a certain potential existence; however, it is its form that provides it with its actual existence to the extent that, were it possible for its form to subsist without the matter, then one could do away with [the matter]? Moreover, it escaped this man's notice that the woodiness is a form and that when there is growth, [this form] is being preserved. So, if the important thing for us to bear in mind when considering the conditions for something's being a nature is that it provide the thing with its substantiality, then the form deserves that [title] most.

---

7. See, for instance, 1.10.7 and *Kitāb fī al-kawn wa-l-fasād* 6–7.
8. Cf. Aristotle, *Physics* 2.1.193a12–17.

وستعلم بعد أنَّ المزاج ما هو، ونرشدك إليه الآن يسيراً فنقول: إنَّ المزاج هو كيفيّة تحصل من تفاعل كيّفيات متضادّة في أجسام متجاورة.

(٦) وقد كان الأقدمون من الأوائل شديدي الشغف بتفضيل المادة والقول بها وتصيّرها طبيعة، ومنهم أنْطيفن الذي يذكره المعلم الأول؛ ويحكي عنه أنَّه أصرّ على أنَّ المادة هي الطبيعة وأنّها هي المقوّمة للجواهر. ويقول لو كانت الصورة هي الطبيعة في الشيء لكان السرير إذا عَفَن وصار بحيث يُفرع غصناً وينْبته فرّع سريراً؛ وليس كذلك، بل يرجع إلى طبيعة الخشبية فينبت خشباً، كأن هذا الرجل رأى أنَّ الطبيعة هي المادة، ولا كلّ مادة، بل المحفوظ ذاتها في كل تغيّر، وكأنه لم يفرّق بين الصورة الصناعية وبين الطبيعة بل لم يفرّق بين العارض وبين الصورة، ولم يعرف أنَّ مقومّ الشيء يجب أنْ لا يكون منه بُدّ عند وجود الشيء، ليس أنَّه الذي لا بُدّ منه عند عدم الشيء ولا انفكاك ويكون ثابتاً، أو يكون ثابتاً عند عدم الشيء. وما يغنينا أنْ يكون الشيء ثابتاً في الأحوال، ووجوده لا يكفي في أنْ يحصل الشيء بالفعل؛ مثل هذا الذي هو الهيولى التي لا تفيد وجود الشيء بالفعل، بل إنّما تفيد قوّة وجوده، بل الصورة هي التي تجعله بالفعل. ألا ترى أنَّ الخشب واللّبن إذا وُجدا كان للبيت وجودٌ ما بالقوة، ولكن وجوده بالفعل مستفادٌ من صورته، حتى لو جاز أنْ تقوم صورته لا في المادة لا ستغنى عنها. وهذا الرجل ذهب عليه أيضاً أنَّ الخشبية صورة، وأنَّها عند الإنْبات محفوظة. فإنْ كان الذي يهمّنا في مراعاة شرائط كون الشيء طبيعة؛ هو أنْ تكون مفيدة للشيء جوهريته، فالصورة أولى بذلك.

(7) Now, since simple bodies are actually what they are through their forms and not through their matter (otherwise they would not differ), then clearly, nature is not the matter. In simple substances, it is the form, and it is a certain form in itself, not a certain matter. As for composites, you are well aware that the defined nature and its definition do not yield their essences but are together with certain additional factors; nevertheless, their perfecting forms are synonymously called *nature,* in which case *nature* is predicated in common of both this case and the first one. As for motion, it is the farthest removed from the nature of things, for, as will become clear,[9] it arises in the case of deficiency and is foreign to the substance.

9. See 4.9.5.

(٧) ولما كانت الأجسام البسيطة هي ما هي بالفعل بصورها ، ولم تكن هي ما هي بموادّها ، وإلاّ لما اختلفت ، فبيّنٌ أنَّ الطبيعة ليست هي المادة ، وأنّها هي الصورة في البسائط ، وأنها في نفسها صورة من الصور ليست مادة من المواد . وأمّا في المركبات فغير خافٍ عليك أنَّ الطبيعة المحدودة وحدّها لا يعطي ماهيّاتها ، بل هي مع زوائد . إلاّ أنْ تسمى صورتها الكاملة طبيعة على سبيل الترادف ، فتكون الطبيعة تقال حينئذ على هذه وعلى الأُولى بالاشتراك . وأمّا الحركة فهي أبعد من أنْ تكون طبيعة للأشياء ، فإنَّها - كما يتضح - طارئة ، في حال النقص ، وغريبة عن الجوهر .

Chapter Seven

Of certain terms derived from nature and an explanation of their status

(1) Here are some terms used: *nature, natural, what has a nature, what is by nature, what is naturally,* and *what follows the natural course. Nature* has already been defined;[1] and as for the *natural,* it is whatever is related to nature. Now, whatever is related to nature is either that in which there is the nature or that which is from the nature. *That in which there is the nature* is either that which is informed by the nature or that which is the nature, like a part of its form. *That which is from the nature* is effects and motions as well as things of the same kind falling under place, time, and the like. That which in itself has something like this principle is *what has a nature*—namely, the body that is moved and is at rest by its natural dispositions. As for *what is by nature,* it is anything whose actual existence or actual subsistence is from the nature, whether existing primarily, like natural individuals, or existing secondarily, like natural species.[2] *What is naturally* is whatever necessarily follows upon the nature, however it might be, whether as resembling the intention (such as the individuals and species of substances) or [as] its necessary concomitant (such as necessary and[3] incidental accidents). *What follows the natural course* is, for example, the motions and rests that the nature of itself necessitates essentially and that do not lie outside of what is proper to it. Now, *what lies outside what is proper to it* sometimes results from some foreign cause and sometimes is from [the nature] itself through some cause receptive to its action, namely the matter. So, [for example], the oversized head and additional finger do not follow the natural course, and yet they [occur] naturally and are by nature, since their cause is the nature, albeit not of itself but only accidentally, namely [because] the matter is in a certain state with respect to its quality and quantity so as to be susceptible to that.

1. See 1.5.5–6.

2. The text's literal "by the first existence" and "by the second existence" are probably references to *Kitāb al-madkhal* 1.2, where Avicenna distinguishes two existences: the first is in concrete particulars and the second is in conceptualization.

3. Reading *wa,* which is omitted in **Y**, with **Z** and **T**; the Latin reads *aut,* which corresponds with the Arabic *au* (or).

<الفصل السابع>

في ألفاظ مشتقة من الطبيعة وبيان أحكامها

(١) ها هنا ألفاظٌ تُستعمل فيقال: الطبيعة، والطبيعي، وما له الطبيعة، وما بالطبع، وما يجري المجرى الطبيعي. فالطبيعة قد عرفتها، وأمّا الطبيعي فهو كل منسوبٍ إلى الطبيعة، والمنسوب إلى الطبيعة هو إمّا ما فيه الطبيعة، وإمّا ما عن الطبيعة، والذي فيه الطبيعة فالمتصوّر بالطبيعة أو الذي الطبيعة كالجزء من صورته. وأمّا ما عن الطبيعة فالآثار والحركات وما يجانس ذلك من المكان والزمان وغيره. وأمّا ما له الطبيعة؛ فهو الذي له في نفسه مثل هذا المبدأ؛ وهو الجسم المتحرّك بطباعه والساكن بطباعه. وأمّا ما بالطبيعة فهو كلّ ما وجوده بالفعل عن الطبيعة، أو قوامه بالفعل عن الطبيعة بالوجود الأول كالأشخاص الطبيعية، أو بالوجود الثاني كالأنواع الطبيعية. وأمّا ما بالطبع؛ فهو كلّ ما يلزم الطبيعة كيف كان - على مُشاكلة القصد كالأشخاص والأنواع الجوهرية، أو لازماً لها كالأعراض اللازمة والحادثة. وأمّا ما يجري المجرى الطبيعي؛ فمثل الحركات والسكونات التي توجبها الطبيعة بنفسها لذاتها لا خارجة عن مقتضاها. والخارج عن مقتضاها ربّما كان بسببٍ غريب، وربّما كان عنها نفسها بسببٍ قابل فعلها وهو المادة. فإنّ الرأس المُسَفَّط والإصبع الزائدة ليسا جاريين على المجرى الطبيعي، ولكنهما بالطبع وبالطبيعة، إذْ سببهما الطبيعة، ولكن ليس لنفسها بل عارضٍ؛ وهو كون المادة بحالٍ - في كيفيتها أو كميّتها - تقبل ذلك.

(2) *Nature* is predicated in the manner of a particular and a universal. That which is predicated in the manner of a particular is the nature proper to each of the individuals, whereas the nature that is predicated in the manner of a universal is sometimes a universal relative to a species and sometimes a universal absolutely. Neither of these has an existence in concrete particulars as subsisting entities, except[4] in conceptualization. In fact, however, only the particular has existence. The first of the two [universals] is what our intellects recognize as a principle proper to the management necessary for the conservation of a species, whereas the second is what our intellects recognize as a principle proper to the management necessary for the conservation of the universe according to its order.

(3) Some had supposed that each of these two is a certain existing power: the first permeating the individuals of the species, and the other permeating the universe.[5] Others supposed that each one of these, [considered] in itself as an emanation from the first principle, is one, but is divided by the divisions of the universe, varying with respect to the recipients.[6] None of this should be listened to, since only the various powers that are in recipients exist, and they were never united and thereafter divided. Certainly, they have some relation to a single thing; but the relation to the single thing, which is the principle, does not eliminate the essential difference resulting from the things, nor do the things that result from the relation [namely, the universals] subsist separately in themselves. In fact, nature in this sense has no existence, neither in

4. Reading *illā* with **Z**, **T**, and the Latin (*nisi*), which has been (inadvertently) omitted in **Y**.

5. It is not clear what the source is for the present position or that of the next sentence. Neither position appears in Aristotle's *Physics* nor in what I have seen in the extant commentaries on the *Physics* available in Arabic. The first position has certain similarities in content and terminology with Alexander of Aphrodisias's *The Principles of the Universe in Accordance with the Opinion of Aristotle;* see ed. and trans., Charles Genequand, *Alexander of Aphrodisias on the Cosmos,* Islamic Philosophy, Theology and Science Texts and Studies 44 (Leiden: E. J. Brill, 2001), §§278ff.

6. This position has certain affinities with that of Plotinus and the Neoplatonists; see Plotinus, *Enneads* 3.2. Concerning PlotinPus as a possible source, it should be noted that a redaction of his Enneads 4–6 was made in Arabic under the title *The Theology of Aristotle,* for a discussion of the *Theology of Aristotle* see Peter Adamson, *The Arabic Plotinus, A Philosophical Study of the "Theology of Aristotle"* (London: Duckworth, 2002).

(٢) والطبيعة تقال على وجه جزئي، وتقال على وجه كلّي؛ فالتي تقال على وجه جزئي هي الطبيعة الخاصّة لشخص شخص، والطبيعة التي تقال بوجه كلّي فربّما كانت كلّية بحسب نوعٍ وربّما كانت كلّية على الإطلاق، وكلاهما لا وجود لهما في الأعيان ذواتاً قائمة إلا في التصوّر، بل لا وجود إلّا للجزئي. أمّا أحدهما فهو ما نعقله من مبدأ مقتضى التدبير الواجب في استحفاظ نوع، والثاني ما نعقله من مبدأ مقتضى التدبير الواجب في استحفاظ الكل على نظامه.

(٣) وقد ظنّ بعضهم أنّ كل واحدٍ منهما قوة موجودة؛ أمّا الأولى فسارية في أشخاص النوع، والأخرى فسارية في الكل. وظنّ بعضهم أنّ كل واحدٍ منهما هو في ذاته، وفيضانه عن المبدأ الأول واحدٌ، ومنقسمٌ بانقسام الكل، ويختلف في القوابل. وليس من هذا شيءٌ يجب أن يُصغى إليه، فإنّه لا وجود إلّا للقوى المختلفة التي في القوابل، ولم تكن البتة متحدة ثم انقسمت. نعم لها نسبة إلى شيء واحد؛ والنسبة إلى الشيء الواحد - الذي هو المبدأ - لا ترفع الاختلاف الذاتي عن الأشياء، ولا تقوّم المنسوبات مجردة بأنفسها، بل لا وجود للطبيعة بهذا المعنى؛ لا في ذات المبدأ الأول - فإنّه من

the First Principle itself (for it is impossible that there be in it itself anything other than it itself, as you will learn)[7] nor in the manner of the procession to other things, as if it were an emanation but has not yet arrived. It has no existence in things as some uniform thing without difference; rather, everything's nature is something either specifically or numerically different. Moreover, the example they give of the Sun's shining is not at all like that, since nothing that subsists separately departs from the Sun, neither a body nor an accident. Quite the contrary, its ray comes to be in the recipient and in every other numerically different recipient. It is neither the case that that ray exists in anything other than the recipient nor that some part of the whole ray of the Sun's substance has sunk down toward and then spread over bits of matter. It is true that, if there were not different recipients, but only one, there would in that case be only one effect. (The confirmation of all of this will be explained to you in another discipline.)[8] If there were a universal nature of this kind, however, it would not be *qua* nature, but, rather, *qua* intelligible object vis-à-vis the first principles from which the management of the universe emanates, or *qua* nature of the first of the heavenly bodies through whose mediation the order [of the universe] is conserved. There simply is no nature of a single essence permeating different bodies. So, it is in this way that you must conceive the universal and particular nature.

---

7. Cf. *Ilāhīyāt* 8.3–5.

8. Cf. *Kitāb al-nafs* 3.2, where he critiques the view that rays of light move and so are bodies.

المحال أنْ يكون في ذاته شيء غير ذاته، كما ستعلم - ولا في طريق السلوك إلى الأشياء كأنَّه فائض، لكنه بعد لم يصل ولا له وجودٌ في الأشياء متحدّاً بلا اختلاف، بل طبيعة كلّ شيء، شيء آخر تقوم بالنوع أو بالعدد. ولا أيضاً ما يمثّلونه من شروق الشمس كذلك، فإنَّ الشمس لا ينفصل عنها شيء يقوم واحداً لا جسم ولا عرض، بل إنّما يحدث شعاعها في القابل، ويحدث في كل قابل آخر بالعدد، وليس لذلك الشّعاع وجودٌ في غير القابل، ولا هو شيء من جملة شعاع جوهر الشمس قد انحدر منه إلى المواد فغشيها. نعم، لو لم يختلف القابل، وكان واحداً، لكان الأثر واحداً بحسبه حينئذ. ويتبيّن لك تحقيق هذا كله في غير هذه الصناعة. لكن إنْ كانت طبيعة كلّية من هذا الجنس؛ فلا تكون على أنَّها طبيعة، بل على أنَّها أمرٌ معقولٌ عند الأوائل والمبادىء التي منها يفيض تدبير الكلّ، أو على أنَّها طبيعة جرم أول من الأجرام السماوية الذي يتوسطه يستحفظ النظام، ولا تكون البتة طبيعة واحدة الماهيّة سارية في الأجسام الأخرى، فهكذا يجب أن تُتصور الطبيعة الكلّية والجزئية.

(4) Next, you know that what lies outside the natural particular course frequently does not lie outside the natural universal course; for even if death is not what is intended with respect to the particular nature that is in Zayd, it is in certain ways what is intended with respect to the universal nature. One of these ways is that [death] frees the soul from the body for the sake of flourishing among the blessed, which is [the soul's] aim and for which the body was created, and should [the soul] fail to achieve that, it is not because of the nature, but owing to evil choice. Another [way that death is something that the universal nature intends] is that other people deserve a share in existence just like this individual; for if the former ones did not die, there would not be space and food to go around for the latter ones. Also, those latter ones—namely, the ones deserving a share of something like this existence—have something [almost] owed them on the part of matter's potential, [since] they no more deserve perpetual nonexistence than the former deserve never to die. So this and others are certain things intended by the universal nature. The same is true of the additional finger, since it is something intended by the universal nature, which requires that any matter that is prepared for some form receive it and that [that form] not be hindered; so when there is excessive matter deserving the form of fingerness, it will not be denied and wasted.

(٤) ثم تعلم أنَّ كثيراً مما هو خارج عن مجرى الطبيعة الجزئية ليس بخارج عن مجرى الطبيعة الكلّية. فإنَّ الموت - وإنْ كان غير مقصودٍ في الطبيعة الجزئية التي في زيد - فهو مقصودٌ في الطبيعة الكلّية من وجوهٍ: منها أنْ تتخلّص النفس عن البدن للسعادة في السعداء وهي المقصودة ولها خُلِقَ البدن، وإذا أخلفت فليس بسببٍ من الطباع، بل لسوء الاختيار. ومنها أنْ يكون لقوم آخرين حالهم في استحقاق الوجود؛ حال هذا الشخص وجوداً، فإنَّه إنْ خَلُدَ هؤلاءً لم يسع الآخرين مكانٌ ولا قوت. وفي قوة المادة فضلٌ للآخرين وهم يستحقون مثل هذا الوجود، وليسوا أولى بالعدم الدائم من هؤلاء بالخلود، فهذه - وغيرها - مقاصد في الطبيعة الكلّية. وكذا الإصبع الزائدة فهي مقصودة في الطبيعة الكلّية التي تقتضي أنْ تكسى كل مادةٍ ما تستعد لها من الصورة ولا تعطّل، فإذا فضلتْ مادة تستحق الصورة الإصبعية لم تُحرم ولم تضيع!

Chapter Eight

On how the science of physics conducts investigation and what, if anything, it shares in common with the other sciences

(1) Since nature and issues related to nature have been defined, it will have become abundantly clear to you which things physics investigates. Now, since delimited magnitude is among the necessary concomitants of this natural body and its essential attributes (I mean the length, breadth, and depth to which one can point) and [since] shape is among the necessary concomitants of magnitude, then shape is also among the accidents of the natural body. Since, however, the subject of the geometer is magnitude, his subject is one of the accidents pertaining to the natural body, and the accidents that he investigates fall under the accidents of this accident [namely, magnitude]. In this way, geometry is, in a certain way, a part of natural science, albeit pure geometry and natural science do not share in common the [same] set of questions.[1] Arithmetic is the least likely to share something in common, [owing to] its greater simplicity. These two, however, have other subalternate sciences, such as the science of weights, music, spherics,[2] optics, and astronomy, all of which are closely related to the science of physics. Spherics is the simplest of them, and its subject matter is the *moving* sphere. Now, on the one hand, motion, on account of its continuity, is closely associated with magnitudes, even if its continuity is not essential but [is so associated] because of distance and time

1. "Pure geometry" would consider only those mathematical factors found, for example, in Euclid's *Elements,* whereas geometry considered as a part of natural science might roughly correspond with engineering, which must include not only geometrical knowledge, but also knowledge relevant to the material and form.

2. Literally, "the science of the moved spheres." See the Introduction to *Ptolemy's Almagest,* trans. G. J. Toomer (Princeton, NJ: Princeton University Press, 1998), 6, which briefly describes this science as dealing "with the phenomena arising from the rotation of stars and Sun about a central, spherical earth, e.g., their risings, settings, first and last visibilities, periods of invisibility, etc., using elementary geometry, but arriving mainly at qualitative rather than quantitative results." This science was considered to be quite basic, which matches Avicenna's own description of it below.

<الفصل الثامن>

في كيّفية بحث العلم الطبيعي ومشاركاته لعلوم أُخرى إنْ كانت تشاركه

(١) وإذْ قد عرفت الطبيعة وعرفت الأُمور الطبيعية، فقد اتضح لك فضْل اتضاحِ أنَّ العلم الطبيعي عن أي الأشياء يبحث. ولمّا كان المقدار المحدود من لوازم هذا الجسم الطبيعي وعوارضه الذاتية؛ أعني الطول والعرض والعمق المشار إليها - وكان الشكل من لوازم المقدار - كان الشكل أيضاً من عوارض الجسم الطبيعي. ولمّا كان المهندس موضوعه المقدار، فموضوعه عارضٌ من عوارض الجسم الطبيعي، والعوارض التي يبحث عنها هي من عوارض هذا العارض. فمن هذه الجهة تصير الهندسة الجزئية بوجهٍ مّا عند العلم الطبيعي، ولكن الهندسة الصرفة لا تشارك العلم الطبيعي في المسائل. وأمّا علم الحساب فهو أبعد من المشاركة وأشدّ بساطة. بل ها هنا علومٌ أخرى تحتهما كعلم الأثقال وعلم الموسيقى وعلم الأُكر المتحرّكة وعلم المناظر وعلم الهيئة. وهذه العلوم أقرب مناسبة إلى العلم الطبيعي، وعلم الأُكر المتحرّكة أبسطها؛ وموضوعه كرة متحركة. والحركة شديدة المناسبة للمقادير لاتصالها، وإنْ كان اتصالها لا لذاتها، بل بسبب مسافةٍ

(as we shall make clear later);[3] whereas, on the other hand, the demonstrations in the science of spherics do not use any physical premises. The subject matter of music is musical notes and intervals, and it has principles from both physics and arithmetic; and the same holds for the science of weights. Likewise, optics, whose subject matter is magnitudes related to a certain point of vision, draws its principles from both physics and geometry. Now, none of these sciences shares the [same] set of questions in common with physics; and although all of them consider the things belonging to them insofar as they possess quantity and have the accidents of quantity, their being conceived as such does not require that we make them some quantity in a natural body in which there is a principle of rest and motion, and neither do we need that.

(2) The subject matter of astronomy is the more significant portions of the subject matter of physics, and its principles are both physical and geometrical. The physical ones are, for example, that the motion of the heavenly bodies must be preserved according to a single system and other such things that are frequently used at the beginning of the *Almagest*.[4] As for the geometrical ones, they are well known. [Astronomy] differs from the other sciences in that it equally shares the [same] questions in common with physics, and so the questions it raises are a subset of the questions physics raises. Likewise, what is referred to in it and in the questions of physics is some accident or other belonging to the natural body—as, for example, that the Earth is a sphere and the Heavens are a sphere and the like. So it is as if this science is a mixture of something physical and something mathematical, as if the purely mathematical is something abstracted and not at all in matter, while that one [namely, what is physical] is an instantiation of that abstract thing in a determinate matter. Still, the astronomer and natural philosopher have different premises

3. The reference appears to be to 3.6 below, where, after having shown that there are no spatial magnitudes that are composed of atomic units, Avicenna argues that, given the mutual relation among distance (a spatial magnitude), time, and motion, none of them can have a discrete or atomic structure, and so all must be continuous.

4. The astronomical treatise of Ptolemy, translated during the reign of the ʿAbbāsid caliph al-Manṣūr (754–775). The *Almagest,* with its system of embedded spheres, deferents, and epicycles, provided the basis for virtually all medieval astronomy until Copernicus. For an account of the Ptolemaic system see Thomas Kuhn, *The Copernican Revolution: Planetary Astronomy in the Development of Western Thought* (Cambridge, MA: Harvard University Press, 1957).

أو زمانٍ - كما نبيّن نحن من بعد ، ثم البراهين في علم الأُكر المتحرّكة لا تستعمل فيها المقدمات الطبيعية البتة. وأمّا علم الموسيقى فموضوعه النَغَم والأزمنة ، وله مبادىء من علم الطبيعين ومباديء من علم الحساب. وكذلك علم الأثقال وعلم المناظر أيضاً موضوعه مقادير منسوبة إلى وَضْع مّا من البصر ، وله مبادىء من الطبيعيات ومن الهندسة. وهذه العلوم كلّها لا تشارك العلمَ الطبيعي في المسائل البتة ، وكلّها تنظر في الأشياء التي لها من حيث هي ذوات كمّ ، ومن حيث لها عوارض الكمّ التي لا توجب تصور عروضها للكمّ أنْ نجعلها كمّاً في جسم طبيعي فيه مبدأ حركة وسكون ، ولا نحتاج إلى ذلك.

(٢) وأمّا علم الهيئة فموضوعه أعظم أجزاء موضوع العلم الطبيعي ، ومبادؤه طبيعية وهندسية ؛ أمّا الطبيعية فمثل أنَّ حركة الأجرام السماوية يجب أنْ تكون محفوظة على نظام واحدٍ وما أشبه ذلك ، ممّا استُعمِل كثيرٌ منه في أول المجسطي. وأمّا الهندسية فممّا لا يخفى ، ويخالف سائر تلك العلوم في أنَّه يشارك الطبيعي في المسائل أيضاً ؛ فيكون موضوع مسائله شيئاً من موضوعات مسائل العلم الطبيعي ، والمحمول فيه أيضاً عارضٌ من عوارض الجسم الطبيعي ومحمولٌ أيضاً في مسائل العلم الطبيعي ؛ مثل الأرض كُرية والسماء كُرية ، وما أشبه ذلك. فهذا العلم كأنه ممتزج من طبيعي ومن تعليمي ، فإنّ التعليمي المحض مجرد لا في مادة البتّة ، وكأن هذا موقع لذلك المجرد في مادةٍ معيّنة. لكن المقدمات المُبرهن بها على المسائل المشتركة لصاحب الهيئة والطبيعي مختلفة ، أمّا مقدمات

by which they construct their demonstrations for the commonly shared questions: mathematical premises involve astronomical observation,[5] optics, and geometry, while physical premises are taken from whatever the nature of the natural body requires. Sometimes the natural philosopher combines [the two] and so introduces mathematical premises into his demonstrations; and the same for the mathematician, when he introduces physical premises into his demonstrations. When you hear the natural philosopher say, "If the Earth were not a sphere, then the remnant left during the Moon's eclipse would not be a crescent," know that he has combined [physical and mathematical premises]; and when you hear the mathematician say, "The noblest body has the noblest shape—namely, that which is circular," and "Portions of [the] Earth are moved rectilinearly," know that he has provided a mixed [demonstration]. Now, consider how the natural philosopher and mathematician differ in demonstrating that a certain simple body is spherical. To prove that, the mathematician uses what he discovers about the states of planets with respect to their rising, setting, elevation on the horizon, and declination, all of which would be impossible unless the Earth is spherical. The natural philosopher, however, says that the Earth is a simple body, and so its natural shape, which necessarily results from its homogeneous nature, cannot be something in which there are dissimilarities, such that part of it is angular and another part rectilinear, or such that part of it has one kind of curve and another its opposite. So you find that the first produced proofs that draw on the relation of oppositions, positions, and conjunctions without needing to turn to some power of nature that is necessary in order to make sense of them, while the second advanced premises drawn from what is proper to the nature of the natural body *qua* natural. The first has provided the *fact that* but not the cause, whereas the second has provided the cause and the *reason why*.[6]

---

5. See Dozy, *Supplément aux Dictionnaires Arabes,* 2 vols. (Leiden: E. J. Brill, 1881; reproduced, Beirut: Librairie du Liban, 1991), s.v. *raṣd* for this reading of *raṣdīya.*

6. In other words, the astronomer provides a demonstration *quia,* whereas the natural philosopher provides the demonstration *propter quid.* For Avicenna's discussion of this Aristotelian distinction, see *Kitāb al-burhan,* 1.7.

التعليمي فرصْدية مُناظرية أو هندســية ، وأمّا مقدمات الطبيعي فمأخوذة ممّا توجبه طبيعة الجســم الطبيعي. وربّما خلــط الطبيعي فأدخل المقدمات التعليميــة في براهينه ، وخلط التعليمــي فأدخل المقدمات الطبيعية في براهينه وإذا ســمعت الطبيعي يقول : لو لَمْ تكن الأرض كُرية ؛ لم يكن فضل الكسوف القمري هلالياً – فأعلم أنّه قد خلط . وإذا سمعتَ التعليمي يقول : وأشــرف الأجرام له أشرف الأشــكال وهو المستدير ، وأنّ أجزاء الأرض تتحرك إليه على الاستقامة – فأعلم أنّه قد خلط . وانظر كيف يختلف الطبيعي والتعليمي في البرهان على أنّ جُرْماً مّا من البســائط كُري ؛ أمّا التعليمي فيســتعمل في بيان ذلك ما يجد عليه حال الكواكب في شروقها وغروبها ، وارتفاعها عن الأفق وانخفاظها ، وأنّ ذلك لا يمكن إلاّ أن تكون الأرض كُرية . والطبيعي يقول إنّ الأرض جرْمٌ بسيطٌ ، فشكله الطبيعي الذي يجب عن طبيعةٍ متشــابهة مســتحيل أنْ يكون مختلفاً فيه ؛ فيكون في بعضه زاوية وفي بعضه خطٌ مســتقيم ، أو يكون بعضه على ضَرْبٍ من الانحناء والآخر على خلافه. فتجد الأول قد أتى بدلائل مأخوذة من مناسبة المقابلات والأوضاع والمحاذيات ، من غير أنْ تكــون محتاجة إلى أنْ يكون فيها تعرضٌ لقوةٍ طبيعيةٍ موجبة فيها لمعنى ، وتجد الثاني قد أتى بمقدماتٍ مأخوذة من مقتضى طبيعة الجسم الطبيعي – بما هو طبيعي . والأول يكون قد أعطى الإنّية ولم يعطِ العِلّة ، والثاني أعطى العِلّة واللّميّة .

(3) Numbers *qua* numbers might exist in natural existents, since one unit and then another is found in them. Now, [the fact] that each one of [the natural existents] is a unit is different from its being water, fire, earth, or the like; rather, the unity is some necessary concomitant belonging to [the natural existent] extrinsic to its essence. The consideration of those two units, insofar as they are together in some manner of existence, is the form of duality in that existence. The same holds for the other numbers as well. This is the numerable number, which also might exist in non-natural existents, which will be shown to have a certain "that-ness" and subsistence. So number is not included in physics, because it is neither a part nor species of the subject matter of [physics], nor is it some accident proper to it. Consequently, its identity does not require any dependence relation upon either natural or non-natural things—where *dependence relation* means that its existence is proper to that of which it is said to be dependent as something requiring it—but, rather, it is distinct from either one of the two in subsistence and definition. It is dependent (if it is and it is necessary) on what exists commonly and so is among the necessary concomitants of it. So the nature of number is fittingly understood by the intellect as something wholly abstracted from matter. Now, on the one hand, the consideration of it *qua* the nature of number and what is accidental to it from that perspective is a consideration of something abstracted from matter, while, on the other hand, it may have certain accidental states that someone who is counting considers, [while] it does not have those accidental states except inasmuch as they are necessarily dependent upon matter to subsist, even though [the nature of number] is necessarily not dependent upon [matter] by definition and is not something properly belonging to a determinate matter. So the consideration of the nature of number, as such, is a mathematical consideration.

(٣) والأعداد ، بما هي أعـــداد ، قد توجد في الموجودات الطبيعة ، إذ يوجد فيها واحدٌ وواحدٌ آخر ، وكون كلّ واحدٍ منها واحداً ؛ ليس كونه ذاته من ماء أو نار أو أرض أو غيــر ذلك ، بــل الوحدة أمرٌ لازمٌ له خارج عن ماهيَّته ، واعتبار ذينك الواحدين - من حيــث همــا في نحوٍ من أنحاء الوجــود معاً - هو صورة الإثْنينيــة في ذلك الوجود ، وكذلــك في غير ذلك من الأعــداد ، وهذا هو العدد المعدود ، وقد يوجد في الموجودات غير الطبيعية التي ســيتضح أنَّ لها إنِّيةً وقواماً . فليس العــدد داخلاً في العلم الطبيعي ؛ لأنــه لا هو جزء ولا هو نوع من موضوعه ، ولا هــو عارض خاصّ به ، فهويته لا تقتضي تعلّقاً لا بالطبيعيات ولا بغير الطبيعيات . ومعنى التعلّق أنْ يكون وجوده خاصّاً بما قيل إنَّه متعلقٌ به مقتضياً إياه ، بل هو مباينٌ لكل واحدٍ منهما بالقوام وبالحدّ ويتعلق - إنْ كان ولا بُــدّ - بالموجود العام ؛ فيكون من الأمور اللازمة له . فطبيعة العدد تصلح أنْ تُعقَل مجرّدة عن المادة أصلاً ، والنظر فيها من حيث هي طبيعة العدد وما يعرض لها من هذه الجهة ، نظرٌ مجرّدٌ عن المادة . ثم قد تعرض لها أحوال ينظر فيها الحاسبو تلك الأحوال لا تعرض لها إلاّ وقد وجب تعلّقها بالقوام بالمادة وإنْ لم يجب تعلّقها بها بالحدّ ، ولم تكون ممّا تُخصّصها بمادةٍ معيّنة ، فيكون النظر في طبيعة العدد - من حيث هي كذلك - نظراً رياضياً .

(4) Magnitudes are common to those things dependent upon matter but distinct from it. They are common to those things dependent upon matter because magnitudes are among the features that absolutely subsist in matter, whereas they are distinct from it in a number of ways. One way is that among the natural forms, there are some that it is immediately obvious cannot belong to just any matter, as chance would have it. An example would be the form that belongs to water *qua* water, for it simply cannot exist in stony matter as such, given its [elemental] mixture—unlike being round, which can inhere in both materials, as well as any other matter. Also, the form and nature of human-ness cannot exist in woody matter. This is not something that requires a great deal of mental effort to confirm, but can be grasped readily. There are other [natural forms] whose chancing to belonging to just any matter would not at first glance seem impossible—as, for example, white, black, and things of that kind (for the mind does not find this repugnant). Despite that, however, intellectual consideration will subsequently affirm that the nature of white and black belongs only to a certain mixture and a specific disposition and that what is disposed to turning black—in the sense of [naturally] becoming colored, not [artificially] being dyed—is not susceptible to turning white, which in the former sense [that is, the sense of naturally becoming colored] is due to something in its [elemental] mixture and inherent disposition. Even if the two are like that, however, neither of them is mentally conceptualized without being joined to a certain thing, which is not [color]. That thing is surface or magnitude, which is distinct from the color in the way it is affected.

(٤) وأمّا المقادير فإنَّها تشـارك المتعلّقات بالمادة وتباينها . أمّا مشـاركتها للمتعلّقات بالمـادة فلأنَّ المقادير هي من المعاني القائمة في المادة لا محالة ، وأمّا مباينتها فمن جهات . من ذلك أنَّ من الصور الطبيعية ما يظهر من أمره في أول الأمر أنَّه لا يصلح أنْ يكون عارضاً لكل مادة اتفقت ، مثل الصورة التي للماء من حيث هي ماء ؛ فإنَّها مستحيلة أنْ توجد في المادة الحجرية من حيث هي على مزاجها ، لا كالتدوير الذي يصح أن يحلّ المادتين جميعاً وأي مادة كانت . والصورة الإنسانية وطبيعتها فإنَّها مستحيلة أنْ توجد في المادة الخشبية ، وهـذا أمرٌ لا يلـزم الذهن في تحقّقه كثير تكلّف ، بل يقرب مناله . ومنها ما لا يسـتحيل ، في بادىء النظر ، أنْ يعرض لأي مادة اتفقت - مثل البياض والسـواد وأشـباء من هذا الجنس - فإنَّ الذهن لا يستوحش من إحلالها أيّة مادة اتفقت ، لكن العقل والنظر يوجبان مـن بعد أُنَّ طبيعة البياض والسـواد غير عارضةٍ إلاّ لمزاج واسـتعدادٍ مخصوص ، وأنَّ المستعدّ للتسوّد - بمعنى التّلون لا بمعنى التصبّغ - ليس قابلاً للبياض الذي بذلك المعنى لأمر في مزاجه وغريزته ، لكنهما ، وإنْ كانا كذلك ، فلا يُتصور ولا واحدٌ منهما في الذهن إلاّ مقارناً لأمرٍ ليس هو هو ، وذلك الأمر هو السـطح أو المقدار المباين للّون في المفعول .

(5) Moreover, the two forgoing divisions [of natural forms] frequently share a certain thing in common: namely, the mind receives one of [those divisions of forms] only when there is attached to it a specific relation to another thing, like the subject, which is joined to the thing itself. So when the mind brings up the form of human-ness, it necessarily brings it up together with its relation to a certain specific matter, appearing in the imagination only like that. Similarly, when it conceptually brings up white, it brings it up together with a certain extension in which [the white] necessarily is, and [the mind] refuses to conceptualize a given white unless it conceptualizes a given amount. Now, it is known that being white is not being a certain amount, but we do make the relation of being white to being an amount similarly to *x*'s relation to *y*, which is *x*'s subject.

(6) Magnitude is distinct from these two classes of things with respect to that which the two share in common, since the mind receives magnitude as something abstract. How could it do so otherwise, when a thorough investigation is needed in order to reveal that magnitude exists only in matter? It is distinct from the first division [of natural forms] by virtue of something peculiar to it—namely, that [even] when the mind finally discovers magnitude's relation to matter, it is not forced to consider its having a specific matter. It is distinct from the second division in that, [as with the first division,] in conceptualizing magnitude, the mind is not forced to make it have a specific matter; but additionally, neither do deduction and the intellect force it to that, since the intellect, in the same act of conceptualizing magnitude, can dispense with conceptualizing it in matter. Similarly, deduction does not require that magnitude have some unique relation to some species-specific matter because magnitude is never separate from [all the various kinds of] material things, and so it is not proper to some determinate [kind] of matter.

(7) Besides that, one has no need of matter in order to imagine and define [magnitude]. Now, it has been supposed that this is the status of white and black as well, but that is not so. [That] is because neither conceptualization in imagination nor descriptions nor definitions are

(٥) ثم قد يتشارك أيضاً هذان القسمان المذكوران في أمرٍ وهو أنَّ الذهن لا يقبل واحداً منهما إلاّ وقد لحقته خاصّية نسبةٍ إلى أمرٍ آخر يقارن ذاته كالموضوع، فإنَّ الذّهن إذا أحضر صورة الإنسانية لزمه أنْ يحضر معها نسبة لها إلى مادةٍ مخصوصةٍ لا تُتخيّل إلاّ كذلك. والبياض أيضاً إذا أحضره التصوّر أحضر معه إنبساطاً هو فيه ضرورة، وأبى أنْ يتصور بياضاً إلاّ تصوّر قدراً، ومعلومٌ أنَّ البياضية غير القدرية. ونجعل نسبة البياضية إلى القدرية شبيهة بنسبة شيءٍ إلى أمرٍ موضوعٍ له.

(٦) ثم المقدار يفارق هذين الصنفين فيما يشتركان فيه، وإذ الذهن يقبل المقدار على أنَّه مجرّد، وكيف لا يقبله وهو محتاجٌ إلى استقصاءٍ في البحث حتى ينكشف أنَّ المقدار لا يوجد إلاّ في مادة؟ ويفارق القسم الأول بشيءٍ يخصّه وهو أنَّ الذهن إذا تكلّف نسبة المقدار إلى المادة لم يضطر إلى أنْ تُعدّ له مادة مخصوصة، ويفارق القسم الثاني بأنَّ الذّهن – وإنْ لم يضْطر في تصوّر المقدار إلى أنْ يجعل له مادة مخصوصة – فالقياس والعقل لا يضطره إليها أيضاً، إذ الذّهن يستغني في نفس تصوّر المقدار عن تصوّره في المادة. والقياس لا يوجب أيضاً أنْ يكون للمقدار اختصاصٌ بمادةٍ نوعيةٍ معيّنة؛ لأنَّ المقدار لا يفارق شيئاً من المواد، فليس ممّا يكون خاصّاً بمادة معيّنة.

(٧) ومع ذلك فهو مستغنٍ في التوهّم والتحديد عن المادة. وقد ظُنَّ أنَّ البياض والسواد هذا حكمهما أيضاً وليس كذلك؛ فإنَّه لا التصوّر التخيّلي ولا الرسوم ولا الحدود

provided for them that are totally free of that when it is thoroughly investigated. The two are abstract only in another sense—namely, that matter is not a part of their subsistence, as it is a part of the subsistence of the composite; however, it is a part of their definition. Many things are part of something's definition while not being part of its subsistence, when its definition includes a certain relation to something external to the thing's existence. This point has been explained in the *Book of Demonstration*.[7]

(8) So, in the construction of demonstrations in the disciplines of arithmetic and geometry, neither discipline needs to turn to natural matter or take premises that refer to matter in any way. In contrast, the disciplines of spherics, music, optics, and astronomy progressively take matter, or some accident or other of matter, more and more [in their premises] because they are involved in investigating its states and so must take it into account. That is because these disciplines investigate either something's number or magnitude or some shape in a thing, where number, magnitude, and shape are accidents of all natural things. Now, occurring together with number and magnitude are also the things that essentially follow upon number and magnitude; and so, when we want to investigate the states in a given natural thing that are accidental to number and magnitude, we must necessarily take into account that natural thing. It is as if the physical sciences were a single discipline and mathematics—which is pure arithmetic and pure geometry—were a single discipline; and what is begot between the two are disciplines whose subject matter is from one discipline, while the things taken up in answering their set of questions are from another. When some of the sciences related to mathematics require that the mind turn toward matter, owing to the relation between them and the objects of physics, then how much greater should be your opinion about physics itself! What utter rot, then, is the ungrounded opinion that in physics we should focus solely on the form to the exclusion of matter!

---

7. The reference appears to be to *Kitāb al-burhān* 4.4.

المعطاة لهما تُغني عن ذلك إذا حُقّق واستُقْصي، وإنّما يتجردان بمعنى آخر؛ وهو أنّ المادة ليس جزء قوامهما كما هي جزء قوام المركّب لكنها جزء حدّهما. وكثيرٌ من الأشياء يكون جزء حدّ الشيء ولا يكون جزءاً من قوامه - إذا كان حدّه يتضمن نسبة مّا إلى شيءٍ خارجٍ عن وجود الشيء - وقد شُرح هذا المعنى في كتاب البرهان.

(٨) فصناعة الحساب وصناعة الهندسة؛ صناعتان لا تحتاجان في إقامتهما البراهين أن تتعرضا للمادة الطبيعية، أو تأخْذا مقدماتٍ تعرض للمادة بوجه. لكن صناعة الكُرة المتحرّكة - وأشدّ منها صناعة الموسيقى وأشدّ منها صناعة المناظر وأشدّ من ذلك صناعة الهيئة - تأخذ المادة أو شيئاً من عوارض المادة. وذلك لأنّها تبحث عن أحوالها، فمن الضرورة أنّها تأخذها؛ وذلك لأنّ هذه الصناعات إمّا أنّ تبحث عن عددٍ لشيء، أو مقدارٍ أو شكلٍ في شيء، والعدد والمقدار والشكل عوارض لجميع الأمور الطبيعية. ويعرض مع العدد والمقدار اللّواحق الذاتية أيضاً بالعدد والمقدار، فإذا أُريد أنْ نبحث عمّا يعرض من أحوال العدد والمقدار في أمرٍ من الأمور الطبيعية، لزم ضرورة أنْ نلتفت إلى ذلك الأمر الطبيعي؛ وكأنّ الصناعة الطبيعية صناعة بسيطة. والصناعة التعلمية، التي هي حسابٌ صِرف وهندسةٌ صِرفة؛ صناعةٌ بسيطة، ويتولّد ما بينهما صنائع موضوعاتها من صناعة؛ ومحمولات المسائل فيها من صناعة. وإذا كان بعض العلوم المنسوبة إلى الرياضة ممّا يحوج الذّهن إلى إلتفاتٍ نحو المادة، لمناسبةٍ بينه وبين الطبيعيات، فكيف ظنّك بالعلم الطبيعي نفسه! وما أفسد ظنَّ مَنْ يظنّ أنَّ الواجب أنْ نشتغل في العلم الطبيعي بالصورة، ونخلّي عن المادة أصلاً.

Chapter Nine

On defining the causes that are of the greatest interest to the natural philosopher in his investigation

(1) Some natural philosophers, of whom Antiphon was one, wholly dismissed form from consideration, believing that matter is what must be acquired and known [as the proper object of physics]; and when it is in fact acquired, what is subsequent to that is an infinite number of accidents and concomitants that are beyond the mastery [of the natural philosopher]. It would appear that this matter to which they restricted their inquiry is corporeal matter [already] impressed with a nature to the exclusion of Prime [Matter], as if they were oblivious to Prime [Matter]. Some of them at times appealed to one of the crafts, and drew a comparison between physics and some menial trade, saying that the iron miner toils to acquire iron without a thought for its form, and the pearl diver toils to acquire the pearl without a thought for its form.

(2) For our part, what makes it obvious that this opinion is wrong is that it would strip us of the opportunity to learn the specific and generic properties of natural things, which are their forms. Also, the very proponent of this school contradicts himself. So, on the one hand, if simply learning about formless material is enough to content him, then with respect to science he should sit perfectly happy in the knowledge of "something" that has no actual existence but, rather, is like some mere potential. On the other hand, in what way will he reach an awareness of it, since he has dismissively turned away form and accidents, whereas it is the forms and accidents themselves that lead our minds to affirm [this potential thing]? If, then, simply learning about formless material

<الفصل التاسع>

في تعريف أشْدّ العلل اهتماماً للطبيعي في بحثه

(١) قد رفض بعض الطبيعيين، ومنهم أنْطيفن، مراعاة أمر الصورة رفضاً كلّياً، واعتقدَ أنَّ المادة هي التي يجب أنْ تحصّل وتعرف. فإذا حصلت تحصيلاً فما بعد ذلك أعراض ولواحق غير متناهية لا تُضبط. ويشبه أنْ تكون هذه المادة التي قصر عليها هؤلاء نظرهم هي المادة المتجسمة المنطبعة دون الأولى، فكأنَّهم عن الأولى غافلون. وربّما احتجّ بعض هؤلاء ببعض الصنائع، وقايس بين الصناعة الطبيعية وبين الصناعة المهنية؛ فقال: إنَّ مستنبط الحديد وكده تحصيل الحديد وما عليه من صورته، والغوّاص وكده تحصيل الدُّرَّة وما عليه من صورتها.

(٢) والذي يُظهر لنا فساد هذا الرأي إفقاده إيانا الوقوف على خصائص الأمور الطبيعية ونوعياتها التي هي صورها، ومناقضة صاحب المذهب نفسِه نفسَه فإنَّه إنْ أقنعه الوقوف على الهيولى غير المصورة فقد قنع من العلم بمعرفة شيء لا وجود له بالفعل، بل كأنَّه أمرٌ بالقوة. ثم من أي الطرق يسلك إلى إدراكه إذ قد أعرض عن الصورة والأعراض صفْحاً، والصور والأعراض هي التي تجر أذهاننا إلى إثباته. فإنْ لم يقنعه الوقوف على

is not enough to content him, and he seeks a form belonging to the material, such as the form of water, air, or the like, then he will not have set aside[1] the inquiry into form. Also, his belief that the iron miner is not forced to consider form is false, for the subject of the iron miner's trade is not iron—which, in fact, is a certain end in his trade—but mineral bodies with which he busies himself through excavation and smelting, where his doing that is the form of his trade. Again, acquiring iron is a certain end of his trade, whereas it is something just taken for granted by the other trades whose master craftsmen are [only] coincidentally concerned with it as a result of their right to dispose of it by imposing a certain form or accident upon it.

(3) Alongside these, another group who were given to physical speculation also arose, but they attached absolutely no importance to matter. They said that its sole purpose for existing is that form may be made manifest through its effects, and that what is primarily intended is form, and that whoever scientifically comprehends form no longer needs to turn to matter, unless to dabble[2] in what is no concern of his. These equally exaggerate on the side of rejecting matter as the first had done on the side of rejecting form.

(4) Aside from the inadequacy of what they say with respect to the physical sciences (at which we already gestured in the previous chapter),[3] they remain happily ignorant of the relations between forms and the [kinds of] matter, since not just any form is conducive to just any matter,

1. Reading *kharaja* with **Z** and **T** for **Y**'s *taḥarraja* (to refrain); the Latin has *praetermisit* (to pass over).

2. Reading with **Z**, **T**, and the Latin *ᶜalá sabīl shurūᶜ*, literally "by way of making an attempt . . . ," for **Y**'s *ᶜalá sabīl mashrūᶜ* (by the attempted way).

3. See 1.8.4–6.

الهيولى غير المصوّرة ورام للهيولى صورة مثل صورة المائية والهوائية أو غير ذلك؛ فما خرج عن النظر في الصورة. وظنّه أنّ مستنبط الحديد غير مضطرٍ إلى مراعاة أمر الصورة ظنٌّ فاسد، فإنَّ مستنبط الحديد ليس موضوع صناعته هو الحديد، بل هو غايةٌ في صناعته، وموضوعه الأجسام المعدنية التي يُكبُّ عليها بالحفْر والتذويب؛ وفعله ذلك هو صورة صناعته، ثم تحصيل الحديد غاية صناعته، وهو موضوعٌ لصنائع أُخرى أربابها لا يُغنيهم مصادفة الحديد عن التصرّف فيه بإعطائه صورةً أو عرضاً.

(٣) وقد قام بإزاء هؤلاء طائفة أخرى من الناظرين في علم الطبيعة، فاستخفوا بالمادة أصلاً وقالوا إنَّها إنّما قصدت في الوجود لتظهر فيها الصورة بآثارها، وأنّ المقصود الأول هو الصورة؛ وأنَّ مَنْ أحاط بالصورة علماً فقد استغنى عن الالتفات إلى المادة إلاّ على سبيل شروعٍ فيما لا يعنيه. وهؤلاء أيضاً مسرفون في جَنْبة إطّراح المادة، كما أولئك كانوا مسرفين في جنْبة إطّراح الصورة.

(٤) وبعد تعذّر ما يقولونه في العلوم الطبيعية – على ما أومأنا إليه قبل هذا الفصل – فقد قنعوا بأنْ يجهلو المناسبات التي بين الصور وبين المواد. إذ ليس كلّ صورة

nor does just any matter conform to just any form. Quite [to] the contrary, in order for the natural species-forms actually to exist in the natures, they need [different kinds] of matter that are species-specific to the forms [and] for the sake of which their being prepared for those forms was completed. Also, how many an accident does the form produce that is only commensurate with its matter? Also [again], when real and complete scientific knowledge comprehends something as it is and what necessarily follows upon it, and the essence of the species-form is something needing a determinate matter or whose existence necessarily follows upon the existence of a determinate matter, then how can our scientific knowledge of form be perfect when this aspect of [the form] is not something that we investigate? Or how can this aspect of it be something that we investigate when we do not take into account the matter, where there is no matter more generally shared in common with it and yet none more further removed from form than Prime Matter?

(5) It is in our scientifically knowing [matter's] nature—namely, that it is potentially all things—that we acquire a scientific knowledge that when the form which is in some particular instance of this matter passes away, there must be the succession of some other [form]; otherwise, there would be some wholly nebulous possible thing. Now, which of the accounts of which we ought to have scientific knowledge is nobler than the state of something with respect to its very existence and that either is fixed or is not? In fact, the natural philosopher is lacking in his demonstration and needs to comprehend fully both the form and the matter in order to complete his discipline. Still, forms provide him with a greater scientific understanding of the actual being of the thing than the matter, whereas the matter more frequently provides him with scientific knowledge of its potential existence, while from both of them together the scientific knowledge of a thing's substance is completed.

مساعدة لكلّ مادّة، ولا كلّ مادة متمهدة لكل صورة؛ بل تحتاج الصور النوعية الطبيعية، في أنْ تحصْل موجودة في الطباع، إلى مواد نوّعيةٍ متخصّصةٍ بصورٍ لأجلها ما استتم استعدادها لهذه الصور؛ وكم من عرضٍ إنّما يحصل عن الصورة بحسب مادتها. وإذا كان العلم التّام الحقيقي هو الإحاطة بالشيء كما هو وما يلزمه، وكانت ماهّية الصورة النوعية أنَّها مفتقرة إلى مادة معيّنة أو لازمٌ لوجودها وجود مادة معيّنة؛ فكيف يستكمل علمنا بالصورة إذا لم يكن هذا من حالها متحقّقاً عندنا؟ أو كيف يكون هذا من حالها متحقّقاً عندنا، ونحن لا نلتفت إلى المادّة، ولا مادة أعمّ اشتراكاً منها وأبعد عن الصورة من المادة الأُولى؟

(٥) وفي علمنا بطبيعتها، وأنَّها بالقوة كل شيء، بكتسب علماً بأنَّ الصورة – التي في مثل هذه المادة – إمّا واجبٌ زوالها بخلافةِ أُخرى غيرها، أو ممكنٌ غير موثوقٍ به. وأي معنى أشرف من المعاني التي من حقّها أنْ نعلم؛ من معنى حال الشيء في وجود نفسه وإنَّه وثيقٌ أو قلقٌ؟ بل الطبيعي مفتقر في براهينه، ومحتاجٌ في استتمام صناعته إلى أنْ يكون محصّلاً للإ حاطة بالصورة والمادة جميعاً؛ لكن الصور تكسبه علم هوية الشيء بالفعل أكثر من المادة، والمادة تكسبه العلم بقوة وجوده في أكثر الأحيان، ومنهما جميعاً يستتم العلم بجوهر الشيء.

Chapter Ten

On defining each of the four kinds of causes

(1) In what preceded, we indicated that natural bodies have certain material, efficient, formal, and final causes. It is now fitting that we define the states of these causes, which in turn facilitates our coming to know natural effects.

(2) [Proving] that there are existing causes for everything that is subject to generation and corruption or undergoes motion or is some composite of matter and form, and that these causes are only four [in number], is not something that natural philosophers undertake—[this] falls [instead] to the metaphysician;[1] however, it is indispensable for the natural philosopher to affirm the essences of [the causes] and to indicate their states as a posit. So we say that the causes that are essential to natural things are four: the agent, matter, form, and end. In natural things, *agent* is often said of the principle of another's motion insofar as it is other.[2] By *motion,* we mean here whatever passes from potency to act in a given matter. This agent is that which is a certain cause for the transition of another, producing within it a motion from potency to act. Similarly, when the physician cures himself, he is a certain principle of motion in another insofar as he is other, because he produces a motion in the patient, and the patient is different from the physician precisely from the perspective of being a patient, whereas he cures only from the perspective that he is what he is (I mean, inasmuch as he is a physician). Now, he undergoes or receives the cure and the cure produces motion in him not from the perspective that he is a physician, but only from the perspective that he is a patient.

1. See *Ilāhīyāt* 6.1–5.
2. Cf. Aristotle, *Metaphysics* 5.12.1019a15ff.

<الفصل العاشر>

في تعريف أصناف علّةٍ علّةٍ من الأربع

(١) قد استعملنا فيما سلف إشاراتٍ دلّتْ على أنَّ للجسم الطبيعي علّة عنصرية وعلّـة فاعلة وعلّـة صورية وعلّة غائية. فحريٌ بنا الآن أنْ نعرف أحوال هذه العلل، فنستفيد منها سهولة سلوك السبيل إلى معرفة المعلولات الطبيعية.

(٢) أمّا أنَّ لكل كائنٍ فاسد؛ أو لكل واقعٍ في الحركة؛ أو لكل ما هو مؤلّف من مادة وصورة عللاً موجودة، وأنَّها هذه الأربع لا غير، فأمرٌ لا يتكفله نظر الطبيعيين، وهو إلى الإلهي، وأمّا تحقيق ماهيّتها والدلالة على أحوالها وضْعاً فأمرٌ لا يستغني عنه الطبيعي؛ فنقول: إنَّ العلل الذاتية للأمور الطبيعية أربعٌ؛ الفاعل والمادة والصورة والغاية. والفاعل في الأمور الطبيعية قد يقال لمبدأ الحركة في آخر غيره من جهة ما هو آخر؛ ونعني بالحركة ها هنا كل خروجٍ من قوةٍ إلى فعلٍ في مادة، وهذا المبدأ هو الذي يكون سبباً لإحالة غيره وتحريكه عن قوةٍ إلى فعل. والطبيب أيضاً إذا عالج نفسه فإنَّه مبدأ حركةٍ في آخَر بأنَّه آخَر، لأنَّه إنما يحرّك العليل؛ والعليل غير الطبيب من جهة ما هو عليل، وهو إنّما يعالج من جهة ما هو هو؛ أعني من جهة ما هو طبيب، فأمّا تَعالُجه وقبوله العلاج وتحرّكه بالعلاج فليس من جهة ما هو طبيب بل من جهة ما هو عليل.

(3) The principle of motion is either what prepares or what completes. What prepares is that which makes the matter suitable, like what moves semen during the preparatory states; whereas what completes is that which gives the form. It would seem that the Giver of that form by which the natural species subsist is outside of the natural order, and it does not fall to the natural philosopher to investigate that, beyond positing that there is that which prepares and there is a Giver of Form.[3] Without doubt, what prepares is a principle of motion, as is what completes, because it is what in fact brings about the emergence from potency to act.

(4) The auxiliary and guiding [principles] also might be numbered among the principles of motion. The auxiliary [principle] is like a part of the principle of motion, as if the principle of motion is the sum of the primary and auxiliary [principles], except that the difference between the primary and auxiliary is that the primary produces motion for a given end, while the auxiliary produces motion either for a certain end that is not for its own sake, but for the sake of the primary [principle], or for some end that itself is not the end that the primary achieves by its producing motion but is for some other end, such as gratitude, pay, or charity. As for the guiding [principle], it is an intermediary principle of motion, for it is a cause of the form characteristic of that soul[4] that is the principle of the first motion of something having volition. So it is the principle of the principle. As far as issues related to physics are concerned, this is the efficient principle; however, when the efficient principle is not concerned with issues of physics, but, instead, with existence itself, the sense is more general than this one, where whatever is a cause of some separate existence is essentially as such separate and as such that existence[5] is not for the sake of that efficient cause.

---

3. For discussions of the role of the "Giver of Forms" (*wāhib al-ṣuwar*) in the processes of generation and corruption, see *Kitāb fī al-kawn wa-l-fasād* 14 and *Ilāhīyāt* 9.5.

4. Literally, "psychic form."

5. Reading *dhālika al-wujūd* with **Z**, **T**, and the Latin (*illud esse*) for **Y**'s simple *dhālika* (that).

(٣) ومبـدأ الحركة إمّا مهيِّء وإمّا متمّم، والمهيـىء هو الذي يصلح المادة كمحرّك النُطْفة في الإحالات المعدّة، والمتمّم هو الذي يعطي الصورة. ويُشبه أن يكون الذي يعطي الصـورة المقوّمة للأنواع الطبيعية خارجـاً عن الطبيعيات، وليس على الطبيعي أنْ يتحقّق ذلـك بعد أنْ يضـع أنَّ ها هنا مُهيئاً، وها هنا معطي صورة. ولا شـك أنَّ المهيىء مبدأ حركة، والمتمّم أيضاً هو مبدأ الحركة، لأنَّه المخرج بالحقيقة من القوة إلى الفعل.

(٤) وقد يُعدّ المعين والمشـير في مبادىء الحركة؛ أمّا المعين فيشـبه أنْ يكون جزءاً مـن مبدأ الحركة، كأنَّ مبدأ الحركة جملة الأصـل والمعين؛ إلاّ أنَّ الفرق بين الأصل والمعين أنَّ الأصْل يُحرّك لغايةٍ ما، والمعين يحرّك لغايةٍ ليسـت له بل للأصْل، أو لغايةٍ ليسـت نفس غاية الأصل الحاصلة بالتحريك، بل <ل> غايةٍ أُخرى؛ كشُـكْرٍ أو أجْرٍ أو برّ. وأمّا المشير فهو مبدأ الحركة بتوسـط؛ فإنَّه سبب للصورة النفسانية التي هي مبدأ الحركة الأولى لأمرٍ إرادي؛ فهو مبدأ المبدأ، فهذا هو المبدأ الفاعلي بحسـب الأمور الطبيعية. وأمّا إذا أُخذ المبدأ الفاعلي لا بحسـب الأمور الطبيعية، بل بحسب الوجود نفسه، كان معنى أعمّ من هذا، وكان كل ما هو سـبب لوجودٍ مباينٍ لذاتـه من حيث هو مباين، ومن حيث ليس ذلك الوجود لأجله عِلّة فاعلة.

(5) Let us now discuss the material principle and say that it has an equivocal meaning—namely, that it is naturally disposed to bearing things foreign to it, having one relation to the thing composed from it and those essences and another to those essences themselves. An example is that the body has a certain relation to what is a composite (namely, to white [for example]) and a certain relation to what is not a composite (namely, to whiteness [for example]). Its relation to the composite is always a relation of a cause, because it is part of what makes the composite subsist, and the part in itself is prior to the whole as well as to that which subsists by it essentially.

(6) As for [the material principle's] relation to those latter things [namely, what is simple and not composite], there are only three logical possibilities. The first is that it is neither prior nor posterior to them in existence; [by this] I mean [that] they do not need some other thing in order to subsist, nor does that other thing [namely, the material principle] need them in order to subsist. The second is that the matter needs something like that [simple and incomposite] thing [call it *F*] in order to actually subsist, where *F* is essentially prior in existence to [the matter]. It is as if *F*'s existence is not dependent upon matter, but on different principles; however, when it exists, it necessarily entails that its matter subsist and [that] it make [the matter] actual, just like many things whose subsistence is through one thing, and after it subsists, it necessarily entails that something else subsist. Still, sometimes what makes it subsist is through something essentially separate from it, and at other times its subsistence[6] is through something essentially mixed with it, an example of which is called *form*, either having a share in making the matter to subsist by essentially being joined with it, or being whatever is a proximate cause of subsistence, which will be explained in metaphysics.[7] The

6. Reading *taqawwum* with **Z** for **Y**'s *taqwīm* (to make subsist); it should be noted that there is an almost equal split among the MSS on this point. The Latin's active indicative *constituet* would suggest that *taqwīm* was in that translator's exemplar.

7. See *Ilāhīyāt* 2.2–4.

(٥) ولننقل الآن في المبدأ المادي؛ فنقول: إنّ المباديء المادية تشترك في معنى وهي أنّها في طبائعها حاملة لأمورٍ غريبة عنها، ولها نسبة إلى المرّكب منها ومن تلك الماهيّات، ولها نسبة إلى تلك الماهيّات نفسها. مثلاً أنّ الجسم له نسبة إلى المركّب؛ أي إلى الأبيض، ونسبة إلى البسيط أي إلى البياض، ونسبته إلى المركّب نسبة علّة أبداً؛ لأنّه جزءٌ من قوام المركب، والجزء في ذاته أقدم من الكلّ ومقوّم لذاته.

(٦) وأمّا نسبته إلى تلك الأمور فلا تُعقل إلاّ على أقسام ثلاثة: إمّا أنْ يكون لا يتقدمها في الوجود ولا يتأخّر عنها؛ أعني لا هي محتاجة إلى الأمر الآخر في التقوُّم، ولا ذلك الأمر محتاج إليها في التقوّم. والقسم الثاني أنْ تكون المادة محتاجة إلى مثل ذلك الأمر في التقوّم بالفعل، والأمر يكون متقدماً عليها في الوجود الذاتي، كأنَّ وجوده ليس متعلقاً بالمادة بل بمباديء أُخرى، ولكنه يلزمه إذا وجد أنْ يقوّم مادته ويجعلها بالفعل. كما أنَّ كثيراً من الأشياء يكون تقوّمه بشيءٍ، ويلزمه بعد تقوّمه أنْ يقوِّم شيئاً آخر، لكنه ربّما كان ما يقوّمه بمفارقةٍ لذاته، وربّما كان تقوّمها بمخالطةٍ من ذاته، ومثل هذا الأمر يسمّى صورة وله قسطٌ في تقويم المادة بمقارنة ذاته، أو هو كل المقوّم القريب؛ وبيان

third is that the matter subsists in itself and is fully actual, being prior to F and making it subsist; and this thing is what we properly call an *accident,* even if we sometimes call all of those dispositions *accidents.* So the first division requires a relation of simultaneity, and the latter two require a relation of priority and posteriority. In the first case, what is in the matter is prior; while in the second case, the matter is prior. The first class does not obviously exist and would (if there is an instance of it at all) be like the soul and Prime Matter when they come together in order to make the human subsist. As for the latter two [namely, form and accidents], we have already spoken about them repeatedly.[8]

(7) There is another way to consider the relation of matter together with what is generated from it, of which [the matter] is a part. This relation can be transferred to the form. [That] is because, on the one hand, the matter alone might be sufficient in that it is the material part of what possesses matter, where that concerns a certain kind of things. On the other hand, it might not be sufficient unless some other matter is united[9] with it, so that from it and the other there is a combination like the single matter that is for the sake of perfecting something's form, where that concerns [another] kind of things, such as drugs for the sake of the poultice and gastric juices for the sake of the body. When something occurs from the matter alone in that [that thing] is together with it and nothing else, then it is according to one of the following. [(1)] It might be according to the combination only—as, for example, individual people for the sake of the military and homes for the city. [(2)] It might be according[10] to only the combination and composition together—as, for example, bricks and wood for the sake of the house. [(3)] It might be according to the combination, composition, and alteration—as, for

8. See, for example, 1.2.16–17.

9. Reading *lam tanḍammi* with **Z**, **T**, and the Latin for **Y**'s *lam tanẓim* (not arranged).

10. Reading *bi-ḥasabi* with **Z**, **T**, and the Latin for **Y**'s *bi-ḥaddi* (by definition).

ذلك في الصناعة الأُولى. والقسم الثالث هو أنْ تكون المادة متقوّمة في ذاتها وحاصلة بالفعل، وأقدم من ذلك الشيء وتقوّم ذلك الشيء؛ وهذا الشيء هو الذي نسميه عَرَضاً بالتخصيص. وإنْ كنّا ربّما سمّينا جميع هذه الهيئات أعراضاً. فيكون القسم الأول يوجب إضافة المعيّة، والقسمان الآخران إضافة تقدم وتأخّر، لكن في الأول منهما التقدّم لما في المادة، وفي الثاني منهما التقدّم للمادة، والقسمُ الأول ليس بظاهر الوجود؛ وكأنّه إنْ كان له مثالٌ فهو النفس والمادة الأُولى إذا اجتمعا في تقويم الإنسان، وأمّا القسمان الآخران فقد أخبرنا عنهما مراراً.

(٧) وللمادة مع المتكوّن عنها، التي هي جزءٌ من وجوده، نوعٌ آخر من اعتبار المناسبة. ويصلح أيضاً أنْ تُنقل هذه المناسبة إلى الصورة، فإنَّ المادة قد تكفي وحدها في أنْ تكون هي الجزء المادي لما هو ذو مادة وذلك في صنْفٍ من الأشياء، وقد لا تكفي ما لم تنضمّ إليها مادة أخرى فتجتمع منها ومن الأخرى كالمادة الواحدة لتمامية صورة الشيء، وذلك في صنفٍ من الأشياء كالعقاقير للمعجون والكيموسات للبدن. وإذا كانت المادة إنّما يحصل منها الشيء بأنْ يكون معها غيرها؛ فإمّا أنْ يكون بحسب الاجتماع فقط؛ كأشخاص الناس للعسكرية والمنازل للمدينة، وإمّا بحسب الاجتماع والتركيب معاً فقط، كاللَّبِن والخشب للبيت، وإمّا بحسب الاجتماع والتركيب والاستحالة، كالأُسْطُقسّات

example, the elements for the sake of the things subject to generation. [That] is because neither the very combination of the elements nor there being a composition from them (be it by touching, meeting, and receiving shape) is sufficient for things to be generated from them. Instead, it is that some of them act upon others, while others are acted upon, becoming stabilized for the sake of the whole as a homogeneous quality, which is called an [*elemental* or *humoral*] *mixture,* in which case it is prepared for the species-form. This is why, when the ingredients of an electuary[11] or the like are mixed, combined, and composed, it is not yet an electuary, nor does it have the form of the electuary until a certain period of time elapses, during which some [of the ingredients], as a result of their various qualities, act upon others and are acted upon by others, after which a single quality stabilizes as something homogeneous with respect to all of them, and by sharing [in all those qualities], a single activity arises from them. So the essential forms of these [ingredients] remains conserved, while the accidents by which they interact so as to bring about an alteration change and undergo alteration such that as any excess that is in any of its individual [ingredients] decreases until the quality of the overpowering [ingredient] stabilizes in it, falling below the point where it overpowers.

(8) The common practice is to say that the relation of the premises to the conclusion resembles the relation of the materials and forms.[12] It seems more likely that the form of the premises is their figure, where the premises through their figure resemble the efficient cause. [That] is because they are like the efficient cause of the conclusion, and the conclusion, *qua* conclusion, is something that emerges from them. Since

11. Literally, a theriaca, which is a paste used in the ancient and medieval periods as an antidote to poison, particularly snake venom. It was made up of some sixty or seventy different ingredients mixed and combined with honey.

12. Cf. Aristotle, *Posterior Analytics* 2.11.94a24–35, *Physics* 2.3.195a16–21, and *Metaphysics* 5.2.1013b20–21. Avicenna briefly touches on this point again in his *Ilāhīyāt* 6.4.

للكائنات. فإنَّ الأُسْطُقُسّات لا يكفي نفس اجتماعها ولا نفس تركيبها - بالتّماس والتلاقي وقبول الشكل - لأنْ تكون منها الكائنات، بل بأنْ يفعل بعضها في بعض، وينْفعل بعضها من بعض، ويستقر للجملة كيْفية متشابهة تسمّى مزاجاً، فحينئذ تستعد للصورة النوعية. ولهذا ما كان الترياق وما أشبهه؛ إذا خُلطت أخلاطه واجتمعت وتركّبت لم تكن ترياقاً بعد، ولا له صورة الترياقية، إلى أنْ تأتي عليه مدّة في مثلها يفعل بعضها في بعض وينفعل بعضها عن بعض بكيْفياتها، فتستقر كيْفية واحدة كالمتشابهة في جميعها، فيصدر عنها فعلٌ واحدٌ بالمشاركة. وهذه؛ فإنَّ صورها الذاتية تكون ثابتة محفوظة، والأعراض التي بها تتفاعل التفاعل الاستحالي فتتغيّر وتستحيل استحالة بأنْ تنتقص كل إفراطٍ يكون في كل مفرد منها إلى أنْ تستقر فيها كيْفية الغالبات أنقص ممّا في الغالب.

(٨) وقد جرت العادة بأنْ يقال إنَّ المقدمات نسبتها إلى النتيجة مشاكلة لمناسبة المواد والصور، والأشبه أنْ تكون صورة المقدمات شكلها، وتكون المقدمات بشكلها تشاكل السبب الفاعل، فإنّها كسببٍ فاعلٍ للنتيجة، والنتيجة - من حيث هي نتيجة - شيء

they found, however, that when the minor and major terms are properly distributed, the conclusion follows, and that the two [terms] before that had been in the syllogism, it was supposed that the subject of the conclusion is in the syllogism;[13] and that, in turn, was taken to the extreme in the belief that the syllogism itself is that in which the conclusion inheres. The fact is that the natures of the minor and major terms are subjects for certain forms, for they are subjects for the form of the conclusion. Now, in the case where there are no minor and major terms and subjects, so as to be a minor term and a major term, there will similarly be no subjects for the conclusion, because when each of them has a certain kind of relation to the other, there is a minor term and major term. That kind is [(1)] that both actually have a determinate relation to the middle [term] and [(2)] that they potentially have a relation to the conclusion. When [they] have another kind [of relation], they are actually subjects of the conclusion. The latter kind is that they stand to one another according to the relation of predicate and logical subject or antecedent and consequent after having had a certain relation. Despite that, what itself is in the syllogism, whether as a major term or minor term, is not also potentially the subject of the conclusion, but is something else of the same species as it; for we cannot say that, numerically, one thing accidentally happens to be a subject for the sake of its being a major and a minor term, while it is a subject for the sake of its being a part of the conclusion. So I simply do not understand how we should make the premises a subject for the conclusion. [That] is because when we compare the matter to what comes to be from it, sometimes the matter is a matter susceptible to generation, and sometimes it is susceptible to alteration, and sometimes it is susceptible to combination and composition, and sometimes it is susceptible to both composition and alteration. So this is what we have to say about the material cause.

13. The Arabic *mawḍūᶜ* is used both in the sense of "logical subject" (as opposed to the logical predicate) and the sense of "[material] subject" or "underlying thing" (cf. 1.2.3, 6). Since the immediate issue is why some thought that the premises stood to the conclusion as matter to form, the sense of *mawḍūᶜ* is almost certainly that of "material subject" or "underlying thing," even if the vocabulary is otherwise that of logic.

خارج عنها. لكنهم لمّا وجدوا الحدّ الأصغر والحدّ الأكبر إذا التأما حصلت النتيجة، وقد كانا قبل ذلك في القياس، وقع الظنّ بأنَّ في القياس موضوع النتيجة، فتخطّى ذلك إلى أنْ ظنّ أنَّ القياس نفسه موضوع النتيجة. لكن الحدّ الأصغر والحدّ الأكبر طبيعتاهما موضوعتان لصور، فإنَّهما موضوعتان لصورة النتيجة، وليستا حينئذ الحدّ الأصغر والحدّ الأكبر، وموضوعتان لأنْ يكونا حدّاً أصغر وحدّاً أكبر، وليستا حينئذ موضوعتين للنتيجة، لأنَّ كل واحدٍ منهما إذا كان على نمطٍ من النسبة إلى الآخر كان حدّاً أكبر وحدّاً أصغر؛ وذلك النمط هو أنْ يُنْسبا معاً بالفعل نسبة معيّنة إلى الأوسط، وأنْ تكون لهما إلى النتيجة نسبة إلى شيءٍ بالقوة. وإذا كانا على نمطٍ آخر؛ كانا موضوعين للنتيجة بالفعل، وذلك النمط هو أنْ ينسب كل واحدٍ منهما إلى الآخر نسبة الحَمْل والوضْع أو التلو والتقدّم بعد نسبةٍ كانت لهما. ومع ذلك فليس أيضاً عين ما هو في القياس حدّاً أكبر أو ⟨حدّاً⟩ أصغر هو بالقوة موضوع النتيجة، بل آخر من نوعه. فليس يمكن أنْ نقول إنَّ شيئاً واحداً بالعدد يعرض له أنْ يكون موضوعاً لكونه حدّاً أكبر وحدّاً أصغر، وموضوعاً لكونه جزء النتيجة! فلستُ أفهم كيف ينبغي أنْ نجعل المقدمات موضوعة للنتيجة، فإذا قسنا المادة إلى ما عنها يحدث؛ فقد تكون المادة مادة لقبول الكون، وقد تكون لقبول الاستحالة، وقد تكون لقبول الاجتماع والتركيب، وقد تكون لقبول التركيب والاستحالة معاً، فهذا ما نقول في العلّة المادية.

(9) The *form* may be said of the essence—which, when it occurs in the matter, makes a species subsist—as well as being said of the species itself. *Form* is sometimes said specifically of the shape and outline. [At] other times, it is said of the combination's disposition, like the form of the army and the form of conjunctive premises, as well as being said of the regulative order, such as the law. *Form* might be said of every disposition, however it might be, as well as being said of any thing, whether a substance or an accident, that is separate in the species (for this is said of the highest genus). *Form* may even be said of the intelligibles that are separate from matter. The form taken as one of the principles is relative to what is composed of it and the matter—namely, that it is a part of it that necessitates its being actual in its instance, whereas the matter is a part that does not necessitate its being actual (for the existence of the matter is not sufficient for the actual generation of something, but only for something's potential generation). So the thing is not what it is through the matter; rather, it is through the existence of the form that something becomes actual. As for the form that makes the matter subsist, it stands above [any] other kind. The formal cause might be related to either a genus or species—that is, the form that makes matter to subsist. It also might be related to the class—that is, it is not the form that makes the matter subsist as a species, but it is coincidental to it, such as the form of the shape belonging to the bed and the whiteness in relation to a white body.

(10) The end is the thing for the sake of which the form occurs in the matter—namely, either the real or apparent good. So any production of motion that proceeds—not accidentally, but essentially—from an agent is one whereby he intends some good relative to himself. Sometimes it is truly good, and at other times it is [only] apparently good, for either it is such or appears to be such.

(٩) وأمّا الصورة؛ فقد يُقال للماهيّة التي إذا حصلت في المادة قوّمتها نوعاً، ويُقال صورة لنفس النوع، ويُقال صورة للشكل والتخطيط خاصّة، ويُقال صورة لهيئة الاجتماع كصورة العسكر وصورة المقدمات المقترنة، ويُقال صورة للنظام المستحفظ كالشريعة، ويُقال صورة لكل هيئة كيف كانت، ويُقال صورة لحقيقة كل شيء كان جوهراً أو عرضاً ويفارق النوع؛ فإنَّ هذا قد يقال للجنس الأعلى، وربّما قيل صورة للمعقولات المفارقة للمادة، والصورة المأخوذة إحدى المبادئ هي بالقياس إلى المركّب منها ومن المادة؛ إنَّها جزء له يوجبه بالفعل في مثله، والمادة جزء لا يوجبه بالفعل. فإنَّ وجود المادة لا يكفي في كون الشيء بالفعل بل في كون الشيء بالقوة، فليس الشيء هو ما هو بمادته، بل بوجود الصورة يصير الشيء بالفعل، وأمّا تقويم الصورة للمادة فعلى نوعٍ آخر. والعلّة الصورية قد تكون بالقياس إلى جنسٍ أو نوعٍ، وهو الصورة التي تقوّم المادةً، وقد تكون بالقياس إلى الصنْف، وهو الصورة التي قد قامتً المادة دونها نوعاً وهو طارئ عليها؛ كصورة الشكل للسرير، والبياض بالقياس إلى جسم أبيض.

(١٠) وأمّا الغاية؛ فهي المعنى الذي لأجله تحصل الصورة في المادة، وهو الخير الحقيقي أو الخير المظنون. فإنَّ كلَّ تحريكٍ يصدر عن فاعلٍ لا بالعرض بل بالذات؛ فإنَّه يروم به ما هو خيرٌ بالقياس إليه، فربّما كان بالحقيقة وربّما كان بالظنّ؛ فإنّه إمّا أنْ يكون كذلك، أو يُظنّ به ظنّاً.

Chapter Eleven

On the interrelations of causes

(1) In a certain respect, the agent is a cause of the end; and how could it be otherwise, when the agent is what makes the end exist? In another respect, however, the end is a cause of the agent; and how could it be otherwise, when the agent acts only for the sake of [the end] and otherwise does not act? So the end moves the agent so as to be an agent. This is why, when it is asked, "Why did he exercise?" and we say, "For the sake of health," then this is an answer, just as when it is asked, "Why are you healthy?" and I say, "Because I exercised," it is an answer. Exercise is an efficient cause of health, and health is a final cause of exercise.

(2) If it is asked, "Why is health sought?" and it is said, "For the sake of exercise," it is not, in fact, an answer resulting from true choice; however, if it is asked, "Why was exercise sought?" and it is said, "In order that I be healthy," it is, in fact, an answer. Now, the agent is neither the cause of the end's becoming an end nor of the end's essence in itself; rather, it is a cause of the end's essence existing concretely in particulars, where there is a difference between essence and existence, as you have learned.[1] So the end is a cause of the agent's being an agent and so is a cause of its being a cause, whereas the agent is not a cause of the end with respect to its being a cause. This is something that will be explained in First Philosophy.[2]

1. A possible reference is *Kitāb al-madkhal* 1.6.
2. See *Ilāhīyāt* 6.5.

<الفصل الحادي عشر>

في مناسبات العلل

(١) الفاعل من جهةٍ سببٌ للغاية، وكيف لا يكون كذلك والفاعل هو الذي يحصّل الغاية موجودة. والغاية من جهةٍ هي سببٌ للفاعل، وكيف لا تكون كذلك؛ وإنّما يفعل الفاعل لأجلها؛ وإلّا لما كان الفعل. فالغاية تحرّك الفاعل إلى أنْ يكون فاعلاً؛ ولهذا إذا قيل: لِمَ يرتاض؟ فنقول: ليصح، فيكون هذا جواباً. كما إذا قيل: لِمَ صَحَحتَ؟ فيقول: لأنّي ارتضتُ، ويكون جواباً. والرياضة سببٌ فاعلي للصحة، والصحة سببٌ غائي للرياضة.

(٢) ثم إنْ قيل: لِمَ تطلب الصحة؟ فقيل: لأرتاض، لم يكن جواباً صحيحاً عن صادق الاختيار. ثم إنْ قيل: لِمَ تطلب الرياضة؟ فقيل: لكي أصح؛ كان الجواب صحيحاً. والفاعل ليس علّة لصيرورة الغاية غاية، ولا لماهيّة الغاية في نفسها، ولكن علّة لوجود ماهيّة الغاية في الأعيان؛ وفرقٌ بين الماهيّة والوجود كما علمته. والغاية علّة لكون الفاعل فاعلاً؛ فهو علّة له في كونه علّة، وليس الفاعل علّة للغاية في كونها علّة، وهذا سيتضح في الفلسفة الأولى.

(3) Next, the agent and end are like nonproximate principles of the caused composite. [That is] because, on the one hand, the agent either prepares the matter and so causes the existence of the effect's proximate matter while not [itself] being a proximate cause of the effect, or it provides a form and so is a cause of the proximate form's existence. On the other hand, the end is a cause of the agent as an agent, and a cause of the form and matter by means of its producing motion in the agent that brings about the composite.

(4) Next, the proximate principles of a thing are the material and form, with no intermediary between them and [that] thing; rather, they are the causes of it as two parts that make it subsist without intermediary (even if [the role] each one plays in making it subsist is different, as if *this* cause is different from the one that is *that*). It might accidentally happen, however, that matter and form are a cause both through an intermediary and without an intermediary from two [different] perspectives. As for the matter, when the composite is not a species but a class (that is, the form is not the form in the proper sense of the term but is an accidental disposition), then the matter is a cause of that accident that causes that class, as a class, to subsist, and so is a certain cause of the cause. Even so, however, insofar as the matter is a part of the composite and a material cause, there is no intermediary between the two. As for the form, when it is a true form (namely, belonging to the category of substance) and is causing the matter actually to subsist (where matter is a cause of the composite), then this form is a cause of the cause of the composite. Even so, however, insofar as the form is a part of the composite and a formal cause, there is no intermediary between the two. So when the matter is the cause of the cause of the composite, it is not, as such, a material cause of the composite; and when the form is the cause of the cause of the composite, then, as such, it is not a formal cause of the composite.

(٣) ثـم الفاعل والغاية كأنهما مبدآن غير قريبين من المركّب المعلول؛ فإنَّ الفاعل إمّا أنْ يكون مهيئاً للمادة فيكون سبباً لإيجاد المادة القريبة من المعلول لا سبباً قريباً من المعلول، أو يكون معطياً للصورة فيكون سـبباً لإيجاد الصورة القريبة. والغاية سبب للفاعل في أنّه فاعل، وسببٌ للصورة والمادة بتوسط تحريكها للفاعل المركّب.

(٤) فالمبادىء القريبة من الشـيء هـي الهيولى والصورة ولا واسـطة بينهما وبين الشـيء؛ بل هما عِلّتاه، على أنّهما جزءآن يقوّمانه بلا واسـطة - وإنْ اختلف تقويم كلّ منهما - وكأنَّ هذا علّة غير العلّة التي هي ذاك، لكنه ربّما عرض أنْ كانت المادة والصورة علّة بواسـطة وبغير واسـطة معاً من وجهين. أمّا المادة؛ فإذا كان المركّب ليس نوعاً بل صِنْفاً، وكانت الصورة لا التي تُخص باسـم الصورة، بـل هيئة عرضية، فحينئذ تكون المـادة مقوّمـة لذات ذلك العرض الذي يقومّ ذلك الصنـف من حيث هو صنف، فيكون علّـة مّا للعلّة، لكـن - وإنْ كان كذلك - فمن حيث المادة جزء من المركّب وعلّة مادية فلا واسـطة بينهما. وأمّا الصورة؛ فإذا كانت الصورة صورة حقيقية ومن مقولة الجوهر، وكانت تقوّم المادة بالفعل - والمادة علّة للمركّب - فتكون هذه الصورة علّة لعلّة المركّب. لكنه، وإنْ كان كذلك، فمن حيث الصورة جزء من المركّب وعلّة صورية، فلا واسـطة بينهمـا. فالمـادة إذا كانت علّة علّـة المركّب فليس من حيث هي علّـة مادية للمركّب، والصورة إذا كانت عِلّة علّة المركّب فليس من حيث هي علّة صورية للمركّب.

(5) Now, it may perchance be that the essence of the agent, form, and end is a single essence,[3] but that it should be an agent, form, and end is accidental to it. For in the father, there is a principle for generating the human form from semen. Now, that is not everything there is to the father, but only his human form, and it is only the human form that exists in the semen. Also, the end toward which the semen is moved is nothing but the human form. Insofar as it[4] makes the human species subsist with the matter, however, it is a form, whereas insofar as the semen's motion terminates at it, then it is an end, and insofar as its composition begins from it, then it is an agent. Again, when it is related to the matter and composite, it is a form. When it is related to the motion, then sometimes it is an end and at other times it is an agent: it is an end with respect to the motion's termination, which is the form that is in the son, while it is an agent with respect to the motion's beginning, which is the form that is in the father.

---

3. Cf. Aristotle, *Physics* 2.7.

4. Reading *lākinhā* with **Z** and **T** for **Y**'s *lākinhumā,* which would make the referent "the two" but would cause the verb "to subsist" to be incorrectly conjugated.

(٥) وقد يتفق أنْ تكون ماهيّة الفاعل والصورة والغاية ماهيّة واحدة، فتكون هي التي يعرض لها أنْ تكون فاعلاً وصورة وغاية. فإنَّ في الأب مبدأ لتكوّن الصورة الإنسانية من النُطْفة، وليس ذلك كل شيءٍ من الأب؛ بل صورته الإنسانية، وليس الحاصل في النُطْفة إلاّ الصورة الإنسانية. وليست الغاية التي تتحرّك إليها النُطْفة إلاّ الصورة الإنسانية؛ لكنها من حيث تُقوّم مع المادة نوع الإنسان؛ فهي صورة، ومن حيث تنتهي إليها حركة النُطْفة فهي غاية، ومن حيث يبتدىء منه تركيبها فهي فاعلة. وإذا قيست إلى المادة والمركّب كانت صورة، وإذا قيست إلى الحركة كانت غاية مرّة وفاعلة مرّة؛ أمّا غاية فباعتبار انتهاء الحركة وهي الصورة التي في الإبن، وأمّا فاعلة فباعتبار ابتداء الحركة؛ وهي الصورة التي في الأب.

Chapter Twelve

On the divisions of causal states

(1) Each one of the causes may be essential or accidental, proximate or remote, specific or general, particular or universal, simple or compound, potential or actual, as well as some combination of these.

(2) Let us first illustrate these states with respect to the efficient cause. So we say that the essential efficient cause is, for example, the physician when he heals, and fire when it heats. That is, the cause is a principle of that very act itself and taken insofar as it is its principle. The accidental efficient cause is whatever is not in keeping with that and is of various sorts. Among these is that the agent performs some action such that that action removes a certain contrary [*x*] that is holding its [opposing] contrary [*y*] in check. In that case [the agent's action] strengthens [*y*] so that the action of [*y*] is attributed to [the accidental efficient cause]—as, for example, scammony when, by purging bile, it cools.[1] Alternatively, the agent might remove something hindering a thing from its natural action, even if it does not require a contrary together with the hindrance—as, for example, one who removes the pillar from some tall building, since it is said that he destroys the building. Another [instance of accidental agency] is when a single thing is considered in various respects because it has varying attributes, and insofar as it has one of them, it is an essential principle of a certain action; but then [the action] is not attributed to [that attribute], but to one joined to it. For example, it is said that the physician builds—that is, what is posited of the physician is the [attribute of] building; however, he builds because he is a builder, not because he is a physician. Or what is posited alone

1. The example may strike modern readers as odd, and so must be understood against the background of ancient and medieval humoral medicine. Health and illness for most early physicians were understood in terms of a balance or imbalance of one of the four basic humors: blood, phlegm, black bile, or yellow bile. Yellow bile was seen as a substance essentially possessing the powers hot and dry. When one ingests scammony, the scammony essentially produces a yellowish, burning diarrhea, which was associated with the purging of yellow bile. Again, since this was believed to be a hot-dry humor, it was thought that there would be an accidental cooling effect.

<الفصل الثاني عشر>

في أقسام أحوال العلل

(١) إنَّ كل واحدٍ من العِلل قد يكون بالذات وقد يكون بالعرض، وقد يكون قريباً وقد يكون بعيداً، وقد يكون خاصّاً وقد يكون عاماً، وقد يكون جزئياً وقد يكون كلّياً، وقد يكون بسيطاً وقد يكون مركّباً، وقد يكون بالقوة وقد يكون بالفعل، وقد يتركب من بعض هذه مع بعض.

(٢) لنصور هذه الأحوال أولاً في العلة الفاعلة فنقول: إنَّ العلّة الفاعلة بالذات هي مثل الطبيب إذا عالج والنار إذا سخّنت، وهو أنْ تكون العلّة مبدأ لذات ذلك الفعل وأُخذت من حيث هي مبدأ له. والعلّة الفاعلة بالعرض ما خالف ذلك وهي على أصناف: من ذلك أنْ يكون الفاعل يفعل فعلاً؛ فيكون ذلك الفعل مزيلاً لضدّ ممانع ضدّه فيقوى الضدّ الآخر فينسب إليه فعل الضدّ الآخر؛ مثل السَّقَمُونْيا إذا برد بإسهال الصفراء. أو يكون الفاعل مزيلاً لمانعٍ شيئاً عن فعله الطبيعي، وإنْ لم يكن يوجب مع المَنْع ضدّاً؛ مثل مزيل الدعامة عن هدفٍ، فإنَّه يقال إنّه هو هادم الهدف. ومنه أنْ يكون الشيء الواحد معتبراً باعتباراتٍ لأنَّه ذو صفات، ويكون من حيث له واحدة منها مبدأ بالذات لفعلٍ فلا ينسب إليها بل إلى بعض المقارنة لها؛ كما يقال إنَّ الطبيب يبني، أي الموضوع الذي للطبيب هو بناءٌ فيبني، لأنَّه بنّاء لا لأنَّه طبيب، أو يؤخذ الموضوع وحده غير مقترنٍ بتلك

can be taken without that attribute, and so it is said that the man builds. Also among [the kinds of accidental efficient causes] is the agent, whether natural or voluntary, which is directed toward a certain end and then either reaches it or does not; however, together with [that end], there is accidentally another end—as, for example, the stone that splits open a head. Now, that is accidental to it precisely because it belongs essentially to [the stone] to fall, in which case it was just by happenstance that a head was passing by and so, through [the stone's] weight, it landed on it and so fractured the head.[2] Something might also be called an accidental efficient cause even if it does not do anything at all, save that its presence is frequently attended by something laudable or dangerous, and so it is recognized by that. So its being close at hand is deemed desirable if there attends it something laudable, which is auspicious; or keeping one's distance is deemed desirable if there attends it something dangerous, which is inauspicious, where it is supposed that its presence is a cause of that good or evil.

(3) As for the proximate agent, there is no intermediary between it and its effect (as, for example, the sinew in moving the limbs), while between the remote [agent] and the effect there is an intermediary (as, for example, the soul in moving the limbs). The specific agent involves only that single thing alone by which precisely one thing is acted upon (as, for instance, the medicine that Zayd ingests), whereas the general agent is that which is common among many things with respect to being acted upon by it (as, for example, the ambient air that produces change in many things).[3] The particular [agent] is either the individual cause of an individual effect (like *this* physician for *this* cure), or the specific

2. The phrase "because it belongs essentially to [the stone] to fall, in which case it was just by happenstance that a head was . . . ," is absent in **Y** but is confirmed by **Z**, **T**, and the Latin.

3. Literally, "air's producing change in many things, even without an intermediary"; however "ambient air" is almost certainly the intended sense, given the preceding medical example and the fact that ambient air played a central role in the maintenance of health in Galenic medicine (a medical system, one might add, that Avicenna adopted in general).

الصفة؛ فيقال: إنَّ الإنسان يبني. ومن ذلك أنْ يكون الفاعل بالطبع أو الإرادة متوجهاً إلى غايةٍ مّا فيبلغها أو لا يبلغها، لكن تعرض معها غاية أُخرى؛ مثل الحجر ليشجّ، وإنّما عرض له ذلك لأنه بذاته يهبط فاتفق أن وقعت هامة في ممّره فأتى عليها بثقله فشجّها. وقد يقال للشـــيء إنَّه فاعل بالعرض، وإنْ كان ذلك الشيء لم يفعل أصلاً، إلاَّ أنَّه يتفق أنْ يكون في أكثر الأمر يتبع حضوره أمرٌ محمودٌ أو مَحْذورٌ فيعرف بذلك، فيستحب قربه إنْ كان يتبعه أمرٌ محمودٌ يُتيمّن به، أو يُســتحب بُعْده؛ إنْ كان يتبعه أمرٌ محذور يُتطيّر منه، ويُظنّ أنَّ حضوره سببٌ لذلك الخير أو لذلك الشر.

(٣) وأمَّا الفاعل القريب، فهو الذي لا واســطة بينه وبين المفعول؛ مثل الوتر لتحريك الأعضـــاء، والبعيد هو الذي بينه وبين المفعول واســطة مثل النفــس لتحريك الأعضاء. وأمّـــا الفاعل الخاص فهو الذي إنّما ينفعل عن الواحد منه وحده شـــيءٌ بعينه؛ مثل الدواء الذي يتناوله زيدٌ في بدنه والفاعل العام فهو مثل الهواء المغيّر لأشـــياء كثيرة، وإنْ كان بلا واسطة. وأمّا الجزئي فهو، إمّا العلّة الشخصية لمعلولٍ شخصي كهذا الطبيب لهذا العلاج،

cause of a specific effect, being equal to it in the degree of generality and specificity (as, for instance, doctor for cure). As for the universal [agent], that nature does not exactly mirror the effect but is more general, (as, for example, physician for *this* cure, or professional for cure). Simple [agency] involves the action's arising out of a single active power (as, for example, pushing and pulling in bodily powers), whereas compound [agency] is that the act arises from a number of powers, whether agreeing in species (like many men who move a ship) or differing in species (like hunger resulting from the faculty of desire and sensation). That which is actual is like the fire in relation to what it is burning, while that which is potential is like the fire in relation to what it has not [yet] burnt, but it is so suited as to burn it. Sometimes the potential is proximate and at other times remote, where the remote [potential] is like the young child's potential to write [someday], while the proximate [potential] is like the potential to write of the one who has [already] acquired the talent of writing. Also, these may be combined in any number of ways, which we leave up to you to imagine.

(4) Let us now present these considerations with respect to the material principle. The matter in itself is that which, of itself, is susceptible to some thing—as, for example, oil's [being susceptible] to burning. As for that which is accidental, it is of various kinds. One of these is that the matter exists together with a certain form that is contrary to some [other] form that passes away with [the certain form's] arrival, and so a certain matter belonging to the present form is considered[4] together with the passing form, just as it is said that water is the subject for air and semen is the subject for human. Here, it is not the case that the semen is a subject *qua* semen, since the semen ceases once there is the human.

4. Reading *ta*ʾ*khudhu* with **Z**, **T**, and the Latin, for **Y**'s *yūjadu* (exists).

أو العلّة النوعية لمعلولٍ نوعي مساوٍ له في مرتبة العموم والخصوص؛ مثل الطبيب للعلاج. وأمّا الكلّي فأنْ تكون تلك الطبيعة غير موازية لما بإزائها من المعلول، بل أعمّ، مثل الطبيب لهذا العلاج، أو الصانع للعلاج. وأمّا البسيط فأنْ يكون صدور الفعل عن قوة فاعلة واحدةٍ مثل الدفع أو الجذب في القوى البدنية. وأمّا المركّب فأنْ يكون صدور الفعل عن عدّة قوى؛ إمّا متفقة النوع كعدّةٍ يحرّكون سفينة، أو مختلفي النوع كالجوع الكائن عن القوة الجاذبة والحاسّة. وأمّا الذي بالفعل؛ فمثل النار بالقياس إلى ما اشتعلت فيه. وأمّا الذي بالقوة؛ فمثل النار بالقياس إلى ما لَمْ تشتعل فيه ويصح اشتعالها فيه. والقوة قد تكون قريبة وقد تكون بعيدة، والبعيدة كقوة الصبي على الكتابة، والقريبة كقوة الكاتب المقتني الملكة الكتابية على الكتابة. وقد يمكنك أنْ تركّب بعض هذه مع بعض، وقد وكلناه إلى ذهنك.

(٤) ولنورد هذه الاعتبارات أيضاً في المبدأ المادي. فأمّا المادة بالذات فهي التي لأجل نفسها تقبل الشيء: مثل الدهن للإشعال، وأمّا التي بالعرض فعلى أصناف: من ذلك أنْ توجد المادة مع صورةٍ مضادّة لصورةٍ تزول بحلولها، فتؤخذ مع الصورة الزائلة مادة للصورة الحاصلة، كما يقال إنَّ الماء موضوعٌ للهواء والنُطْفة موضوعةٌ للإنسان، والنُطْفة ليست موضوعة بما هي نُطْفة؛ لأنَّ النُطْفية تبطل عند كون الإنسان. أو يؤخذ الموضوع مع صورة

Alternatively, the subject may be considered together with a form that does not enter into the subject's being a subject; and even if it is not a certain contrary of the final intended form, it is taken as a subject. An example would be our saying that the physician is cured, for he is not cured inasmuch as he is a physician, but only inasmuch as he is a patient, and so the subject of the cure is [the individual *qua*] patient, not physician. The proximate subject is, for example, the body's limbs, while the remote [subject] is like the [humoral and elemental] mixtures and, really, the four underlying elements. The specific subject of the human form, for example, is the human body with its humoral mixture, while the general [subject] is like the wood of the bed, chair, and the like. There is a difference between the proximate and the specific [subject], for the material cause might be proximate while being general—as, for instance, the wood of the bed. An example of the particular subject is *this* wood for *this* chair, or *this* substance for *this* chair. The[5] universal [subject] is like wood for this chair, or substance for chair. The simple subject is, for example, the material belonging to all things and perceptible wood to wooden things, while the composite [subject] is like the humoral mixtures of the living body and the drugs of the electuary. The actual subject of the human form, for example, is the human body, whereas its potential [subject] is like the semen or like the unworked wood for *this* chair. Here, again, the potential is sometimes proximate and sometimes remote.

---

5. Omitting the phrase *wa-l-ʿāmm mithla hādhā al-khashab li-hādhā al-kursī,* (the general [subject] is, for example, *this* wood of *this* chair), which **Y** includes, noting that it was omitted in at least one manuscript and should probably not be retained. The phrase is not found in **Z**, **T**, or the Latin.

ليست داخلة في كون الموضوع موضوعاً ، وإنْ لم يكن ضدّاً للصورة الأخرى المقصودة ، فيُجعل موضوعاً ؛ مثل قولنا : إنَّ الطبيب يتعالج ، فإنَّه ليس إنَّما يتعالج من حيث هو طبيب ولكن من حيث هو عليل ، فالموضوع للعلاج هو العليل لا الطبيب . وأمّا الموضوع القريب ؛ فمثل الأعضاء للبدن ، والبعيد مثل الأخلاط بل الأركان ، والموضوع الخاص فمثل جسم الإنسان بمزاجه لصورته ، والعام مثل الخشب للسرير والكرسي ولغيره . وفرقٌ بين القريب والخاص ؛ فقد يكون السبب المادي قريباً وعامّاً مثل الخشب للسرير ، والموضوع الجزئي مثل هذا الخشب لهذا الكرسي ، أو هذا الجوهر لهذا الكرسي ، والكلّي مثل الخشب لهذا الكرسي أو الجوهر للكرسي . والموضوع البسيط ؛ فمثل الهيولى للأشياء كلّها أو الخشب عند الحسّ للخشبيات ، والمركّب مثل الأخلاط للبدن ومثل العقاقير للترياق . والموضوع بالفعل مثل بدن الإنسان لصورته ، وبالقوة مثل النُطفة لها ، أو الخشب غير المصور بالصناعة لهذا الكرسي ، وهاهنا أيضاً قد تكون القوة قريبة وقد تكون بعيدة .

(5) As for these considerations on the part of the form, the form that is essential is, for instance, the chair shape belonging to the chair, while that which is accidental is like its whiteness or blackness. [The accidental form] might be something useful with respect to what is essential—as, for example, the wood's hardness in order that it receive the shape of the chair. Also, the form might be accidental because of vicinity, like the motion of one standing still on the boat, for it is said of the one standing still on the boat that he is moved and being carried along accidentally. The proximate form is, for example, the squareness of *this* square, while the remote [form] is its possessing angles (for instance). The specific form is no different from the particular (namely, for example, the definition, species difference, or property of something), nor is the general form different from the universal (namely, for example, the genus[6] of the property). The simple form is like the form of water and fire, which is form whose subsistence does not result from a combination of a number of forms, while the composite [form] is like the form of human, which does result from the combination of a number of powers and forms. The actual form is well known, whereas the potential form is, in a certain way, the potential together with the privation.

---

6. Following **Z**, **T**, and Latin (*genus*) that have *jins* for **Y**'s *khashab* (wood).

(٥) وأمـا هـذه الاعتبارات من جهة الصـورة؛ فالصورة التي بالذات مثل شـكل الكرسـي للكرسي، والذي بالعرض فمثل البياض أو السواد له. وربّما كان نافعاً في الذي بالذات، مثل صلابة الخشب لقبوله شكل الكرسي. وربّما كانت الصورة بالعرض وبسبب المجاورة؛ كحركة الساكن في السفينة؛ فإنَّه يقال لساكن السفينة متنقلٌ ومتحركٌ بالعرض. والصـورة القريبة؛ فمثل التربيع لهـذا المربع، والبعيدة مثل ذي الزاوية له. والصورة الخاصّة لا تخالف الجزئية؛ وهو مثل حدّ الشيء أو فصل الشيء أو خاصّة الشيء، و<الصورة> العامة فلا تفارق الكلّية، وهو مثل الجنس للخاصّة، والصورة البسـيطة فمثل صورة الماء والنـار الذي هو صورة لم تتقوّم عن عِدّة صور مجتمعة. والمركّبة؛ مثل صورة الإنسـان التـي تحصل مـن عِدّة قوى وصورةٍ تجتمع. والصورة بالفعـل معروفة، والصورة بالقوة من وجهٍ مّا فهي القوة مع العدم.

(6) As for considering these accounts from the vantage point of the end, the essential end is that toward which natural or voluntary motion tends for its own sake, not for another's—as, for example, the health due to medicine. The accidental end is, again, of varying sorts. One of them is what is intended, but not for its own sake, such as pulverizing the medicine for the sake of drinking, which is in turn for the sake of health. This might be either what is beneficial or what appears to be beneficial. The former is the good, the latter what appears to be good. Another sort is what the end either necessarily or accidentally entails. An example of what the end necessarily entails would be that the end of eating is a bowel movement, where that necessarily follows owing to the end but is not the end, which instead is to stave off hunger. What the end accidentally entails is, for example, the beauty that results from exercise, for, although beauty may be accidental to health, beauty is not what is intended by exercise. Another sort is when motion is not directed toward something but meets up with it on the way—as, for example, the head's being fractured owing to the falling rock, and whoever shoots at a bird and [instead] hits a man. Sometimes the essential end is found together with [the accidental end],[7] and sometimes it is not.

(7) The proximate end is like the [immediate] health owing to the medicine, whereas the remote [end] is like the [life of] flourishing on account of the medicine. The specific end is like Zayd's meeting his friend so-and-so, while the general [end] is like the purging of bile owing to drinking camelthorn,[8] since it is the end of drinking violet [root][9] as well. The particular end is like Zayd's collecting money from some debtor whom [Zayd] traveled to find, while the universal end is like his seeking justice from the unjust absolutely. As for the simple end, it is like eating to satisfy one's appetite, whereas the composite [end] is like wearing silk for the sake of beauty and to do away with lice,[10] which are really two [different] ends. The actual and potential ends are like the actual and potential forms.

---

7. Alternatively, the pronoun *hā* might refer to "motion."

8. That is, *alhagi maurorum* or "Persian manna."

9. Large doses of the violet (*viola odorata*) root contain an alkaloid called violine, which is a purgative.

10. It was recognized that lice cannot hold onto the slippery surface of silk.

(٦) وأمّا اعتبار هذه المعاني من جهة الغاية، فالغاية بالذات هي التي تنحوها الحركة الطبيعية أو الإرادية لأجل نفسها لا غيرها؛ مثل الصحة للدواء. والغاية بالعرض على أصناف: فمن ذلك ما يُقصد ولكن لا لأجله؛ مثل دق الدواء لأجل شرب الدواء لأجل الصحة، وهذا هو النافع أو المظنون نافعاً، والأول هو الخير أو المظنون خيراً. ومن ذلك ما يلزم الغاية أو يعرض لها؛ أمّا ما يلزم الغاية فمثل الأكل غايته التغوّط، وذلك لازمٌ للغاية لا غاية؛ بل الغاية هو كفّ الجوع. وأمّا ما يعرض للغاية؛ فمثل الجمال للرياضة، فإنّ الصحة قد يعرض لها الجمال وليس الجمال هو المقصود بالرياضة. ومن ذلك ما تكون الحركة متوجهة لا إليه فيعارضها هو، مثل الشجّة للحجر الهابط، ومثل مَنْ يرمي طيراً فيصيب إنساناً، وربّما كانت الغاية الذاتية موجودة معها وربّما لم توجد.

(٧) وأمّا الغاية القريبة فكالصحة للدواء، والبعيدة فكالسعادة للدواء. وأمّا الغاية الخاصّة فمثل لقاء زيد صديقه فلاناً، وأمّا العامة فكأسهال الصفراء لشرب الترنجبين، فإنّه غاية له ولشرب البنفسج أيضاً. وأمّا الغاية الجزئية فكقبض زيد على فلان الغريم المقصود كان في سفره، وأمّا ‹الغاية› الكلّية فكأنتصافه من الظالم مطلقاً. وأمّا الغاية البسيطة فمثل الأكل للشبع، والمركّبة مثل لُبس الحرير للجمال ولقتل القُمّل، وهذا بالحقيقة غايتان. وأمّا الغاية بالفعل والغاية بالقوة فمثل الصورة بالفعل والصورة بالقوة.

(8) Know that the potential cause mirrors the potential effect; and so, as long as the cause is potentially a cause, the effect is potentially an effect. Also, each one of them might be essentially something else. An example would be that the cause is a human, while the effect is wood, since the man is potentially a carpenter and the wood is something potentially worked by the carpenter. What is not possible is that the effect itself should exist while the cause is entirely absent. What [seems] to throw doubt upon this is the case of a building and its remaining after the builder [departs]. So you must know that the building *qua* the effect of the builder does not remain after the builder [departs], for the effect of the builder is to move the parts of the building until they form an integral whole—[an action] that does not continue after he departs. As for the persistence of the integral whole and the presence of the shape, it persists as a result of certain existing causes that when they are destroyed, bring about the destruction of the building. The independent verification of this account and what in the preceding was like it will be deferred until first philosophy, and so wait until then.[11]

11. See *Ilāhīyāt* 6.2.

(٨) واعلـم أنّ العلة بالقوة بإزاء المعلـول بالقوة، فما دامت العلّة بالقوة عِلّة فالمعلول بالقوة معلولٌ. ويجوز أنْ يكون كل واحدٍ منهما بالفعل ذاتاً أخرى، مثل أنْ تكون العلّة إنساناً والمعلول خشباً، فيكون الإنسان نجاراً بالقوة، والخشب منجوراً بالقوة، ولا يجوز أنْ تكون ذات المعلول موجودة والعلّة معدومة البتة. والذي يشكل في هذا من أمر البناء وبقائه بعد البانـي. فيجب أنْ تعلم أنَّ البناء ليس يبقى بعد الباني، على أنَّ البناء معلول الباني؛ فإنَّ معلول الباني هو تحريك أجزاء البناء إلى الاجتماع وهو لا يتأخّر عنه. وأمّا ثبات الاجتماع وحصول الشـكل فيثبت عن عللٍ موجودة إذا فسـدتْ فسد البناء. وتحقيق هذا المعنى وما يجري مجراه – ممّا سلف – موكول إلى الفلسفة الأولى، فليتربَّص به إلى ما هناك.

Chapter Thirteen

Discussion of luck and chance: The difference between them and an explanation of their true state

(1) Since we have been discussing causes and it is supposed that luck, chance, and what happens spontaneously are among the causes, we ought not to neglect considering these accounts and whether they are to be considered causes or not, and if they are, then what their manner of causality is.

(2) Now, the earliest Ancients differed concerning luck and chance. One group denied that luck and chance could be included among the causes and, in fact, denied that there was any sense at all in which they exist. They said that it is absurd that we should discover and observe that things have necessitating causes and then turn our backs upon [those causes], dismissing them as causes and going for unknown "causes" such as luck and chance. So, when someone who is digging a well stumbles across a treasure, the ignorant say with absolute conviction that good luck attended him, whereas if he slips into it and breaks a leg, they say with equal conviction that bad luck attended him. [This group said in response] that no luck attends him at all here [either good or bad]; but, rather, whoever digs where a treasure is buried will acquire it and whoever leans over a slippery edge will slip over it. Also, they say that when someone goes to market to tend shop and sees someone who owes him money and so collects what he is owed, that [might seem] to be an act of luck, but it is not; rather, it is because he went someplace where his debtor was and, having good eyesight, saw him.[1] They also said that even if his end in going to the market was not this one, it does not necessarily follow that going to the market was not a real cause of his collecting the money that was owed him, for a single action might have various ends. In fact, most actions are like that; however, the one who performs that action just happened to stipulate that one of those ends is [his] end and

1. Cf. Aristotle, *Physics* 2.5.196b33ff.

<الفصل الثالث عشر>

في ذِكْر البخْت والاتفاق والاختلاف فيهما وإيضاح حقيقة حالهما

(١) وإذ قد تكلمنا على الأسباب، وكان البخْت والاتفاق وما يكون من تلقاء نفسه؛ قد ظُنَّ بها أنَّها من الأسباب، فحريٌ بنا أنْ لا نُغْفل أمر النظر في هذه المعاني وأنَّها هل هي في الأسباب أو ليست في الأسباب، وإنْ كانت، فكيف هي في الأسباب؟

(٢) وأمَّا القدماء الأقدمون فقد كانوا اختلفوا في أمر البخْت والاتفاق، ففرقة أنكرت أنْ يكون للبخْت والاتفاق مدخلٌ في العلل، بل أنكرت أنْ يكون لهما معنى في الوجود البتة. وقالت إنَّه من المحال أنْ نجد للأشياء أسباباً موجبة ونشاهدها فنعدل عنها ونعْزلها عن أنْ تكون عللاً، ونرتاد لها عللاً مجهولة من البخْت والاتفاق. فإنّ الحافر بئراً إذا عثر على كنْز جزم أهل الغباوة القول بأنَّ البخْت السعيد قد لحقه، وإنْ زلق فيه فانكسر جزموا القول بأنَّ البخْت الشقي قد لحقه، ولم يلحقه هناك بخْتٌ البتة، بل كل مَنْ يحفر إلى الدفين يناله، ومَنْ يميل على زلقٍ في شفير يزلق عنه. ويقولون إنَّ فلاناً لمّا خرج إلى السوق ليقعد في دكانه لمح غريماً له فظفر بحقه، فذلك من فعل البخْت، وليس كذلك، بل ذلك لأنّه قد توجّه إلى مكانٍ به غريمه، وله حُسْن بصرٍ فرآه. قالوا وليس - وإنْ كان غايته في خروجه غير هذه الغاية - يجب أنْ لا يكون الخروج إلى السوق سبباً حقيقياً للظفر بالغريم، فإنَّه يجوز أنْ يكون لفعلٍ واحدٍ غايات شتى، بل أكثر الأفعال كذلك. لكنه يعرض أنْ يجعل المستعمل لذلك الفعل إحدى تلك الغايات غاية فتتعطل الأخرى، بوضعه لا في

so renders the others ineffectual by stipulation, though not with respect to the thing itself (that is, the thing itself is [still] an end suitably disposed to being set up as an end and the others set aside). If this person were aware of the debtor's presence there and so went running after him so as to collect from him, why don't we say that that occurs by luck, when we say of other cases that they are by luck or chance? So, you see that to stipulate [only] one of the things to which *going out* leads as an end strips *going out* of being a cause in itself of whatever else it causes; but how can it be supposed that that changes simply by stipulation? This is one side.

(3) Another side, with many splinter groups, rose up in mirror opposition to them, which touted the significance of luck. Some said that luck is a divine, hidden cause that is beyond the grasp of our intellects. Some who believed this opinion took it to the point of setting up luck as something to draw near to or, by worshiping it, to draw near to God, being something for which a temple was built and in whose name an idol was made that was worshipped in the way that idols are worshipped.

(4) Another group went so far as [to make] luck like the natural causes in a certain way, and so they made the world come to be through luck. This is Democritus and those who followed him, for they believe the following: [(1)] the principles of the universe are atoms that are indivisible owing to their solidity and absence of [interstitial] void [space]; [(2)] [these atoms] are infinite in number and scattered throughout an infinitely extended void; [(3)] with respect to the nature of their substance, [atoms] are generically alike, whereas they differ by means of their shapes; [(4)] [these atoms] are in constant motion within the void, and so a group of them chanced to collide and so combine according to some configuration, from which then the world comes to be; and [(5)] there are an infinite number of worlds just like this one, arranged throughout an infinite void. Despite that, [Democritus further] thinks that [(6)] particular things like animals and plants do not come to be according to chance.

نفس الأمر، وهو في نفس الأمر غاية يصلح أنْ ينصبها غاية ويرفض ما سواها. أليس لو كان هذا الإنسان شاعراً بمقام الغريم هناك؛ فخرج يرومه فظفر به لم يقل أنَّ ذلك واقعٌ منه بالبخْت، بل قيل لما عداه أنَّه بالبخْت أو بالاتفاق؟ فترى أنَّ جَعْلَه أحد الأمور التي يؤدّي إليها خروجه غاية يُصرْف الخروج عن أنْ يكون في نفسه سبباً لما هو سببه، فكيف يُظن أنَّ ذلك يتغيّر بجعْل جاعل؟ فهؤلاء طائفة.

(٣) وقد قام بإزائهم طائفة أخرى عظموا أمر البخْت جداً وتشعبوا فرقاً، فقائل منهم إنَّ البخْت سببٌ إلهي مستور يرتفع عن أنْ تدركه العقول، حتى أنَّ بعض من يرى رأي هذا القائل؛ أحلَّ البخْت محل الشيء الذي يتقرب إليه أو إلى الله بعبادته، وأمر فبُنيَ له هيكلٌ، واتُّخذ باسمه صنمٌ يعبد على نحو ما تعبد عليه الأصنام.

(٤) وفرقة قدمت البخْت من وجهٍ على الأسباب الطبيعية، فجعلت كون العالم بالبخْت؛ وهذا هو ديموكريتس وشيعته؛ فإنَّهم يرون أنَّ مبادىء الكلّ هي أجرامٌ صغارٌ لا تتجزأ لصلابتها ولعدمها الخلاء، وأنَّها غير متناهية بالعدد ومبثوثة في خلاء غير متناهي القدر، وأن جوهرها في طباعه متشاكلٌ وبأشكالها مختلفٌ، وأنَّها دائمة الحركة في الخلاء، فيتفق أنْ تتصادم منها جملة فتجتمع على هيئةٍ، فيكون منه عالمٌ، وأنَّ في الوجود عوالم مثل هذا العالم غير متناهية بالعدد مرتبة في خلاء غير متناه. ومع ذلك فيرى أنَّ الأمور الجزئية مثل الحيوانات والنبات كائنة لا بحسب الاتفاق.

(5) Yet another group [namely, Empedocles and those following him] did not go so far as to make the world in its entirety come to be by chance, but they did make the things subject to generation come to be by chance from the elemental principles; and so the disposition of whatever's combination [that] is in some way suited by chance to survive and to reproduce survives and reproduces, whereas the one that by chance is not so suited, does not reproduce. [They also held] that at the beginning of evolution, there were engendered animals of various limbs of different kinds—as, for example, an animal that was half stag and half goat—and that the limbs of animals were not as they now are with respect to magnitude, natural disposition, and accidental qualities but, rather, were such as chance would have it. For example, they said that the incisors are not sharp for the sake of cutting, nor are molars wide for the sake of grinding; rather, the matter chanced to combine according to this form, and this form chanced to be useful in suiting one for survival, and so the individual derives the benefit of survival from that. Sometimes, through the mechanizations of reproduction, [that individual] chanced to have an offspring—not for the sake of the preservation of the species, but simply by chance.

(6) We ourselves say that some things always occur, while others occur for the most part[2]—as, for example, fire for the most part burns wood when put into contact with it, and whoever heads from his house to his garden for the most part reaches it—whereas other things [occur] neither always nor for the most part. The things that [occur] for the most part are those that are not seldom [in their occurrence]; and so their coming to be, when they do, is either the result of a certain regularity in the nature of the cause alone ordered toward them, or it is not. If, on the one hand, it is not [due to the cause considered alone], then either the cause needs to be joined with some [other] cause (whether a

2. The Arabic *fī akthar al-amr,* and its equivalents, corresponds with the Greek *hos epi to polu,* which is technical vocabulary within the Aristotelian and Galenic systems of thought. In normal parlance it simply means "usually"; however, modern scholarship has fixed upon the somewhat cumbersome phrase, "for the most part," which is adopted here.

(٥) فرقـةٌ أخرى لم تقدم على أنْ تجعل العالم بكلّيته كائناً بالاتفاق، ولكنها جعلت الكائنات متكونة عن المباديء الأُسْطُقسّية بالاتفاق، فما اتفق أن كان هيئة اجتماعه على نمطٍ يصلح للبقاء والنسْـل بقي ونسـل، وما اتفق أنْ لم يكن كذلك لم ينسل. وأنّه قد كان في ابتداء النشـوء؛ ربّما تتولد حيوانات مختلطة الأعضاء من أنواع مختلفة، وكأنْ يكون حيوانٌ نصفه أيّلٌ ونصفه عنْز، وأنَّ أعضاء الحيوان ليست هي على ما هي عليه من المقادير والخِلَق والكيفيّات لأغراضٍ؛ بل اتفقت كذلك. مثلاً قالوا: ليست الثنايا حادّة لتقطع، ولا الأضـراس عريضة لتطحن، بل اتفقت أنْ كانت المادة تجتمع على هذه الصورة، واتفق أنْ كانت هذه الصورة نافعة في مصالح البقاء؛ فاستفاد الشخص بذلك بقاء، وبما اتفق له من آلات النسل نَسَلَ، لا ليستحفظ به النوع؛ بل إتفاقاً.

(٦) فنقـول: إنَّ الأُمـور منها ما هي دائمة، ومنها ما هي فـي أكثر الأمر، مثل أنَّ النار في أكثر الأمر تُحرق الحطب إذا لاقته، وأنَّ الخارج من بيته إلى بستانه، في كثير من الأمـر، يصـل إليه. ومنها ما ليس دائماً ولا في أكثر الأمـر؛ والأمور التي تكون في أكثر الأمـر هي التي لا تكون في أقل الأمر، فكونها - إذا كانت - لا يخلو إمّا أنْ تكون عن اطرادٍ في طبيعة السـبب إليها وحده، أو لا بكون كذلـك. فإنْ لم تكن كذلك، فإمّا أنْ يحتاج السبب إلى قرينٍ من سبب أو شريكٍ أوزوال مانعٍ أولا يحتاج. فإنْ لم يكن كذلك،

cooperative cause or the removal of some obstacle), or it does not. If it is not like that (namely, the cause does not need to be conjoined [with some other factor]), then the coming to be of [those things that occur for the most part] is no more apt to result from the cause than not, since neither the thing considered in itself and alone nor [the thing] considered along with what is joined to it selectively determines the coming to be from the not coming to be. So *x* is no more apt to result from *y* than not, and so it is not something subject to generation for the most part. If, on the other hand, it does not need the aforementioned cooperative cause, then it must in itself be something regularly ordered, unless some impediment hinders and opposes it and, owing to its being opposed, it occurs seldom.[3] From that, it is necessary that, when no impediment hinders and opposes [the cause], and its nature is unimpaired to continue along its course, then the difference between what always [occurs] and what [occurs] for the most part is that *what always [occurs]* never encounters opposition, while *what [occurs] for the most part* does encounter opposition. Also, following on that is that what [occurs] for the most part is necessary, on the condition that the obstacles and opposition have been removed. That is obvious with respect to natural things. It is equally the case with respect to volitional things; for, when the volition is firm and completely made up and the limbs are prepared to move and submit and there is no hindering cause or cause that undermines the resolve, and [moreover] the intended thing can be achieved, then it is clearly impossible that it not be achieved. Now, since what occurs always is not said to come to be by luck insofar as it always comes to be, then, likewise, it should not be said that what [occurs] for the most part comes to be by luck, for the two are alike in kind and in status. Certainly, when it is opposed, it turns away, and so it might be said that its being turned from its course comes to be by chance or luck; but you also know that people do not say that what results for the most part from one and the same cause or [occurs] always comes to be by chance or luck.

3. Literally, "it no longer fails to be seldom."

ولم يحتج السبب إلى قرينٍ، فليس كونها عن السبب أولى من لا كونها ؛ إذ ليس في نفس الأمـر ، لا فيه وحده، ولا فيه، ولا في مقارنٍ له مـا يرجح الكون على اللاكون، فيكون كون هذا الشيء عن الشيء ليس أولى من لا كونه، فليس كائناً على الأكثر. فإذن - إنْ لم يحتج إلى الشـريك المذكور - فيجب أنْ يكون مطرداً بنفســه إليه، إلاّ أنْ يعوق عائقٌ ويعـارض معارضٌ ولمعارضته ما تخلّف في الأقـل. ويجب من ذلك أنّه إذا لم يعق عائقٌ ولم يعارض معارضٌ وسـلمت طبيعته أنْ يسـتمر إلى ما ينحوه، فحينئذ يكون الفرق بين الدائـم والأكثري أنَّ الدائم لا يعارضه معارضٌ البتة، وأنَّ الأكثري يعارضه معارضٌ. ويتبع ذلك أنَّ الأكثري - بشـرط رفع الموانع وإماطة العـوارض - واجبٌ، وذلك في الأمور الطبيعيــة ظاهرٌ وفي الأمور الإرادية أيضاً. فإنَّ الإرادة إذا صحّتْ وتمّتْ وآتتْ الأعضاء للحركة والطاعة، ولم يقع سـببٌ مانعٌ أو سببٌ ناقضٌ للعزيمة، وكان المقصود من شأنه أُنْ يوصل إليه؛ فبيّنٌ أنّه يسـتحيل أنْ لا يوصل إليه. وإذا كان الدائم، من حيث هو دائم، لا يقال إنَّه كائن بالبخْت، فالأكثري أيضاً لا يقال إنَّه كائنٌ بالبخْت، فأنه من جنســه وفي مثل حكمه. نعم إذا عورض فصرف فربّما قيل انصرافه عن وجهته كائن بالبخت أو بالاتفاق. وأنـت تعلم؛ أنَّ الناس لا يقولون لما يكون كثيراً عن سـببٍ واحدٍ بعينه أو دائماً إنه كائنٌ اتفاقاً أو بالبخْت.

(7) What remains for us to do is to consider what comes to be [and does not come to be] equally, and what comes to be seldom. Now, the issue concerning what comes to be equally seems to be whether or not to say it just chanced to be by chance or luck. Now, modern Peripatetics made it a condition of being by chance and luck that it only concerns things that seldom come to be from their causes,[4] whereas the one who worked out this course for them [namely, Aristotle] did not make that a condition and instead only made it a condition that it not come to be always and for the most part.[5] What incited the moderns to associate chance with things that seldom come to be to the exclusion of what comes to be equally was the form present in voluntary affairs, for these moderns say that eating and not eating, walking and not walking, and the like are things that proceed equally from their principles; and yet, when one voluntarily walks or eats, we do not say that that was a matter of chance.

(8) As for ourselves, we do not approve of any addition to the condition that their teacher made and shall lay bare the error of their position by something well known—namely, that in one respect and from one perspective, one and the same thing might come to be for the most part (in fact, be necessary), while in another respect and from another perspective, it comes to be equally. In fact, when certain conditions are made about what seldom comes to be and certain states are taken into account, it [too] becomes necessary. An example would be that during the coming to be of the embryonic palm, it is made a condition that the matter exceed that which is reserved for five fingers and that the divine power emanating into the bodies encounters a perfect preparedness in a given matter whose nature deserves a certain form, and also, having encountered that, [the divine power] does not forgo providing [the matter] with [the form]. In that case, an additional finger will necessarily be created.[6] This class, even if it is uncommon and the possibility is quite seldom in relation to the universal nature, it is not uncommon and seldom

4. Cf. Ibn al-Samḥ, who made this point clearly, and John Philoponus and Abū Bishr, who strongly suggested it in their comments to Aristotle's *Physics* 2.5.196b10ff. The commentaries of all these expositors can be found in *Arisṭuṭālīs: al-Ṭabīʿī*, ed. ʿA. Badawī, 2 vols. (Cairo: The General Egyptian Book Organization, 1964–65; henceforth, the Arabic *Physics*).

5. Aristotle, *Physics* 2.5.196b10–13.

6. Reading *yatakhallaqu* with **Z**, **T**, and the Latin (*creetur*) for **Y**'s *yatakhallafu* (to lag behind or be absent).

(٧) وقد بقي لنا ما يكون بالتساوي وما يكون على الأقل، والأمر مشتبه في الكائن بالتساوي أنَّه هل يقال فيه أنَّه اتفق إتفاقاً وكان بالبخْت، أو لا يقال؟ وقد اشترط متأخرو المشائين أنَّ ما يكون بالاتفاق والبخت فإنَّما يكون في الأمور الأقلّية الكون عن أسبابها. والذي رسم لهم هذا النهج لم يشترط ذلك، بل اشترط أنْ لا يكون دائماً ولا أكثرياً. وإنَّ ما دعا المتأخرين إلى أنْ جعلوا الاتفاق متعلقاً بالأمور الأقلّية دون المتساوية صورة الحال في الأمور الإرادية. فإنَّ هؤلاء المتأخرين يقولون إنَّ الأكل واللاآكل والمشي واللامشي، وما أشبه ذلك، هي من الأمور المتساوية الصدور عن مبادئها، ثم إذا مشى ماش أو أكل آكل بإرادته لم يقل إنَّه اتفق ذلك.

(٨) وأمّا نحن فلا نستصوب زيادة على ما اشترطه معلّمهم، ونبيّن بطلان قولهم بشيءٍ يسير وهو: أنَّ الشيء الواحد قد يكون بقياس واعتبارٍ أكثرياً بل واجباً، وبقياس واعتبارٍ آخر متساوياً، بل الأقلي إذا أُشترطت فيه شرائطُ واعتُبرت أحوالٌ صار واجباً؛ مثل أنْ يُشترط أنَّ المادّة في كَوْن كف الجنين فضلت عن المصروف منها إلى الأصابع الخمس، والقوة الإلهية الفائضة في الأجسام صادفت استعداداً تاماً في مادة طبيعية لصورة مستحقة، وهي أيضاً صادفت ذلك، لم تعطّلها عنها، فيجب هناك أنْ يتخلّق إصبعٌ زائدة فيكون هذا الباب - وإنْ كان هو أقلّي الإمكان ونادراً بالقياس إلى الطبيعة الكلّية - فليس أقلّياً ونادراً بالقياس إلى الأسباب التي ذكرناها، بل هو واجبٌ. ولعل الاستقصاء

in relation to the causes we mentioned, but necessary. Perhaps a thorough investigation would reveal that the thing does not exist necessarily from its causes and does not come to exist from the nature of possibility; however, the explanation of this and what is like it will have to wait until first philosophy.[7]

(9) When the situation is like this, it is not improbable that a single nature in relation to one thing is for the most part, while in relation to another thing it is equal, for the gap between what is for the most part and what is equal is narrower than that between what is necessary and what is seldom. Again, when eating and walking are related to the will, which is assumed to be fully determined, then the two shift from coming to be equally to coming to be for the most part, and [once they] have so shifted, it is not at all correct to say that they are matters of chance and came to be by luck. When, however, they are not related to the will, but are considered in themselves at some time when eating and not eating are equal,[8] then it is correctly said, "I visited him, and as chance would have it he was eating," where that is related to the visiting [and] not to the will. The same holds should someone say, "I bumped into him while he chanced to be walking" or "I met him while he chanced to be sitting," for all of this is recognized and accepted and yet true. In general, when the thing that comes to be is considered in itself and is neither the object of attention nor what is expected (since, in that case, it would not be always and for the most part), it is correct to say of the cause leading to it that it is either by chance or luck—namely, when [the cause] is of such a character to lead to it, but does not lead to it, always and for the most part. When it absolutely and necessarily never leads to it, such as someone's sitting during a lunar eclipse, then we don't say that so-and-so's sitting chanced to be a cause of the lunar eclipse. It is correct, however, to say that [the two] occurred together by chance, in which case the sitting is not a cause of the eclipse, but it is an accidental cause of occurring together with the eclipse. To occur together with the eclipse, however, is not the same as [causing] the eclipse.

---

7. See *Ilāhīyāt* 1.6 and 6.1–2.

8. Reading *yatasāwā* with **Z**, **T**, and the Latin (*aeque*), which seem to have been inadvertently omitted in **Y**.

في البحث يبيّن أنَّ الشيء ما لم يجب أنْ يوجد من أسبابه ولم يخرج عن طبيعة الإمكان، لم يوجد عنها؛ ولكن بيان هذا وأمثاله مؤخّر إلى الفلسفة الأُولى.

(٩) وإذا كان الأمر على هذا، فغير بعيد أنْ تكون طبيعة واحدة بالقياس إلى شيءٍ أكثرية، وبالقياس إلى شيءٍ آخر متساوية، فإنَّ البُعْد ما بين الأكثري والمتساوي أقرب من البُعْد ما بين الواجب والأقلي. ثم الأكل والمشي إذا قيسا إلى الإرادة وفرضت الإرادة حاصلة؛ خرجا عن حدَّ الإمكان المتساوي إلى الأكثري، وإذا خرجا من ذلك لم يصح البتة أنْ يقال إنَّهما اتفقا أو كانا بالبخْت. وإمّا إذا لم يضافا إلى الإرادة ونظر إليهما نفسيهما في وقت يتساوى كوْن الأكل ولا كونه، فصحيح أنْ يقال: دخلتُ عليه واتفق أنْ كان يأكل، وذلك بالقياس إلى الدخول لا إلى الإرادة. وكذلك قول القائل: صادفته واتفق أنْ كان يمشي، ولقيته واتفق أنْ كان قاعداً، فإنَّ هذا كلّه متعارفٌ مقبولٌ، ومع ذلك صحيح. وبالجملة إذا كان الأمر الكائن في نفسه غير متطلّع ولا متوقع – إذ ليس دائماً ولا أكثرياً – فصالحٌ أنْ يقال للسبب المؤدي إليه أنَّه اتفاقٌ أو بختٌ، وذلك إذا كان من شأنه أنْ يؤدي إليه، وليس مؤدياً إليه ولا دائماً ولا أكثرياً. وأمّا إذا لم يكن مؤدياً إليه البتة ولا موجباً له؛ مثل قعود فلان عند كسوف القمر، فلا يقال إنَّ قعود فلان اتفق أنْ كان سبباً لكسوف القمر، بل يصلح أنْ يقال: اتفق أنْ كان معه فيكون القعود لا سبباً للكسوف بل سبباً بالعرض للكون مع الكسوف، وليس الكون مع الكسوف هو الكسوف.

(10) To sum up, when *x* is not at all of the character so as to lead to *y*, then it is not a chance cause of *y*; *x* is a chance cause of *y* only when *x* is of such a character as to lead to *y*, but not always and for the most part. Taken at its extreme, if the agent were aware of the course of the universe's motions and he truly intended to and chose to, then he would, in fact, make [that course] a given end. It is just as if someone were going to the market [and] were aware that somebody who owes him money was on the way [there as well]; he would, in fact, make [going to the market] an end. In this case, there is a shift away from occurring equally and seldom, since to take a course that one knows the debtor is presently taking does for the most part lead to encountering him, whereas inasmuch as he does not know [his debtor's whereabouts], going out might or might not lead [to encountering him]. It is by chance only in relation to going out without any additional condition, while it is not by chance when a certain additional condition [namely, knowing the whereabouts of the debtor] is added to going out. From this, it should be clear that when there are chance causes, they are for the sake of something, except that they are their efficient causes accidentally, and the ends are accidental ends and are included among the causes that are accidental. So chance is a cause of natural and volitional things accidentally, necessitating neither [their occurrence] always nor for the most part. In other words, it concerns what is for the sake of something whose cause does not necessitate it essentially. Also, something might happen to be neither by intention nor by chance—as, for example, leaving footprints on the ground when going to overtake the debtor, for even if that was not intended, it is a necessary effect of what was intended.

(١٠) وبالجملة إذا كان الشيء ليس من شأنه أنْ يؤدي لشيءٍ البتة، فليس سبباً اتفاقياً له. إنّما يكون سبباً اتفاقياً له؛ إذا كان من شأنه أنْ يؤدي إليه وليس دائماً ولا في أكثر الأمر، حتى لو فطن الفاعل بما تجري عليه حركات الكلّ، وصحّ أنْ يريد ويختار؛ لصحّ أنْ يجعله غاية. كما لو فطن الخارج إلى السوق أنَّ الغريم في الطريق؛ لصحّ أنْ يجعله غاية، وكان حينئذ خارجاً عن حدّ التساوي والأقلي، لأنَّ خروج العارف بحصول الغريم في جهة مخرجه؛ يؤدي فى أكثر الأمر إلى مصادفته، وأمّا خروج غير العارف من حيث هو غير عارف؛ فربّما أدى وربّما لم يؤدِ، وإنّما يكون إتفاقاً بالقياس إلى الخروج لا بشرطٍ زائد، ويكون غير إتفاقٍ بالإضافة إلى خروج بشرط زائد. ويتبيّن من هذا أنَّ الأسباب الاتفاقية تكون حيث تكون من أجل شيءٍ، إلاّ أنّها أسبابٌ فاعلة لها بالعرض، والغايات غايات بالعرض، فهي داخلة في جملة الأسباب التي بالعرض. فالاتفاق سببٌ من الأمور الطبيعية والإرادية بالعرض، ليس دائم الإيجاب ولا أكثري الإيجاب، وهو فيما يكون من أجل شيء ليس له سببٌ أوجبه بالذات. وقد تعرض أمورٌ لا بقصدٍ وليست بالاتفاق؛ مثل تخطيط القدم على الأرض عند الخروج إلى أخذ الغريم، فإنَّ ذلك - وإنْ لم يُقصد - فضروري في المقصود.

(11) Now, one could claim that frequently we say that such-and-such happened by chance even though it occurs for the most part. An example would be one who says, "I sought Zayd for the sake of some need or other and I chanced to find him at home," which does not prevent him from saying, "Zayd is at home for the most part." The response is that this person says the latter not only by considering the thing in itself, but also by considering his belief about [Zayd]. So [for example] when his overwhelming opinion is that Zayd should be at home, then he would not say that [his being at home] is by chance, but, rather, he would say it was by chance if he did *not* find him at home. He says the former, however, only when it seems in his opinion, at that time and in that situation, that [Zayd] is equally either at home or not. So at *that* time his opinion is to judge [the two] as being equal and not as being for the most part or necessary, even if, in relation to the time generally, it is for the most part.

(12) Concerning many of the natural things whose existence is rare—such as the gold vein whose [amount of gold] defies being weighed or the sapphire of unprecedented magnitude—it might be supposed that they exist by chance, since they are seldom. That is not so. What seldom occurs enters into the ranks of what is by chance *not only* when it is considered in relation to existence generally, *but also* when it is considered in relation to its efficient cause, and so its existence seldom results from [that efficient cause]. That gold vein and that sapphire, however, proceeded from their efficient cause precisely because of its power and the two instances of a wealth of abundant matter. Given that that is the case, something like this action would proceed essentially and naturally, either always or for the most part.

(١١) لكن لقائل أنْ يقول إنَّا ربّما قلنا إنَّ كذا كان بالاتفاق، وإنْ كان الأمر أكثرياً كقول القائل: إنَّ فلاناً قصدته لحاجة كذا فاتفق أنْ وجدته في البيت، ولا يمنعه عن هذا القول كون زيد في أكثر الأمر في البيت. فالجواب أنَّ هذا القائل إنّما يقول ذلك لا بحسب الأمر في نفسه؛ بل بحسب اعتقاده فيه. فإنَّه إذا كان أغلب ظنّه أنَّ زيداً ينبغي أنْ يكون في البيت، فلا يقول إنَّ ذلك اتفق؛ بل إنْ لم يجده يقول إنَّ ذلك اتفق. ولكن إنّما يقول هذا إذا كان يتساوى عنده في ظنّه في ذلك الوقت وفي تلك الحال أنّه كائن في البيت أو غير كائنٍ، فيكون ظنّه في ذلك الوقت يحكم بالتساوي دون الأكثري والواجب، وإنْ كان بالقياس إلى الوقت المطلق أكثرياً.

(١٢) وقد يُظّنّ في كثيرٍ من الأمور الطبيعة النادرة الوجود - مثل الذهب الثابت على وزْنٍ من الأوزان أو الياقوته المجاوزة للمقدار المعهود - أنّه موجودٌ بالاتفاق لأنّه أقلّي، وليس كذلك. فإنَّ كون الشيء في الأقل إنّما يدخل الشيء في الاتفاق، لا إذا قيس إلى الوجود المطلق، بل إذا قيس إلى السبب الفاعل له، فكان وجوده عنه أقلّياً، والسبب الفاعل لهذا الذهب والياقوت إنّما صدر عنه ذلك لقوته، ولوجدان المادة الوافرة. وإذا كان كذلك فيصدر مثل هذا الفعل عن ذاته دائماً أو في الأكثر صدوراً طبيعياً.

(13) We say that the chance cause might sometimes lead to its essential end and sometimes might not. For example, when the man leaves, headed for his shop, and then by chance comes across one who owes him money, that might interrupt his essential goal [of going to his shop], or it might not and instead he continues along his way until he arrives. Also, should a falling rock fracture a head, it might either lodge there or it might continue downward to its place of descent. So, if it reaches its natural end, it is an essential cause relative to it, while relative to the accidental end, it is a chance cause. If it does not reach it, it is [still] a chance cause relative to the accidental end; but relative to the essential end, it is in vain (just as they say, "he drank medicine in order to be purged, but he was not purged, and so he drank in vain"), while the accidental end relative to it is by chance. It might be supposed that certain things are and come to be, not for some end, but on a whim—though not by chance—such as a desire for a beard and the like. This is not so, and in first philosophy we shall explain the real state of affairs about such cases.[9]

(14) Now, chance is more general than luck in our language, for every instance of luck is an instance of chance, but not every instance of chance is an instance of luck. So it is as if *luck* is said only of what leads to something of account, where its principle is a volition resulting from rational and mature individuals having a choice. If [*luck*], then, is said of something other than one such as that—as, for example, it is said of the piece of wood that is split and whose one half is used for a mosque while its other half is used for a public lavatory, that its one half is fortunate, while its other half is unfortunate[10]—then it is said metaphorically. Anything whose principle is a nature is not said to come to be by luck and instead might be designated more properly as coming to be *spontaneously*, unless it is related to some other voluntary principle.

9. Cf. *Ilāhīyāt* 6.5.

10. Reading *shaqīy* with **Z**, **T**, and the Latin *infortunata* for **Y**'s *shay*ʾ (thing).

(١٣) ونقول إنَّ السبب الاتفاقي قد يجوز أنْ يتأدى إلى غايته الذاتية وقد يجوز أنْ لا يتأدى؛ مثل أنَّ الرجل إذا خرج متوجهاً إلى متجره فلقي غريمه اتفاقاً، فربّما انقطع بذلك عن غايته الذاتية وربّما لم ينقطع، بل توجّه نحوها ووصل إليها. والحجر الهابط إذا شجَّ رأساً فربّما وقفَ وربّما هبط إلى مهبطه، فإنّه إذا وصل إلى غايته الطبيعية فيكون بالقياس إليها سبباً ذاتياً، وبالقياس إلى الغاية العرضية سبباً اتفاقياً. وأمّا إنْ لم يصل إليها فإنّه يكون بالقياس إلى الغاية العرضية سبباً اتفاقياً، وبالقياس إلى الغاية الذاتية باطلاً؛ كقولهم شرب الدواء ليسْهل فلم يسْهل، فكان شربه باطلاً، والغاية العرضية بالقياس إليها تكون اتفاقاً. وقد يظن أنَّه قد يكون وتحدث أمورٌ لا لغاية بل على سبيل العبث، ولا تكون اتفاقاً كالولوع باللحية وما أشبه ذلك، وليس كذلك - وسنبيّن في الفلسفة الأولى حقيقة الأمر فيها.

(١٤) ثم الاتفاق أعمّ من البخْت في لغتنا هذه؛ فإنَّ كل بخْتٍ اتفاقٌ، وليس كل اتفاقٍ بخْتاً، فكأنّهم لا يقولون بخْت إلاّ لما يؤدي إلى شيء يُعتدّ به، ومبدؤه إرادة عن ذي اختيار من الناطقين البالغين. فإنْ قالوا لغير ذلك؛ كما يقال للعود الذي يُشقّ نصفه لمسجدٍ ونصفه لكنيفٍ إنَّ نصفاً منه سعيد ونصفاً منه شقي؛ فهو مجاز. وأمّا ما مبدؤه طبيعي فلا يقال إنَّه كائن بالبخْت؛ بل عسى أنْ يُخصّ باسم الكائن من تلقاء نفسه، إلاّ إذا قيس

So chance events proceed according to various interactions that occur between two or more things. Now, with respect to each member of the interaction, either they both move until they collide and interact, or one of them is at rest and the other is moving toward it; for if they are both at rest in some noninteracting state as they were, then no[11] interaction between them will result. Consequently, there can be two motions from two principles (one of which is natural and the other volitional) that chance to interact vis-à-vis a single end, which in relation to the voluntary [agent] is either accounted good or evil, and so it has [either good or bad] luck, but in relation to the natural motion, it is not luck.

(15) There is a difference between bad luck and bad planning. Bad planning is to choose some cause that, for the most part, leads to some blameworthy end, whereas bad luck is such that the cause, for the most part, does not lead to some blameworthy end but, unfortunately, in the case of the one being held responsible for it, it did so [on this occasion]. A streak of good luck is that which, when it occurs, [brings about], as luck would have it, the repeated occurrence of a number of fortunate causes, whereas a streak of bad luck is that which, when it occurs, [brings about], as luck would have it, the repeated occurrence of a number of unfortunate causes. So, in the first case, one begins to expect the continual repetition of the good that had repeatedly occurred, while in the second case, one expects the continual repetition of the evil that had repeatedly occurred. Now, a single chance cause might have any number of chance ends; and thus, one does not guard against chance the way one guards against essential causes but [instead] seeks refuge in God against misfortune.

---

11. **Y** seems to have inadvertently omitted the negation *lam*, which is verified in **Z**, **T**, and the Latin.

إلى مبدأ آخر إرادي. فإنَّ الأمور الاتفاقية تجري على مصادماتٍ شتّى تحصل بين شيئين أو أشياء، وكل مصادمةٍ فإمّا أنْ يكون فيها كلا المتصادمين متحركين إلى أنْ يتصادما، أو يكون أحدهما ساكناً والآخر متحرّكاً إليه. فإنَّه إذا سكن كلاهما على حالٍ غير التصادم الذي كانا عليه لم ينتج ما بينهما تصادم. وإذا كان كذلك، فجائز أنْ تتفق حركتان من مبدأين أحدهما طبيعي والآخر إرادي، يتصادمان عند غاية واحدة تكون بالقياس إلى الإرادي خيراً يُعتدّ به أو شرّاً يعتدّ به، فيكون حينئذ بخْتاً له ولا يكون بالقياس إلى حركة الطبيعي بخْتاً.

(١٥) وفرقٌ بين رداءة البخْت وسوء التدبير فإنَّ سوء التدبير هو اختيار سببٍ في أكثر الأمر يؤدي إلى غايةٍ مذمومة. ورداءة البخْت هو أنْ يكون السبب في أكثر الأمر غير مؤدٍ إلى غايةٍ مذمومة ولكن تكون عند متولّيها السيء البخْت تؤدي إليه. والشيء الميمون هو الذي قد تكرّر حصول أسبابٍ مُسْعِدة بالبخْت عند حصوله، والشيء المَشْؤوم هو الذي تكرّر حصولِ أسبابٍ مُشْقِية بالبخْت عند حصوله فيستشْعر من حضور الأول عَوْد ما اعتيد تكرّره من الخير، ومن حضور الثاني عَوْد ما اعبيد تكرّره من الشّر. وقد يكون للسبب الواحد الاتفاقي غايات اتفاقية غير محدودة، ولذلك لا يتحرّز عن الاتفاق التحرّز عن الأسباب الذاتية. ويُستعاذ بالله من الشقاوة.

Chapter Fourteen

Some of the arguments of those who were in error concerning chance and luck and the refutation of their views

(1) Since we have explained the essence and existence of chance, we should indicate some of the arguments upon which rest the false views about the class of chance things, even if this explanation might more fittingly wait until metaphysics and first philosophy, since most of the premises that we adopt in explaining this simply have to be asserted [here]. Be that as it may, we shall accommodate tradition in this and certain other analogous cases.

(2) The view that denies chance outright and requires that everything have some known cause, and that we are not forced to contrive some cause that is [called] *chance,* rests upon a proof that does not strictly lead to the desired conclusion. [That] is because it does not follow that when everything has a cause, chance does not exist; rather, the existing cause of something that is not necessitated either always or for the most part is itself the chance cause inasmuch as it is such. Their claim that frequently a single thing has many simultaneous ends involves a fallacy of equivocation concerning the term *end,* for *end* might be said of whatever something ends at, however it might be, and it might be said of that which is actually intended, where both what is intended by nature and what is intended by volition are something definite. We mean here by *essential end* the latter. They say that the end does not cease to be an end by stipulation, such that when one stipulates *catching up with the debtor* as an end, it is not by luck, while if one stipulates *reaching the shop* as an end, it is by luck. The response is not to grant the claim that stipulation does not change the state of this class, unless you believe that the stipulation will make the thing in the one case occur for the most part, whereas in the other case [you believe that] it will occur seldom. Now, certainly, going out in order to catch up with a debtor whose whereabouts are known does, as such, lead for the most part to catching up with him, while going to the shop without such knowledge does not

<الفصل الرابع عشر>

في بعض حُجج مَنْ أخطأ في باب الاتفاق والبخْت ونقْض مذاهبهم

(١) وإذْ قد بينّا ماهيّة الاتفاق ووجوده، فحريٌ بنا أنْ نشير إلى ⟨بعض⟩ حجج المذاهب الفاسدة في باب الاتفاق؛ وإنْ كان الأحرى أنْ يؤخّر هذا البيان إلى ما بعد الطبيعة وإلى الفلسفة الأولى؛ فإنَّ المقدمات التي نأخذها في هذا البيان أكثرها مصادرات لكنّا ساعدنا في هذا الواحد، وفي بعض الأشياء الأُخرى، مجرى العادة؛

(٢) فنقول: أمّا المذهب المبطل للاتفاق أصلاً المحتجّ بأنَّ كل شيءٍ يوجد له سببٌ معلومٌ ولا نضطر إلى اختلاق سببٍ هو الاتفاق؛ فإنَّ احتجاجه ليس ينتج المطلوب، لأنَّه ليس إذا وجد لكل شيءٍ سببٌ لم يكن للاتفاق وجود، بل كان السبب الموجود للشيء الذي لا يوجبه على الدوام أو الأكثر هو السبب الاتفاقي نفسه من حيث هو كذلك. وأمّا قوله إنَّه قد يكون لشيءٍ واحد غايات كثيرة معاً، فإنَّ المغالطة فيه لاشتراك الاسم في الغاية؛ فإنَّ الغاية تقال لما ينتهي إليه الشيء كيف كان، وتقال لما يُقصد بالفعل. والمقصود بالحركة الطبيعية محدودٌ، والمقصود بالإرادة أيضاً محدودٌ، ونحن نعني بالغاية – ها هنا – الذاتية هذا. وقوله: إنه ليس يجب أنْ تصير الغاية غير غاية بالجعْل حتى إذا جعل الظفر بالغريم غاية صار الأمر غير بخْتي، وإنْ جعل الوصول إلى الدكان غاية صار الأمر بخْتياً! فإنَّ الجواب عنه إنَّ قوله إنَّ الجعْل لا يغيّر الحال في هذا الباب هو غير مسلّم؛ ألا ترى أنَّ، الجعْل يجعل الأمر في أحدهما أكثرياً وفي الآخر أقلّياً؛ فإنَّ الشاعر بمقام الغريم الخارج إليه به من حيث هو كذلك، فإنَّه في أكثر الأمر يظفر به، وغير الشاعر الخارج

lead for the most part to catching the debtor. So, if different stipulations can bring about a different status with respect to something's being for the most part as well as not being so, then, likewise, it can bring about a different status with respect to the thing's being by chance or not.

(3) One of the ways to expose the error of Democritus's view—who makes the world come to be by chance, while believing that the things subject to generation are such by nature—is for us explain the essence of chance to him—namely, that it is an accidental end for the sake of something that is either natural, volitional, or forced. Now, what is forced ultimately terminates at some nature or volition, and so it should be obvious that what is forced cannot form an infinite series of forced [causes]. So nature and volition are, in themselves, prior to chance, in which case the ultimate cause of the world is either a nature or volition. Now, he assumes that the bodies that he professes are solid, substantially alike, differing [only] in their shape, and are moved essentially in a void, and then suddenly combine and touch. He also assumes that there is no power or form, but only shape; and, indeed, that the combination of [these bodies] and what their shapes require does not permanently fuse one to another, but, rather, [that] they can separate and continue on with their essential motion. [Given these assumptions], then, [those bodies] must be moved essentially so as to become separated without the continuous combination formed from them remaining. If that were the case, however, the Heavens would not continue to exist according to a single configuration during successive astronomical observations covering a long period of time. Now, if he said that among these bodies, there are different strengths in relation to their substances' chance collisions as well as what is compressed between them, and [that] when one of the weaker ones stands between two compressing agents whose power of compression is exactly balanced, then [the weaker one] will remain like that, it might appear as if he said something of consequence—[that is], until we explain that there is nothing to this and that it is not by chance, which we shall do later.[1] What is truly amazing is the chances that one

1. The reference might be to 4.11.10, but see also *Kitāb al-najāt*, ed. Muhammad Danishpazhuh (Tehran: Dānishgāh-yi Tihrān, 1985; henceforth, the *Najāt*), 298–99, where Avicenna explicitly denies that all bodies naturally descend, which certainly has Democritus as its target, and where much of the language—for example, to compress (variations of *Ḍ-Gh-T*)—is identical to the present passage.

إلى الدكان، من حيث هو كذلك، فإنَّه ليس في أكثر الأمر يظفر بغريمه. فإنَّ كان الجعْل المختلف يختلف له حكم الأمر في أكثريته وغير أكثريته، فكذلك يختلف له حكم الأمر في أنَّه اتفاقي أو غير اتفاقي.

(٣) وأمّا ‹ديموكريتس› الذي يجعل تكوّن العالم بالاتفاق، ويرى أنَّ الكائناتِ كائناتٌ بالطبيعة، فممّا يكشف فساد رأيه هو أنْ نبيّن له ماهيّة الاتفاق وأنَّه غاية عرضية لأمر طبيعي أو إرادي أو ‹قسري› والقَسْر ينتهي إلى طبيعةٍ أو إرادة. فإنَّه سيظهر أنَّه لا يستمر قَسْرٌ على قَسْر إلى غير النهاية، فتكون الطبيعة والإرادة في ذاتهما أقدم من الاتفاق، فيكون السبب الأول للعالم طبيعة أو إرادة. على أنَّ الأجرام التي يقول بها ويراها صلبة ويراها متفقة الجواهر مختلفة بالأشكال، ويراها متحركة بذاتها في الخلاء إذا اجتمعت وتماست، ولا قوة عنده ولا صورة إلاّ الشكل فقط، فإنَّ اجتماعها ومقتضى أشكالها لا يلصق بعضها ببعض، بل يجوز لها الانفصال واستمرار حركتها التي لها بذاتها. فيجب لذاتها أنْ تتحرّك فتنفصل ولا يبقى لها الاتصال. ولو كان كذلك؛ لما وجدت السماء مستمرة الوجود على هيئةٍ واحدةٍ في أرصادٍ متتابعةٍ بين طرفي زمان طويل! ولو كان يقول: إنَّ في هذه الأجرام قوى مختلفة في جواهرها، يتفق لها أنْ تتصادم ويضغط ما بينها، ويقف الضعيف منها بين الضاغطين ويتكافأ ميل الضاغطين بحسب القوتين فيبقى كذلك، لكان ربّما أوهم أنَّه يقول شيئاً؛ إلى أنْ نبيّن أنَّ هذا لا يكون ولا يتفق – وسنشير إليه بعد. والعجب أنَّه يجعل الأمر الدائم، الذي لا يقع فيه خروجٌ عن نظام واحدٍ ولا

would make something that is eternal, in which there is no deviation from a single order and that does not come to be in time, to be by luck or to involve chance, while making particular things that appear to involve chance be for the sake of some end!

(4) Empedocles and those following him made the particulars occur by chance but have confused chance with necessity and so made the material occur by chance, while its being informed with the form it has is by necessity and not for the sake of some end.[2] For example, they said that incisors are not sharp for the sake of cutting; but, rather, a certain matter that is susceptible to only this form chanced to occur, and so they are necessarily sharp.[3] They were inclined to pretty feeble arguments concerning this topic, saying the following: How can nature act for the sake of something when it cannot deliberate? Also, if nature were to act for the sake of something, then there would not be any deformities, additional appendages, and death in nature at all, since these are unintended states. The fact is, [they maintain,] that the matter chanced to be in a certain state, upon which these states followed. The same would hold for the rest of the natural things if they, in some way, chanced to possess something beneficial, and yet [that benefit] was not associated with chance and the necessity of matter, but, rather, was supposed to proceed only from some agent acting for the sake of some thing. Now, if that were the case, there would only be things that are eternal, perpetual, and invariant. This is like the rain, which, we know with absolute certainty, comes to be on account of the necessity of matter, since when the Sun causes evaporation and then the vapors reach the cooler air, they are cooled and the water becomes heavy and then necessarily falls. As chance would have it, certain benefits result, and so it is supposed that the rain was intended by nature for those benefits, but, they add, no notice is taken of [the rain's] destroying [crops] on the threshing-floor.

---

2. Cf. *Physics* 2.8.99a10ff. and *Generation and Corruption* 2.6.333b4ff.

3. Reading *istaḥddat* with **Z**, **T**, and the Latin (*fuerunt acuti*) for **Y**'s *istaḥdatha* (to renew).

أمـر حادث، كائنٌ ببخْتٍ أو اتفـاقٍ فيه؛ إتفاقياً. ويجعل الأمـور الجزئية لغايةٍ وفيها ما يرى الاتفاق.

(٤) وأمّا <أمبيذُكْلِس> ومَنْ جرى مجراه، فإنَّهم جعلوا الجزئيات تكون بالاتفاق، بل خلطوا الاتفاق بالضرورة، فجعلوا حصول المادة بالاتفاق وتصوّرها بصورتها بالضرورة، لا لغاية. مثلاً قالوا: إنَّ الثنايا لم تستحد للقطع؛ بل اتفق أنْ حصلت هناك مادة لا تقبل إلاّ هذه الصورة فاستحدّت بالضرورة. وقد أخلدوا في هذا الباب إلى حجج واهية قالوا: كيف تكون الطبيعة تفعل لأجل شيءٍ وليس لها رويّة، ولو كانت الطبيعة تفعل لأجل شيءٍ لما كانت التشويهات والزوائد والموت في الطبيعة؛ البتة. فإنَّ هذه الأحوال ليست بقصد، ولكن يتفق أنْ تكون المادة بحالةٍ تتبعها هذه الأحوال، فكذلك الحكم في سائر الأمور الطبيعية التي اتفقت أنْ كانت على وجهٍ يتضمن المصلحة فلم تُنسب إلى الاتفاق وإلى ضرورة المادة، بل ظُنَّ أنّها إنّما تصدر عن فاعلٍ يفعل لأجل شيءٍ، ولو كان كذلك، لما كان إلاّ أبداً دائماً لا يختلف. وهذا كالمطر الذي يعلم يقيناً أنَّه كائنٌ لضرورة المادة؛ لأنَّ الشّمس إذا بخّرت فخلص البخار إلى الجو البارد برد فصار ماء ثقيلاً، فنزل ضرورة، فاتفق أنْ يقع في مصالح؛ فظُن أنَّ الأمطار مقصودة في الطبيعة لتلك المصالح، وقالوا – ولم يُلتفت إلى فسادها للبيادر.

(5) They also said something else that was misleading about this topic—namely, that the order found in [both] the generation and passing of natural things follows what the necessity in the materials requires; and this is something about which there should be no mistake. So, if it is conceded that development and generation are ordered, then, indeed, reverting and corruption will be ordered no less than that former order—namely, the order of deteriorating from [the natural thing's] beginning to its end, which is just the reverse of the order of development. In that case, however, it also ought to be supposed that deterioration is for the sake of some thing—namely, death. Moreover, if nature acts for the sake of some thing, then the question remains with respect to that thing itself, namely, "Why did it naturally act the way it did?" and so on *ad infinitum.*[4] Also, they asked: How can nature act for the sake of some thing, while one and the same nature produces different actions on account of material differences, like heat's melting certain things, such as wax, while congealing others, such as egg and salt? Now, what would truly be remarkable is that the heat produces burning for the sake of something. That fact is that that [burning] follows on [the heat] by necessity precisely because the matter is in a certain state, with respect to which it must burn when placed in contact with something hot; and the same holds for the rest of the natural powers.

(6) What we should say and believe about this topic for the nonce is that there is not much dispute about including chance in the generation of natural things—namely, in relation to their individual instances. So neither that *this* clod of dirt occurs at *this* part of the world, nor that *this*

---

4. For example, one asks, "Why are incisors sharp?" and the response is "For the sake of cutting," to which it is asked, "Why are they for the sake of cutting?" and it might be responded, "For the sake of facilitating digestion." Such "why"-questions might either go on without end—which, in fact, undermines the position that nature acts for some end—or they terminate at some end that chances to be beneficial, which is the position of Empedocles and his followers.

(٥) قالوا: وقد عرض في هذا الباب أمرٌ آخر وهو النظام الموجود في تكوّن الأمور الطبيعية وسلوكها إلى ما توجبه الضرورة التي في المواد، وليس ذلك مما يجب أنْ يُغْتر به. فإنّه - وإنْ سُلّم أنَّ للنشوء والتّكون نظاماً - فإنَّ للرجوع والسلوك إلى الفساد نظاماً ليس دون ذلك النظام؛ وهو نظام الذبول من أوله إلى آخره، بعكس من نظام النشوء، فكان يجب أيضاً أنْ يُظنّ أنَّ الذبول لأجل شيءٍ هو الموت. ثم إنْ كانت الطبيعة تفعل لأجل شيء؛ فالسؤال ثابتٌ في ذلك الشيء نفسه؛ وأنَّه لِمَ فِعْلٌ في الطبيعة على ما هو عليه؟ وتستمر المطالبة إلى غير نهاية. قالوا: وكيف تكون الطبيعة فاعلة لأجل شيء؛ والطبيعة الواحدة تختلف أفعالها لاختلاف المواد، كالحرارة تحل شيئاً كالشمع، وتعقد شيئاً كالبيض والملح؟ ومن العجائب أنْ تكون الحرارة تفعل الإحراق لأجل شيء، بل إنَّما يلزمها ذلك بالضرورة؛ لأنَّ المادة بحال يجب لها فيها عند مماسّة الحار الاحتراق، وكذلك حكم سائر القوى الطبيعية.

(٦) والذي يجب علينا أنْ نقوله في هذا الباب ونعتقده هو أنَّه لا كثير مناقشة الآن في أنَّ للاتفاق مدخلاً في تكّون الأُمور الطبيعية، وذلك بالقياس إلى أفرادها. فإنَّه ليس حصول هذه المَدَرة عند هذا الجزء من الأرض، ولا حصول هذه الحبّة من البُرّ في هذه

grain of wheat occurs in *this* plot of land, nor that *this* semen occurs in *this* womb is something that is always or for the most part; and indeed, let us happily grant [Empedocles this point] and what is analogous.

(7) What we should focus on is the generation of the spike of grain from the wheat through the aid of the Earth's matter and the fetus from the semen through the aid of the womb's matter. Is that considered to be by chance? Now, we shall discover that it is not by chance but is something that nature necessitates and some power elicits. Likewise, let them also cheer on their claim that the matter belonging to incisors is susceptible to only this form; but we know not [only] that *this form* determinately belongs to this matter because it is susceptible to only this form, but [also] that *this matter* determinately belongs to this form because it is susceptible to only this form— [that] is, because in a house the stones are on bottom and the wood on top not only because stone is heavier and wood lighter, but also [because] here there is the work of a craftsman who could not do it if he had not related the materials that [his craft] uses in this way and so, through [his craft], produces this relation. Sound reflection reveals the truth of what we say—namely, that when *either* a grain of wheat *or* a grain of barley falls onto one and the same plot of land, then *either* a spear of wheat *or* a spear of barley respectively grows. It is absurd to say that the earthy and watery particles move by themselves and penetrate the substance of the wheat and cause it to grow, for it will become clear that they do not move from their proper places owing to themselves, where the [downward] motions that do essentially belong to the [earth and water] are well known. So the two must move only through some attractive powers latent within the grains, which bring about the attraction, God willing.

البقعة من الأرض، ولا حصول هذه النُطْفة في هذا الرحم أمراً دائماً ولا أكثرياً، بل لنسامح أنَّه وما جرى مجراه؛ اتفاقي.

(٧) ولنمْعن النظر في مثل تكّون السنبلة عن البُرّة باستمداد المادة من الأرض، والجنين عن النُطفة باستمداد المادة من الرحم؛ هل يُعدّ ذلك بالإتفاق؟ فنجده ليس باتفاقي، بل أمراً توجبه الطبيعة وتستدعيه قوة. وكذلك ليساعدوا أيضاً على قولهم أنَّ المادة التي للثنايا لا تقبل إلاّ هذه الصورة، بل حصلت هذه المادة لهذه الصورة لأنَّها لا تقبل إلا هذه الصورة. فإنَّه ليس البيت إنّما رسب فيه الحجر وطفا الخشب لأنَّ الحجر أثقل والخشب أخف، بل هناك صنعة صانع لم يصلح لها إلاّ أنْ تكون نسب مواد ما بفعله هذه النسبة، فجاء بها على هذه النسبة. والتأمل الصادق يُظهر صدق ما قلناه وهو أنَّ البقعة الواحدة إذا سقطت فيها حبّة بُرّة أنبتت سنْبلة بُرّة، أو حبّة شعير أنبتت سنْبلة شعير. ويستحيل أنْ يقال أنَّ الأجزاء الأرضية والمائية تتحرّك بذاتها وتنفذ في جوهر البُرّة وتربيّه، فإنَّه سيظهر أنَّ تحرّكهما عن مواضعهما ليس لذاتهما. والحركات التي لذاتهما معلومة، فيجب أنْ يكون تحركهما إنّما هو بجذْب قوى مستكنّة في الحبّات جاذبة بإذن اللّه.

(8) Furthermore, with respect to that plot of land, one or the other of the following must be the case: either some parts are suited to generating wheat, while others are suited to generating barley; or whatever is suited to the generation of wheat is suited to the generation of barley. On the one hand, if one and the same parts are suited to both, then the necessity associated with matter falls to the wayside, and the issue comes back to the fact that the form coincidentally belongs to the matter from some agent that provides that form specific to it and moves it toward that form and that it does that always or for the most part. So, clearly, whatever is like that is an action that proceeds from that very thing toward which it is directed, either always (in which case it is not impeded) or for the most part (in which case it is impeded). Now, with respect to natural things, this is what we mean by the *end.* On the other hand, if the parts are different, then it is because of a certain affinity between the power in the wheat and that [particular] matter that there is that which always or for the most part attracts that very matter and moves it to some specific place, in which case there is a cause of [that matter's] acquiring a certain form. So, again, it is the power that is in the wheat that essentially moves this matter to that form of substance, quality, shape, and, where;[5] and that will not be on account of the necessity of the matter, even though the matter with that description will inevitably be borne along to that form. So let us posit that the natural characteristics of the matter are, for instance, either suited to this form or are not susceptible to any other. Is it not inevitable that its being borne along to the place where it acquires this form after not having it is not because of some necessity in it, but, rather, is the result of some other cause that moved [the matter] to [the form], such that either [the matter] comes to have what it is suited to receive or it is not suited to receive anything else? It follows clearly from all this that the nature causes the materials to move toward some definite terminus according to a natural intention belonging to [the nature], and that continues always or for the most part, but that is what we mean by the term *end.*

5. *Al-aynu* refers to the category of *pou* (where) in Aristotle's *Categories,* which, as such, may also be understood as "place."

(٨) ثم لا يخلو إمّا أنْ تكون في تلك البقعة أجزاء تصلح لتكوّن البُرّة وأُخرى تصلح لتكوّن الشعير، أو يكون الصالح لتكوّن البُرّة صالحاً لتكوّن الشعير. فإنْ كان الصالح لهما أجزاء واحدة فقد سقطت الضرورة المنسوبة إلى المادة؛ ورجع الأمر إلى أنَّ الصورة طارئة على المادة من مصوّرٍ يخصّها بتلك الصورة ويحرّكها إلى تلك الصورة؛ وأنَّه دائماً أو في أكثر الأمر يفعل ذلك. وقد بانَ أنَّ ما كان كذلك فهو فعل يصدر عن ذات الأمر متوجهاً إليه، إمّا دائمٌ فلا يُعاق، وإمّا أكثريٌ فيُعاق، وهذا هو مرادنا بالغاية في الأُمور الطبيعية. وإنْ كانت الأجزاء مختلفة فلمناسبة ما بين القوة التي في البُرّة وبين تلك المادة ما يجذب تلك المادة بعينها ويحرّكها إلى حيّزٍ مخصوصٍ في الدوام أو الأكثر؛ فهناك تكسبها صورة مّا فتكون أيضاً القوة التي في البُرّة تحرّك بذاتها هذه المادة إلى تلك الصورة من الجوهر والكيف والشكل والأين، ولا يكون ذلك لضرورة المادة - وإن كان لا بُدّ من أنَّ تكون المادة على تلك الصفة لتنقل إلى تلك الصورة. فلنضع أنَّ طباع المادة صالحة لهذه الصورة أو غير قابلة لغيرها مثلاً. فلا بُدّ من أنْ يكون انتقالها إلى حيث تكتسب هذه الصورة بعد ما لَمْ تكن لها، ليس لضرورةٍ فيها، بل عن سببٍ آخر يحرّكها إليها؛ فيحصل لها ما هي صالحة لقبوله، أو لا تصلح لقبول غيره. فيستبين من هذا كلّه أنَّ تحريكات الطبيعة للسواد هي على سبيل قصدٍ طبيعي منها إلى حدّ محدود، وأنَّ ذلك مستمر على الدَّوْام أو على الأكثر؛ وذلك ما نعنيه بلفظ الغاية.

(9) Furthermore, it is obvious that all ends proceeding from nature, in the case where there is neither opposition nor obstacle, are goods and perfections, and that when [nature] results in some disadvantageous end, that result is not always or for the most part from [nature]. The fact of the matter is that our soul does not immediately grasp some accidental cause about it and so asks, "For what purpose did this seedling wither?" and "For what purpose did this woman miscarry?" As it is, nature is moved for the sake of the good. This goes not only for the development of plants and animals, but also for simple bodies and the actions that proceed from them naturally; for they tend toward certain ends to which they are always directed (as long as there is no obstacle) and are so according to a definite order from which they do not deviate unless there is some opposing cause. The same holds for the instincts to build, weave webs, and store up food that belong to animal souls, for they seem to be quite natural and are ends.

(10) Now, if things happen by chance, then why doesn't wheat produce barley, and why is there no fig-olive progeny, as they think there was goat-stag progeny?[6] Why isn't it that these rare things frequently occur, rather than the species' being continually preserved for the most part? Another proof that natural things are for the sake of some end is that when we see some opposition to or some weakness in the nature, we aid the nature by art—just as the physician does who believes that when the offending opposition is removed or [the patient's] strength restored, the nature will tend toward health and well-being.

(11) It does not follow from the fact that nature lacks deliberation that we must judge that the action proceeding from it is not directed toward some end; for deliberation is not in order to make the action have some end, but in order to designate the action that is chosen from among actions that might be chosen, each one of which has some end proper to it. So deliberation is for the sake of specifying the action, not

6. Cf. Aristotle, *Physics* 2.8.199b9–13.

(٩) ثم من الظاهر أنَّ الغايات الصادرة عن الطبيعة - حال ما تكون الطبيعة غير معارضة ولا معوِّقة - كلّها خيرات وكمالات. وأنَّه إذا تأدت إلى غايةٍ ضارّة؛ كان ذلك التأدّي ليس عنها دائماً ولا أكثرياً، بل في حالٍ تفقد النفس مِنّا فيها سبباً عارضاً؛ فيقال: ماذا أصاب هذا الفسيل حتى ذوى؟ وماذا أصاب هذه المرأة حتى أسقطت؟ وإذا كان كذلك، فالطبيعة تتحرّك لأجل الخيرية، وليس هذا في نشوء الحيوان والنبات فقط، بل وفي حركات الأجرام البسيطة وأفعالها التي تصدر عنها بالطبع، فإنَّها تنحو نحو غاياتٍ تتوجه إليها دائماً - ما لم تُعق - توجهاً على نظامٍ محدودٍ لا تخرج عنه إلاّ بسببٍ معارض. وكذلك الإلهامات التي للأنفس الحيوانية البانية والناسجة والمُدّخرة، فإنها تُشبه الأمور الطبيعية؛ وهي لغاية.

(١٠) وإنْ كانت الأمور تجري اتفاقاً؛ فلِمَ لا تنبت البُرّة شعيرة، ولا تتولد شجرة مركّبة من تين وزيتون، كما يتولّد عندهم بالاتفاق عَنْز أيّل، ولَم لا تتكرّر هذه النوادر، بل تبقى الأنواع محفوظة على الأكثر؟ ومّما يدل على أنَّ الأمور الطبيعية لغايةٍ؛ أنّا إذا أحسسنا بمعارضٍ أو قصورٍ من الطبيعة أعنّا الطبيعة بالصناعة؛ كما يفعله الطبيب معتقداً أنَّه إذا زال العارض المعارض أو اشتدت القوة، توجهت الطبيعة إلى الصحة والخير.

(١١) وليس إذا عدمت الطبيعة الرّوية وجب من ذلك أنْ نحكم بأنَّ الفعل الصادر عنها غير متوجهٍ إلى غاية؛ فإنَّ الرّوية ليست لتجعل الفعل ذا غاية، بل لتعيّن الفعل الذي يُختار من بين سائر أفعالٍ جائز اختيارها، لكل واحد منها غاية تخصّه. فالرّوية لأجل

for the sake of making it an end; and were the soul spared of the various opposing[7] likes and dislikes, an identical and uniform action would proceed from it without deliberation. If you want to become clear on this point, consider closely the case of art, for undoubtedly, it is for the sake of some end. Once it becomes a habit, however, doing it no longer requires deliberation, and it even becomes such that when deliberation is present, it is nigh on impossible to do, and even the one well versed in its performance becomes befuddled in its execution. An example would be a writer or lute player, for when they deliberate about the choice of one letter after another or one note after another and intentionally become preoccupied with their instruments, then they become befuddled and perform haltingly. They continue to do what they do uniformly only by not deliberating about each of the successive things they continue to do, even if that action and its intention initially occurred only through deliberation. As for what provides the initial basis and starting point for that [deliberation], it is not an object of deliberation.[8] The same holds in the case where someone grabs something to catch his balance or uses his hand to scratch an itch, which are done without thought, deliberation, or trying to imagine the form of what one is doing. A case even clearer than this one is when the faculty of the soul self-consciously chooses to move some external limb. Now, it is not the case that it moves the [external limb] itself without an intermediary; rather, it in fact moves only the muscles and tendon, and then they, in their turn, move that limb. Now, the soul is not conscious of moving the muscle, despite the fact that that action is chosen and first.

(12) Concerning what was said about deformities and what is analogous to them, some of them involve a deficiency, malformation, and weakness of the natural course, while others involve some addition. Whatever involves a deficiency or malformation involves a certain privation

---

7. Reading *mutafanninah* with **Z** and **T** for **Y**'s (inadvertent) *mutaqannina* (legislated). The Latin's *desideriis* (ardent desire) may be translating some derivation from the root F-T-N, (to be infatuated), but it would be difficult to say exactly what.

8. Avicenna seems to be making the Aristotelian point that one does not deliberate about what is in fact good or even appears to be good to one, but only about the means to acquire that good; cf. Aristotle, *Nicomachean Ethics* 3.3.1112b11ff.

تخصّص الفعل لا لجعله غاية. ولو كانت النفس مسلّمة عن النوازع المختلفة والمعارضات المتفننة، لكان يصدر عنها فعل متشابه على نَهْج واحد من غير روية. وإنْ شئت أنْ تستظهر في هذا الباب فتأمل حال الصناعة، فإنَّ الصناعة لا يُشك في أنّها لغاية، إذا صارت ملكة لم يحتج في استعمالها إلى الروية، وصارت بحيث إذا أُحضرت الروية تعذّرت وتبلّد الماهر فيها عن النفاذ فيما يزاوله، كمَنْ يكتب أو يضرب بالعود؛ فإنّه إذا أخذ يروّي في اختيار حرف حرف أو نغمة نغمة، وأراد أنْ يقف على عدده تبلّد وتعطّل. وإنّما يستمر على نَهْج واحدٍ فيما يفعله بلا روية في كل واحد واحد ممّا يستمر فيه – وإنْ كان ابتداء ذلك الفعل وقصده إنّما وقع بالروية. وأمّا المبني على ذلك الأول والابتداء فلا يروّى فيه. وكذلك حال اعتصام الزالق بما يعصمه، ومبادرة اليد على حكّ العضو المستحكّ من غير فكر ولا روية ولا استحضار لصورة ما يفعله في الخيال. وأوضح من هذه القوة النفسانية إذا حرّكت عضواً ظاهراً تختار تحريكه وتشعر بتحريكه، فليس تحريكه بالذات وبلا واسطة، بل إنّما تحرّك بالحقيقة العضل والوتر فيتبعه تحريك ذلك العضو، والنفس لا تشعر بتحريكها للعضلة؛ مع أنَّ ذلك الفعل اختياري وأوّل.

(١٢) وأمّا حديث التشويهات وما يجري مجراها، فإنَّ بعضها هو نقصٌ وقبحٌ وقصورٌ عن المجرى الطبيعي، وبعضها زيادة، وما كان نقصاً وقبحاً فهو عدم فعل لعصيان

produced because of the recalcitrance of matter. Now, we ourselves never promised that the nature could move every matter to the end, nor that there even are ends for the privations of [nature's] actions; rather, we promised that its actions in the materials that are compliant to [the nature] are ends, and this latter [claim] is not at all at odds with the former. Death and deterioration are on account of the weakness of the bodily nature to impose its form onto the matter and to preserve it as such by replacing what is lost. Now, it is simply not the case that the order of deterioration equally leads to some end, for the order of deterioration has a certain cause different from the nature charged with the care of the body. That cause is heat; and as a cause,[9] it is the nature, albeit accidentally. Now, each one of the two has an end. On the one hand, heat's end is to dissolve and transform moisture and so regularly drives matter toward [dissolution], where that is a given end. On the other hand, the end of the nature that is in the body is to preserve the body as long as it can by replenishing it; however, all replenishing will eventually cease, since the replenishing that [the body] receives later on will become less than what it received at first, owing to certain causes (which we shall note in the particular sciences).[10] So that replenishing is an accidental cause for the order of deterioration. Therefore, deterioration, inasmuch as it has an order, is directed toward a certain end, and so it is some action owing to a nature, even if it is not the action of the nature of the body. We again, however, never promised that every state belonging to natural things must be some end of the nature that is in them. All we said, in fact, is that every nature does its action and that it does it only for the sake of its end, whereas the action of some other [second nature] might not be for the sake of [the first nature's] end. Now, death, dissolution, deterioration, and all of that, even if it is not some beneficial end in relation to the body of Zayd, is a necessary end with

9. Reading *sababan* (acc.) with **Z** and **T** for **Y**'s *sababun* (nom.). The Latin suggests a significant variant and reads *et quae causa est calor, sed causa caloris est natura; ergo causa eticae est natura, sed accidentaliter* (that cause is heat, but the cause of heat is a nature; therefore, the cause of deterioration is a nature, but [only] accidentally).

10. Cf., for instance, *Kitāb al-nafs* 5.4, where Avicenna mentions changes of the humoral temperament as one of the causes for the human body's corruption.

المادة. ونحن لم نضمن أنَّ الطبيعة يمكنها أنْ تحرّك كل مادة إلى الغاية، ولا ضمنّا أنَّ لإعدام أفعالها غايات، بل ضمنّا أنَّ أفعالها في المواد الطبْعية التي لها هي غايات، وهذا لا يُزاحم ذاك. والموت والذبول هو لقصور الطبيعة البدنية عن إلزام المادة صورتها وحفظها إياها عليها بإدخال بدل ما يتحلّل، ونظام الذبول ليس أيضاً غير متأدّ إلى غايةٍ البتة؛ فإنَّ لنظام الذبول سبباً غير الطبيعة الموكلة بالبدن؛ وذلك السبب هو الحرارة، وسبباً هو الطبيعة ولكن بالعرض. ولكلّ واحدٍ منهما غاية، فالحرارة غايتها تحليل الرطوبة وإحالتها فتسوق المادة إليه على النظام وذلك غاية. فالطبيعة التي في البدن غايتها حفظ البدن ما أمكن بإمدادٍ بعد إمدادٍ، لكن كل مددٍ يأتي فإنَّ الاستمداد منه أخيراً يقع أقلَّ من الاستمداد منه بدّياً لعللٍ نذكرها في العلوم الجزئية. فيكون ذلك الإمداد بالعرض سبباً لنظام الذبول، فإذن الذبول - من حيث هو ذو نظامٍ ومتوجّهٍ إلى غاية - فهو فعلٌ لطبيعةٍ وإنْ لم يكن فعل طبيعة البدن. ونحن لم نضمن أنَّ كلّ حالٍ للأمور الطبيعية يجب أنْ تكون غاية للطبيعة التي فيها، بل قلنا إنَّ كلّ طبيعة تفعل فعلها فإنّما تفعله لغايةٍ لها، وأمّا فعل غيرها فقد لا يكون لغايةٍ لها. والموت والتحليل والذبول وكل ذلك - وإنْ لم يكن غاية

respect to the order of the universe. We already alluded to that earlier,[11] and your own knowledge about the state of the soul will draw your attention to the necessary end with respect to death, as well as the necessary ends with respect to frailty.

(13) Additional appendages also come to be for some end. [That] is because when the matter is excessive, the nature moves the excess of it toward the form that it deserves, owing to the preparedness in [the matter], and [the nature] does not forgo giving [the form]; and so the nature's acting on [matter] is for the sake of some end, even if the end toward which it is urged on chances to be a non-natural end.

(14) As for the case of rain and what was said about it, one simply should not concede it. Quite to the contrary, we say that the Sun's proximity and remoteness and the occurrence of warmth and coolness owing to its proximity and remoteness respectively (as you will learn later)[12] are an orderly cause of most of the things in nature that have particular[13] ends; and it is the proximity of the Sun during its motions along the incline that is the very cause that results in the evaporation that brings about the upward motion [of the moisture], where it is cooled and then necessarily falls. The necessity of the matter is not sufficient to account for that [namely, the proximity and remoteness produced by the Sun's motions along the incline]; but, rather, this divine action takes charge of the matter until the necessity of [matter] is reached. In this case, the end is imposed upon it, for either every end or the weighted majority of ends do impose some necessity on a given matter. Yet it is the cause producing motion that seeks out the matter and makes it so as to join the necessity that is in it (if there is) with the intended end. That should be the considered view about all the arts.

---

11. See 1.7.4.

12. The reference would appear to be to *Kitāb fī al-samāʾ wa-l-ʿālam* 2, and *Kitāb al-maʿādin wa-l-āthār al-ʿulwīya* 2.

13. Reading *al-juzʾīyah* with **Z** and **T** for **Y**'s and the Latin's *al-khayrīyah* (good).

نافعة بالقياس إلى بدن زيد - فهي غاية واجبة في نظام الكلّ، وقد أومأنا إلى ذلك فيما سلف، وعِلْمك بحال النفس سينبّهك على غايةٍ في الموت واجبة، وغايات في تناسب الضعف واجبة.

(١٣) وأمّا الزيادات فهي أيضاً كائنة لغايةٍ؛ فإنَّ المادة إذا فضلت حرّكت الطبيعة فضْلها إلى الصورة التي تستحقها بالا ستعداد الذي فيها ولا تعطّلها. فيكون فعل الطبيعة فيها لغايةٍ؛ وإنْ كان المستدعي إلى تلك الغاية إتفاق سببٍ غير طبيعي.

(١٤) وأمّا أمر المطر وما قيل فيه؛ فليس ينبغي أنْ نسلّم ما قيل فيه، بل نقول إنَّ قرب الشمس وبُعْدها وحدوث السخونة بقربها والبرودة ببُعْدها - على ما تعلمه بعد - سببٌ ذو نظام لأمور كثيرة من الغايات الجزئيّة في الطبيعة. ووقوع الشمس مقربة في حركاتها المائلة سببٌ يصدر عن ذاته التبخّر المصعّد إلى حيث يبرد فيهبط للضرورة. وليس يكفي في ذلك ضرورة المادة؛ بل هذا الفعل الإلهي المستعمل للمادة، إلى أنْ ينتهي إلى ضرورتها فتلزمها الغاية. فإنَّ كلّ غاية، أو جُلّ الغايات، تلزم ضرورة في مادة، ولكن العلّة المحرّكة ترتاد المادة وتجعلها بحيث تتصل، بالضرورة التي فيها - إنْ كانت - بما هو الغاية المقصودة؛ تأمّل ذلك في الصناعات كلّها.

(15) We also say to them that it is not the case, when motion and action have an end, that every end must have an end and that there will be no end to the question "Why?" for the true end is what is intended for its own sake, and the rest of the things are intended for the sake of it. Now, the why-question that requires the end as its answer is properly asked of whatever is intended for the sake of something else, whereas when what is intended is the thing itself, the question, "Why is it intended?" is not at all appropriate. It is because of this that it is not asked, "Why do you seek health?" and "Why did you seek the good?" or "Why do you want to escape illness?" and "Why do you shun evil?" Now, if motion and transformation were required in order that the end be found or be an end, then every end would necessarily have an end; but as it is, it requires that [only] in the cases of cessation and renewal that proceed from some natural or volitional cause.

(16) You should also not be at all amazed that heat acts for the sake of burning something and that, in fact, heat truly acts in order to burn and to consume what is burnt and to cause its transition into either something like itself or something like the substance in which there is [the heat]. Chance and the accidental end arise in the case of, for example, the burning of a poor man's cloak precisely because [the burning] is not an essential end, for it neither burns it because it is a poor man's cloak nor is the power to burn that is in the fire for the sake of *this* one instance. Quite [to] the contrary, [fire burns] in order to transform whatever it touches into its own substance and in order to melt what is in a certain state and congeal what is in a certain state. In the present case, it chanced to touch *this* cloak, and so there is a certain end on account of the natural activity of fire, even if its bumping into *this* thing that caught on fire is only accidental. Now, the existence of the accidental end does not preclude the existence of the essential; rather, the essential end is prior to the accidental end.

(١٥) ونقول لهم أيضاً ؛ وليس إذا كان للحركة غاية وللفعل غاية وجب أنْ يكون لكل غايةٍ غاية ، وأنْ لا تقف <المسألة> عن لِمَ . فإنَّ الغاية في الحقيقة تكون مقصودة لذاتها وسائر الأشياء تقصد لها ، وما يُقصد لأجل شيء آخر فحريٌ أنْ <يُسأل> عنه باللّم المقتضي للجواب بالغاية . وأما ما يقصد لذاته فإنَّه لا يليق به السؤال عن أنَّه لِمَ قُصِدَ ، ولهذا لا يقال : لِمَ طلبت الصحة ، ولَمَ طلبت الخيرية ، أو لِمَ هربت عن المرض ، ولَم نفرت عن الشر ؟ ولو كانت الحركة والإحالة تقتضي الغاية ، لأنَّها موجودة أو لأنَّها غاية ؛ لكان يجب أنْ يكون لكل غاية غاية ، لكنها تقتضي ذلك من حيث هناك زوالٌ وتجدّدٌ صادرٌ عن سببٍ طبيعي أو إرادي .

(١٦) وليس يجب أنْ تتعجب من أنَّ الحرارة تفعل لأحراق شيء ، بل حقاً إنَّ الحرارة تفعل لتحرِق وتُفْني المُحْرَق وتحيله إلى مشاكلتها أو مشاكلة الجوهر الذي هي فيه . إنَّما يكون الاتفاق والغاية العرضية في مثل أنْ تحرِق ثوب فقير ؛ وذلك ليس له بغايةٍ ذاتيةٍ ؛ فإنّها ليست تحرقه لأجل أنّه ثوب فقير ، ولا في النار هذه القوة المحرقة لأجل هذا الشأن ، بل لكي تحيل ما تماسّه إلى جوهرها ، ولكي تحلّ ما يكون بحال وتعقد ما يكون بحال ، وقد اتفق الآن أنْ ماسّها هذا الثوب . فلفعل النار في الطبيعة غاية ، وإنْ لم تكن مصادفتها هذا المشتعل إلاّ بالعرض ، ووجود الغاية بالعرض لا يمنع وجود الغاية بالذات ، بل الغاية بالذات متقدمة على الغاية بالعرض .

(17) From all this, it is clear that matter is for the sake of the form and that its purpose is to exist determinately and so have the form exist determinately in it, whereas the form is not for the sake of the matter, even if there inevitably is matter in order for the form to exist in it. Also, whoever closely considers the usefulness of the animal's limbs and the parts of plants will have no doubt that natural things are for the sake of some end, and you will get a whiff of that at the end of our discussion about natural things.[14] Now, despite all this, we do not deny that among natural things there are necessary things, some of which are needed for the sake of the end and some of which impose the end.

---

14. The reference is probably to the whole of *Kitāb al-ḥayawān.*

(١٧) فبيّنٌ من هذا كله؛ أنَّ المادة لأجل الصورة، وأنَّها تتوخى لتحصل فتحصل فيها الصورة، وليست الصورة لأجل المادة، وإنْ كان لا بُدّ من المادة حتى توجد فيها الصورة. ومَنْ تأمّل منافع أعضاء الحيوان وأجزاء النبات؛ لم يبق له شكٌ في أنَّ الأمور الطبيعية لغايةٍ، وشتشمّ من ذلك شيئاً في آخر كلامنا في الطبيعيات، ومع هذا كلّه فلا ننكر أنْ يكون في الأمور الطبيعية أمورٌ ضروريةٌ؛ بعضها يُحتاج إليها للغاية، وبعضها تلزم الغاية.

Chapter Fifteen

How causes enter into investigating and seeking the why-question and the answer to it

(1) Since we have explained the number of causes and their states for you, we should add that the natural philosopher must be interested in comprehending all of them, and especially the form, so that he completely comprehends the effect.

(2) Now, no principle of motion is included among the objects of mathematics, since they cannot move; and for the same reason, motion's end and matter are not at all included in them, and in fact, the only things considered about them are formal causes.

(3) Concerning material things, however, know that the question "Why?" might involve any one of the causes. So if it involves the agent (as, for example, asking, "Why did so-and-so fight so-and-so?") the answer might be the end (such as saying, "In order to avenge himself"). The answer might also be either the advisor[1] or someone who did something to him earlier (namely, the one inciting him to act)—as, for example, to say, "Because so-and-so advised him to" or "Because [so-and-so] robbed him of his property"—where this one is an agent on account of the form of choice that originates from him that provokes the ultimate action.

(4) Whether to provide the form or the matter as an answer is an open question. In the case of form, it will be the form of the action—that is, fighting. When the question concerns precisely the cause of [the form's] existence from the agent,[2] however, it is incorrect to provide [the form]

1. The Arabic *mushīr* is also Avicenna's preferred term for the "guiding principle"; see 1.10.4

2. Here and below the cause in question is what is the cause of the efficient cause's acting.

<الفصل الخامس عشر>

في دخول العلل في المباحث وطلب اللِم والجواب عنه

(١) وإذْ قد بانَ لنا عدّة الأسباب وأحوالها فنقول: أنَّه يجب أنْ يكون الطبيعي معْنيّاً بالإحاطة بكلّيتها، وخصوصاً بالصورة، حتى تتمّ إحاطته بالمعلول.

(٢) وأمّا الأمور التعليمية فلا يدخل فيها مبدأ حركة؛ إذ لاحركة لها، وكذلك لا تدخل فيها غاية حركة ولا مادة البتة، بل تتأمل فيها العلل الصورية فقط.

(٣) واعلم أنَّ السؤال عن الأمور المادية باللّم ربما تضمّن علّة من العلل. فإنْ تضمّن الفاعل؛ كقولهم: لِمَ قاتل فلانٌ فلاناَ؟ فيجوز أنْ يكون جوابه الغاية، كقولهم لكي ينتقم منه، ويجوز أنْ يكون جوابه المشير، <أ>و الفاعل المتقدم للفاعل؛ وهو الداعي إلى الفعل؛ مثل أنْ يقال لأنَّ فلاناً أشار عليه، أو أنَّه غصبه حقّه، وهذا هو الفاعل لصورة الاختيار الذي ينبعث منه الفعل الأخير.

(٤) وأمّا أنَّه هل يجيب بالصورة أو هل يجيب بالمادة، ففيه نظر. أمّا الصورة فإنَّها صورة الفعل وهو القتال، وليس السؤال إلاّ عن علّة وجودها عن الفاعل فلا يصلح أنْ

as an answer, for it is not the cause of its very own existence from the agent, except in the case where that form is the ultimate end—such as the good, for example. In that case, it is what moves the agent to be an agent without itself being caused, as we alluded to when making clear the relation between the agent and the end.[3] Additionally, [the cause in question] is not the proximate cause of the [form's] existing in that matter as a result of the agent; but, rather, it is the cause of the agent's existing as an agent. And so it is not the cause of the agent as something existing in matter, but as an essence or account. So when the question is about [the form's] coming to be an existing thing, then it is incorrect to provide [the form] *qua* something existing as an answer, but only *qua* an account or essence. The form in question might itself be a certain account that is included within [that form] or [an account] that is broader in scope than [the form], being an idea that embraces the idea of [the form]. In that case, that account would be a correct answer, just as it is said, "Why did so-and-so act justly?" and it is said, "Because acting justly is admirable." Here, *being admirable* is included in *acting justly* and is an answer analogous to the form, and [yet] the answer is not the form in question, but another form. [That] is because *being admirable* is either a part of its definition or broader in scope than it, since *being admirable* is more general than *acting justly,* whether necessarily broader in scope or a constitutive part of its definition. When it is correct to provide the form as the answer, then it is so inasmuch as it is included within the whole set [of factors] inciting the mover to choose. The very same thing can be judged about the matter. So, when it is said, "Why did so-and-so turn "*this* wood into a bed?" saying, "Because it was the wood he had" is not enough unless one adds, "It was the good solid wood that he had suitable for being turned into a bed, and he did not need it for anything else."

3. See 1.11.1–2, where he discusses how the final cause is the cause of the causality of the efficient cause.

يجاب بها ، فإنَّها ليست علّة لوجود نفسها عن الفاعل ؛ إلاَّ أنْ تكون تلك الصورة هي غاية الغايات كالخير مثلاً ، فتكون لذاتها لا لسببٍ مّا هي محرّكة للفاعل إلى أنْ يكون فاعلاً ؛ على النحو الذي أومأنا إليه في بيان نسبة ما بين الفاعل والغاية . ومع ذلك فلا تكون علّة قريبة لوجودها في تلك المادة عن الفاعل ، بل علّة لوجود الفاعل فاعلاً ، فلا تكون - من حيث هي موجودة في المادة علّة للفاعل - بل من حيث هي ماهيّة ومعنى . فإذا كان السؤال عن كونها موجودة ، لم يصلح الجواب بها من حيث هي موجودة ؛ بل من حيث هي معنى وماهيّة . وربّما كانت الصورة المَسْئُول عنها ذات معنى داخل فيها أو عارض لها ، ذاهب مذهبها ، فيكون يصلح أنْ يكون ذلك المعنى جواباً ؛ كما يقال : لِمَ عدل فلانٌ؟ فيقال لأنَّ العدل حَسَنٌ ، فيكون الحسن معنى في العدل وجارياً مجرى الصورة ، ولا تكون الصورة المَسْئُول عنها جواباً ، بل صورة غيرها . فإنَّ الحسن هو جزء حدّ أو عارضٍ لها ، فإنَّ الحسن معنى أعمّ من العدل ؛ إمّا عارض لازم ، وإمّا جزء حدّ له مقوّم . وإذا صلحت الصورة أنْ يجاب بها ها هنا فقد دخلت من حيث هي كذلك في جملة الداعي المحرّك للاختيار . وحكم المادة هذا الحكم بعينه ، فإنَّه إذا قيل لِمَ نجرَ فلانٌ هذا الخشب سريراً؟ فقيل لأنَّه كان عنده خشب ، لَمْ يكن مقنعاً ؛ إلاّ أنْ يُزاد فيقال : كان عنده خشب صلب صالح لأنْ يُنجز منه سريرٌ ، وكان لا يحتاج إليه في أمرٍ آخر .

(5) Concerning issues involving volition, however, it is difficult to produce the cause completely, for the will is incited to act [only] after a number of factors are fulfilled, the enumeration of which is not easy. Also, one might not even be conscious of many of them so as to include them in the account.

(6) Concerning issues involving nature, the preparedness of the matter and its encountering the active power is enough, and so the determinate occurrence of matter's relation with respect to [the active power] is by itself an answer, once the presence of the agent is mentioned in the question. When the question involves the end, such as saying, "Why did so-and-so recover?" it is correct to provide the efficient principle as an answer and so say, "Because he drank the medicine." Also, it is correct to provide the material principle as an answer in addition to the agent and so say, "His body's humoral temperament is naturally strong"; but it is not enough to mention the matter alone. Also, mentioning the form alone, as in saying, "Because his humoral temperament is well balanced," rarely will be enough to put an end to the questioning; and, in fact, it will require some other question that will lead to a certain matter and agent. When the question concerns the matter and its preparedness—as, for example, in saying, "Why is the human body mortal?"—one might give the final cause as an answer and so say, "It was made such in order that the soul, once perfected, could free itself from the body." The material cause might also be given as an answer, in which case it is said, "Because it is a composite of contraries." It is not permitted to give the agent as an answer in the case of preparedness that is unlike the form, because it is impossible that the agent provide the matter with the preparedness such that, if it does not provide it, then [the matter] would not be prepared—that is, unless by *preparedness* we mean *to have the disposition completed*. In that case, the agent might provide [the preparedness], just as it is said of the mirror when it is asked, "Why does it receive the image?" and it is said, "Because someone polished it," whereas the original preparedness belongs necessarily to the matter. Also, one might give the form as an answer when it is what completes the preparedness—and so, for example, it is said about the mirror, "Because it is smooth and polished."

(٥) لكن الأمور الإرادية يصعب أنْ تؤدي العلّة بتمامها فيها ، فإنَّ الإرادة تنبعث بعد توافي أمور لا يسهل إحصاؤها ، وربّما لم نشعر بكثير منها فيخبر عنه.

(٦) وأمّا الأمور الطبيعية فيكفي فيها من المادة الاستعداد والملاقاة الفاعلة ، فيكون حصول نسبة المادة فيها جواباً وحده؛ إذا ذكر في السؤال حضور الفاعل. وأمّا إذا تضمن السؤال الغاية كما يقال: لِمَ صحَّ فلانٌ؟ فيصلح أنْ يجاب بالمبدأ الفاعلي؛ فيقال لأنَّه شرب الدواء. ويصلح أنْ يجاب بالمبدأ المادي مضافاً إلى الفاعل، فيقال: لأنَّ مزاج بدنه قوي الطبيعة، ولا يكفي ذكر المادة وحدها. وأمّا الصورة فقلّما تُقْنع ويقطع السؤال ذكرها وحدها ، بأنْ يقال لأنَّ مزاجه اعتدل ، بل تحوج إلى سؤال آخر يؤدي إلى مادة أو فاعل. وأمّا إذا كان السؤال عن المادة واستعدادها ؛ بأنْ يقال مثلاً: لَمِ بدن الإنسان قابل للموت؟ فقد يجوز أنْ يُجاب بالعلّة الغائية؛ فيقال جُعل كذلك لتتخلّص النفس – عند الاستكمال – عن البدن. وقد يجوز أنْ يجاب بالعلّة المادية فيقال لأنَّه مركبٌ من الأضداد ، ولا يجوز أنْ يُجاب بالفاعل في الاستعداد الذي ليس كالصورة؛ لأنَّ الفاعل لا يجوز أنْ يعطي المادة الاستعداد ، كأنّه إنْ لم يعطِ لم تكن مستعدة ، اللّهم إلاّ أنْ نعني بالا ستعداد التهيؤ التام ، فقد يعطيه الفاعل ، كما يقال للمرآة إذا سُئل أنَّها لِمَ تقبل الشبح؟ فيقال لأنَّ الصاقل صقلها ، وأما الاستعداد الأصلي فلازم للمادة. ويجوز أنْ يجاب بالصورة إذ كانت هي المتمّمة للاستعداد ؛ فيقال في المرآة مثلاً لأنَّها مَلْساء صقيلة.

(7) In summary, the question is not directed at the matter unless it is considered along with the form, in which case the question is about the cause of the form's existing in the matter. When the question involves the form, then providing the matter alone as an answer is not enough, and instead, a certain preparedness, which is related to the agent, must be added to it. Both the end and the agent do provide answers. Now, if you want to separate out what is said metaphorically and mention the true state of affairs, then the true answer is to mention all the causes, [even if] they were not included in the question; and when they are mentioned and sealed with the true end, then the question comes to a stop.

◆

(٧) وبالجملة؛ السـؤال لا يتوجه إلى المادة إلاّ وقد أُخذت مع صورة، فيسـأل عن علّـة وجود الصورة في المادة. وأمّا إذا تضمن السـؤال الصورة، فالمادة وحدها لا تكفي أنْ يجاب بها، بل يجب أن يضاف إليها إستعدادٌ ويُنسب إلى الفاعل، والغاية يجاب بها، والفاعل يجاب به. وإذا شئت أنْ ترفض ما يقال على سبيل المجاز وتذكر الأمر الحقيقي، فإنَّ الجواب الحقيقي أنْ تذكر جميع العلل التي لم تتضمنها المسـألة، فإذا ذكرت وختمت بالغاية الحقيقية؛ وقف السؤال.

SECOND BOOK: ON MOTION AND THAT WHICH FOLLOWS IT

Chapter One

On Motion

(1) Having completed the discussion of the general principles of natural things, we appropriately turn to their general accidents. Now, there are none more general than motion and rest, where rest, as we shall explain when treating it, is a privation of motion. So we should start by discussing motion.

(2) We say that some things exist as actual in every respect, while others are actual in one respect but potential in another. It is impossible, however, that there be something that is potential in every respect, itself having no actuality whatsoever. Let this [for now] be accepted and set down as an axiom, although an inquiry into it will be taken up soon.[1]

1. See *Ilāhīyāt* 4.2.

المقالة الثانية

في الحركة وما يجري مجراها

<الفصل الأول>

في الحركة

(١) لقد ختمنا الكلام في المباديء العامة للأمور الطبيعية؛ فحريٌ بنا أنْ ننتقل إلى الكلام في العوارض العامة ‹التي› لها ، ولا أعمّ لها من الحركة والسكون. والسكون، كما سنبيّن من حاله، عدم الحركة، فحريٌ بنا أنْ نقدّم الكلام على الحركة؛

(٢) فنقول: إنَّ الموجودات بعضها بالفعل من كل وجه، وبعضها من جهةٍ بالفعل ومن جهةٍ بالقوة، ويستحيل أنْ يكون شيءٌ من الأشياء بالقوة من كل جهةٍ لا ذات له بالفعل البتة. ليُسلّم هذا، وليُوضع وضْعاً، مع قرب تناول الوقوف عليه. ثم من شأن كلّ ذي قوة

Next, everything possessing potency characteristically passes from it into its corresponding actuality; and whenever it is actually impossible to pass to [some actuality]; there is no [corresponding] potentiality. Now, the passage from potency to actuality is sometimes all at once and sometimes not, whereas [*passage* itself] is more general than either of the two. As most general, it belongs to every category, for there is no category in which there is not some passage from a certain potency belonging to it to a certain actuality belonging to it. With respect to substance, it is like a human's passage into actuality after being in potentiality; with respect to quantity, it is like the growing thing's passage from potency to actuality; and with respect to quality, it is like the passage of blackness from potency into actuality. Concerning what is in [the category] of relation, it is like the father's passage into actuality from potency. In [the category of] where,[2] it is like actually going upward that results from the potency; and in [the category of] when,[3] it is like evening's passing from potency to actuality. As for position, it is like the passing from potency to actuality of the one who stands. The same holds with respect to [the categories of] possession, action, and passion. The technical sense among the Ancients concerning the use of *motion,* however, is not common to all of these kinds of passages from potency to actuality; rather, it is what does not pass all at once, but only [does so] gradually. Now, this happens only in a few categories—as, for instance, quality, for what has a potential quality may advance little by little toward actuality until it reaches it, and likewise what has a potential quantity. Later we shall explain in which of the categories this [type of] passage from potency into actuality may and may not occur.[4]

---

2. The Arabic *aina* corresponds with Aristotle's category of *pou,* which is also frequently identified with the category of place.

3. The Arabic *matá* corresponds with Aristotle's category of *pote,* which is also frequently identified with the category of time.

4. See *Physics* 2.3.

أنْ يخــرج منها إلى الفعل المقابل لهــا ، وما امتنع الخروج إليه بالفعل فلا قوة عليه . والخروج إلــى الفعل عن القوة قد يكون دفعْة ، وقد يكون لا دفعْــة ؛ وهو أعمّ من الأمرين جميعاً ؛ وهــوّ - بما هو أعمّ - أمر يعرض لجميع المقولات ، فأنَّه لا مقولة إلاّ وفيها خروجٌ عن قوةٍ لها إلى فعلٍ لها . أما في الجوهر ؛ فكخروج الإنســان إلى الفعل بعد كونه بالقوة ، وفي الكمّ فكخروج النامي إلى الفعل عن القوة ، وفي الكيف فكخروج الســواد إلى الفعل عن القوة ، وفــي المضاف فكخروج الأب إلــى الفعل عن القوة ، وفي الأين فكالحصول فوق بالفعل عن القوة ، وفي متى فكخروج العشي إلى الفعل عن القوة ، وفي الوضع فكخروج المنتصب إلى الفعــل عن القوة ، وكذلك في الجدة ، وكذلك في الفعل وفي الانفعال . لكن المعنى المتصالح عليــه عند القدماء في اســتعمال لفظ الحركة ليس ما يشــترك فيــه جميع أصناف هذه الخروجات عن القوة إلى الفعل ، بل ما كان خروجاً لا دفعْة بل متدرجاً . وهذا لا يتأتى إلاّ في مقولات معدودة مثلاً كالكيف ؛ فإنَّ ذا الكيف بالقوة يجوز أنْ يتوجه إلى الفعل يسيراً يســيراً إلى أنْ ينتهي إليه ، وكذلك ذو الكمّ والقوة . ونحن ســنبيّن من بعد أنَّ أي المقولات يجوز أنْ يقع فيه هذا الخروج من القوة إلى الفعل ، وأيّها لا يجوز أنْ يقع فيه ذلك .

(3) Now, were it not the case that, in defining *time,* we must take motion in its definition, and that time frequently is taken in the definition of the *continuous* and *gradual,* and likewise in the definition of *all at once*—for *instant*[5] is taken in its definition (for it is said to be what is in an instant) and *time* is taken in the definition of the *instant,* since it is [time's] limit[6]—and [instead] time were taken in *motion*'s definition, then it would be easy for us to say that motion is a passage from potency to actuality either with respect to time, or continuously, or not all at once. As it stands, however, all of these are descriptions that include a hidden circular explanation. Thus, the one [namely, Aristotle] who provided us with this discipline was forced to take another course concerning that. He considered the state of what is being moved when it is being moved in itself and the manner of existence proper to motion in itself.[7] He found that motion in itself is a perfection and actuality—that is, actually being—as long as there is a potency corresponding with [the motion], since something might be moved either potentially or actually and perfectly, where its actuality and perfection are motion. In this way, on the one hand, motion is common to the rest of the perfections, while, on the other, it is different from them in that, when the rest of the perfections are determinate, something actually comes to be therein, and afterwards there is nothing potential in [the thing] associated with that actuality. So [for example] when black actually becomes black, no part of the blackness that it can be remains potentially black and when the square[8] actually becomes a square, no part of the squareness that it can be remains potentially a square. [In contrast], when something movable is actually being moved, it is thought that some part of the continuous motion by which it is moved is still something potentially movable.

5. *Al-āna,* which here is translated as "instant," can also be translated "now" and so, in this respect, is much like the Greek *to nun.*

6. Correcting **Y**'s *ṭarfah* (twinkling), which is clearly a typographical error, to *ṭarafahu* with **Z**, **T**, and the Latin.

7. Cf. Aristotle, *Physics* 3.1.

8. Reading *murabba*ᶜ with **Z**, **T**, and the Latin (*Liber primus naturalium tractatus secundus: De motu et de consimilibus,* ed. S. Van Riet, J. Janssens, and A. Allard Avicenna Latinus [Leuven: Peeters, 2006]) for **Y**'s *mutaḥarrik* (mobile).

(٣) ولولا أنَّ الزمان ممّا نضطر في تحديده إلى أنْ تؤْخذ الحركة في حدّه، وأنَّ الاتصال والتدريج قد يؤْخذ الزمان في حدّهما والدفْعة أيضاً فإنَّها قد يؤْخذ الآن في حدّها؛ فيقال: هــو ما يكون في آن. والآن يؤْخذ الزمان في حــدّه لأنَّه طرفه، والحركة يؤخذ الزمان في حدّهــا ليســهل علينا أنْ نقــول: إنَّ الحركة خروجٌ عن القوة إلى الفعــل في زمانٍ أو على الاتصال أو لا دفْعة؛ لكن جميع هذه الرسوم تتضمن بياناً دورياً خفياً. فاضطر مفيدنا هذه الصناعة أنْ ســلك في ذلك نَهْجاً آخر؛ فنظر إلى حال المتحرّك عندما يكون متحرّكاً في نفسه، ونظر في النحو من الوجود الذي يخصّ الحركة في نفسها فوجد الحركة في نفسها كمالاً وفعلاً، أي كوناً بالفعل، إذ كان بإزائها قوة. إذ الشيء قد يكون متحركاً بالقوة، وقد يكون متحركاً بالفعل وبالكمال، وفعله وكماله هو الحركة. فالحركة تشارك سائر الكمالات من هذه الجهة، وتفارق ســائر الكمالات من جهة أنّ سائر الكمالات، إذا حصلت، صار الشــيء بها بالفعل ولم يكن بعد فيه، ممّا يتعلق بذلك الفعل، شيء بالقوة. فإنَّ الأسود إذا صار بالفعل أسود، لم يبقَ بالقوة أسود من جملة الأسود الذي له، والمربّع إذا صار بالفعل مربّعــاً لم يبقَ بالقوة مربّعاً من جملة المربّع الــذي له، والمتحرّك إذا صار بالفعل متحرّكاً؛ فيظــن أنَّه يكون بعد بالقوة متحرّكاً من جملة الحركة المتّصلة التي هو بها متحرّك. ويوجد

Additionally, [the moved thing] potentially exists as something else different from what is moved. [That is] because, as long as the moved thing itself is not the potential thing toward which it is being moved[9] (but which it will reach through the motion), then its current state and relation during the motion to that which it can be potentially is not like what it was before the motion, since, in the state of rest before the motion, it is that potential thing absolutely. In fact, [what is moved] possesses two potentialities one of which is for the thing [it will potentially become] and the other [of which] is for being directed toward it. In that case, it has two perfections at that time for which it also has two potencies. At that time [namely, during the motion], it will have realized the perfection of one of the two potentialities and yet still have remained in potency to that thing that is the intended object of the two potentialities. In fact, with respect to both of them, even if one of the two perfections actually occurred (namely, the first of them), [what is moved] is still not free of what is in potency with respect to both things together, one of which is that toward which it is directed through the motion and the other [of which] is with respect to the motion, since obviously it has not undergone motion to such a degree that no potentiality whatsoever for [motion] remains. So motion is the first perfection belonging to what is in potency, though not in every respect. [That is because] some other perfection can belong to whatever is in potency—like the perfection of humanity or equinity—where that is not associated with its being in potency insofar as it is in potency. How could it be so associated, when it does not preclude the potency as long as it exists nor the perfection when it occurs? So motion is the first perfection belonging to what is in potency from the perspective of what is in potency.[10]

---

9. Hasnawi suggests emending the text's *mā lam yakun* to *mā lam yaskun* <*yakūnu baʿdu*>, in which case the sense would be "for as long as the moved thing itself is not at rest, <it is still> potentially something toward which it moves." See Ahmad Hasnawi, "La définition du mouvement dans la *Physique* du *Shifāʾ* d'Avicenne," *Arabic Sciences and Philosophy* 11 (2001): 219–55, esp. 242.

10. Cf. Aristotle, *Physics* 3.7.201a10–11 and 201b4–5.

أيضاً بالقوة شيئاً آخر غير أنّه متحرّك. فإنَّ ذات المتحرّك ما لم تكن بالقوة شيئاً مّا يتحرك إليه وأنّه بالحركة يصل إليه، فإنّه لا يكون حاله وقياسه عند الحركة إلى ذلك الذي هو له بالقوة كما كان قبل الحركة، فإنّه – في حال السكون قبل الحركة – يكون هو ذلك الشيء بالقوة المطلقة، بل يكون ذا قوتين إحداهما على الأمر والأخرى على التوجّه إليه، فيكون له في ذلك الوقت كمالان وله عليهما قوتان. ثم يحصل له كمال إحدى القوتين، ويكون قد بقي بعد بالقوة في ذلك الشيء الذي هو المقصود بالقوتين بل في كليهما، وإنْ كان أحدهما حصل بالفعل الذي هو أحد الكمالين وأولهما. فهو بعد لم يتبرأ عمّا هو بالقوة في الأمرين جميعاً: أحدهما المتوجّه إليه بالحركة، والآخر في الحركة. فإنَّ الحركة – في ظاهر الأمرّ – لا تحصل له بحيث لا تبقى قوتها البتة، فتكون الحركة هي الكمال الأول لما بالقوة لا من كل جهة، فإنّه يمكن أنْ يكون لما بالقوة كمالٌ آخر؛ ككمال إنسانية أو فرسية، لا يتعلق ذلك بكونه بالقوة بما هو بالقوة، وكيف يتعلق وهو لا ينافي القوة ما دامت موجودة ولا الكمال إذا حصل؟ فالحركةُ كمالٌ أولٌ لما بالقوة من جهة ما هو بالقوة.

(4) [Motion] has been defined in various obscure ways owing to its obscure nature, since it is a nature whose states do not exist as actually enduring and [since] its existence involves seeing that something [that existed] before [the motion] has ceased, while something new comes to exist. So some of [the Ancients] defined it in term of *otherness,* since it requires that the state become otherwise and is evidence that something is other than what it was.[11] [They] were unaware that what necessarily evidences otherness does not in itself have to be otherness, for not everything that provides evidence of something is [that thing]. Also, if otherness were motion, then everything that is other would be moved, which is not the case. One group said that [motion] is an indefinite nature, and this is appropriate if it is some attribute belonging to it other than a property unique to the species, since there are things such as that other than motion, such as the infinite and time. It is also said that [motion] is a passage from sameness, as if a thing's being the same is for a single attribute to persist with respect to each instant that elapses, whereas the relation of motion's parts and states to something at various times is not the same. So [for example] what is moved has a different place at each instant, and what undergoes alteration has a different quality at each instant. Only exigency and short-sightedness prompted these descriptions; and there is no need for us to go to great lengths refuting and contradicting them, since what we have said is enough for the sound mind to declare them false. It is also said about the definition of motion that it is a process from one state to another or a procession from potency to act.[12] That is mistaken, because the relation of procession and traversal to motion is not like the relation of a genus or what is similar to a genus, but like the relation of synonymous terms, since both of these terms, as well as the term *motion,* apply primarily to change of place and then are extended to states.

---

11. The Arabic *ghayrīyah* is clearly translating the Greek *heterotētes,* which Aristotle mentions at *Physics* 3.2.201b20.

12. See, for instance, Themistius, *In Aristotelis physica paraphrasis,* ed. H. Schenkl, vol. 5 (Berlin: George Reimer, 1900), 70.5–13, who made motion a kind of *poreia*—a passage, procession, or traversal.

(٤) وقد حُدّت بحدود مختلفة مشتبهة، وذلك لاشتباه الأمر في طبيعتها؛ إذْ كانت طبيعة لا توجد أحوالها ثابتة بالفعل، ووجودها فيما يرى أنْ يكون قبلها شيء قد بطل وشيء مستأنف الوجود. فبعضهم حدّها بالغَيْرية؛ إذ كانت توجب تغيّراً للحال وإفادة لغير ما كان، ولم يعلم أنّه ليس يجب أنْ يكون ما يوجب إفادة الغَيْرية؛ فهو في نفسه غَيْرية، فإنّه ليس كلّ ما يفيد شيئاً يكون هو. ولو كانت الغَيْرية حركة لكان كل غَيْر متحرّكاً، ولكن ليس كذلك. وقال قومٌ إنّها طبيعة غير محدودة، والأحرى أنْ يكون هذاً - إنْ كان صفة لهاً - صفة غير خاصّة، فغير الحركة ما هو كذلك كاللانهاية والزمان. وقيل إنّها خروج عن المساواة، كأنَّ الثبات على صفةٍ واحدةٍ مساواة لأمرٍ بالقياس إلى كلّ وقت يمرّ عليه؛ وأنَّ الحركة لا تتساوى نسبة أجزائها وأحوالها إلى الشيء في أزمنة مختلفة، فإنَّ المتحرّك في كلّ آنٍ له أيْنٌ آخر، والمستحيل له في كلّ آنٍ كيْفٌ آخر. وهذه رسومٌ إنّما دعا إليها الاضطرار وضيق المجال، ولا حاجة بنا إلى التطويل في إبطالها ومناقضتها، فإنَّ الذّهن السليم يكفيه في تزيفها ما قلناه. وأمّا ما قيل في حدّ الحركة أنّها زوالٌ من حالٍ إلى حال، أو سلوكٌ من قوة إلى فعل فذلك غلطٌ؛ لأنَّ نسبة الزوال والسلوك إلى الحركة ليس كنسبة الجنس أو ما يشبه الجنس، بل كنسبة الألفاظ المرادفة إياها، إذ هاتان اللفظتان ولفظة الحركة وضعت أولاً لاستبدال المكان ثم نقلت إلى الأحوال.

(5) At this point, one should know that when one fully understands motion as it should be, then it is seen to be a [single] term having two senses, one of which cannot actually subsist in concrete particulars, whereas the other one can. So, if by *motion* one means the continuous thing intellectually understood to belong to that which undergoes motion, [stretching] from start to end, then what is being moved simply does not have that while it is between the starting and end points. Quite to the contrary, supposedly it has occurred in some way only when what is moved is at the end point; but this continuous intelligible thing has ceased to exist there, and so how can it have some real determinate existence? The fact is that this thing is not really something that itself subsists in concrete particulars. It leaves an impression on the imagery faculty only because its form subsists in the mind by reason of the moved thing's relation to two places: the place from which it departs, and the place at which it arrives. Alternatively, it might leave an impression on the imagery faculty because the form of what is moved, which occurs at a certain place and has a certain proximity and remoteness to bodies, has been imprinted upon it; and thereafter, by [the moving thing's] occurring at a different place and having a different proximity and remoteness, it is sensibly perceived that another form has followed [the first]; and so one becomes aware of two forms together as a single form belonging to motion. [Motion so understood], however, does not determinately subsist in reality as it does in the mind, since it does not determinately exist at the two limits together, and the state that is between the two has no subsistent existence.

(٥) وممّا يجب أنْ يُعلم في هذا الموضع؛ أنَّ الحركة إذا حصل من أمرها ما يجب أنْ يُفهم؛ كان مفهومها إسماً لمعنيين: أحدهما لا يجوز أنْ يحصل بالفعل قائماً في الأعيان، والآخر يجوز أنْ يحصل في الأعيان. فإنَّ الحركة إنْ عُني بها الأمر المتصل المعقول للمتحرّك من المبدأ والمنتهى؛ فذلك لا يحصل البتة للمتحرّك – وهو بين المبدأ والمنتهى – بل إنّما يظنّ أنَّه قد حصل نحواً من الحصول إذا كان المتحرّك عند المنتهى، وهناك يكون هذا المتّصل المعقول قد بطل من حيث الوجود، فكيف يكون له حصولٌ حقيقيٌ في الوجود؟ بل، وهذا الأمر بالحقيقية ممّا لا ذات له قائمة في الأعيان؛ وإنّما ترتسم في الخيال، لأنَّ صورته قائمة في الذّهن بسبب نسبة المتحرّك إلى مكانين: مكان تركه، ومكان أدْركه، أو يرتسم في الخيال لأنَّ صورة المتحرّك – وله حصولٌ في مكانٍ وقربٍ وبُعْدٍ من الأجسامّ – تكون قد انطبعت فيه. ثم تلحقها من جهة الحسّ صورة أُخرى بحصولٍ له آخر في مكانٍ آخر، وقربٍ وبُعْدٍ آخرين، فيشعر بالصورتين معاً على أنّها صورة واحدة لحركة، ولا يكون لها في الوجود حصولٌ قائمٌ كما في الذّهن، إذ الطرفان لا يحصل فيهما المتحرّك في الوجود معاً، ولا الحالة التي بينهما لها وجود قائم.

(6) The thing that exists in actuality and to which the name is appropriately applied—namely, the motion that exists in the mobile—is the intermediate state of [the mobile], when neither is it at the first limit of the traversed distance, nor has it reached the end. Instead, it is at an intermediate limiting point in such a way that at no instant that occurs during the period [that] it passes into actuality is it found occurring at that limiting point such that its occurrence would be as something traversing a certain distance (that is, some interval in the traversal), whatever the period of time you stipulate. This is the form of motion existing in the moved thing—namely, an intermediacy between the posited starting and end points inasmuch as at any limiting point at which it is posited, it did not previously exist there nor will it exist there afterwards, unlike [its state at] the points of the two extreme limits. So this intermediacy is the form of motion and is a single description that necessarily entails that the thing is being moved and simply does not change as long as there is something being moved, [although] certainly the points of intermediacy may, by supposition, change. Now, what is being moved is not something intermediate because it is at one limiting point to the exclusion of another; rather, it is something intermediate because it has the aforementioned description, namely that, inasmuch as at any limiting point that you care to choose, neither was it there before [that point], nor will it be there after [that point]. Its having this description is a single state that always follows upon it at any limiting point whatsoever, not being so described at one limiting point to the exclusion of another. This, in fact, is the first perfection, whereas the second perfection happens once it has made the traversal. This form is found in what is being moved and at an instant, because it is correctly said of it at any instant one cares to choose that [the thing being moved] is at some intermediary limiting point before which it was not there and after which it is not there.

(٦) وأمّــا المعنى الموجود بالفعل، الذي بالحري أن يكون الاســم واقعاً عليه، وأنْ يكون الحركة التي توجد في المتحرّك؛ فهي حالته المتوســطة حين يكون ليس في الطرف الأول من المسافة ولم يحصل عند الغاية، بل هو في حدّ متوسط بحيث ليس يوجد ولا في آنٍ من الآنات التي تقع في مدة خروجه إلى الفعل حاصلاً في ذلك الحدّ، فيكون حصوله في أي وقت فرضته قاطعاً لمسافةٍ مّا وهو بعد في القطع، وهذا هو صورة الحركة الموجودة في المتحرّك؛ وهو توسّط بين المبدأ المفروض والنهاية، بحيث أيّ حدّ يفرض فيه لا يوجد قبله ولا بعده فيه؛ لا كحدّي الطرفين، فهذا التوسّط هو صورة الحركة، وهو صفة واحدة تلزم المتحرّك ولا تتغيّر الْبتَّة ما دام متحركاً. نعم قد تتغيّر حدود التوسط بالفرض، وليس المتحرّك متوسّــطاً لأنَّه في حدّ دون حدّ، بل هو متوسطٌ لأنّه بالصفة المذكورة؛ وهي أنَّه بحيث أيّ حدّ تفرضه لا يكون قبله ولا بعده فيه، وكونه بهذه الصفة؛ أمرٌ واحدٌ يلزمه دائماً في أي حدّ كان، ليس يوصف بذلك في حدٍ دون حدّ. وهذا بالحقيقة هو الكمال الأول، وأمّا إذا قُطع فذلك الحصول هو الكمال الثاني، وهذه الصورة توجد في المتحرّك وهو في آنٍ لأنَّه يصح أنْ يقال له في كل آنٍ يفرض أنَّه في حدٍ وسطٍ لم يكن قبله فيه ولا بعده يكون فيه.

(7) The claim that every motion is in time [may be taken in either one of two ways].[13] On the one hand, by *motion* one may mean the state that belongs to something between a given starting point and end point that it reached and at which it then either stops or does not, where this extended state is in time. The existence of this state is in one way like the existence of things in the past, while in another way it is distinct from them. [That] is because, on the one hand, things existing in the past had an existence at some past instant that was present, while, on the other hand, it is not like this, for one understands by this motion *traversal*. On the other hand, one might mean by *motion* the first perfection that we previously mentioned. In this case, its being in time does not mean that it must map onto a period of time—the fact is that it won't lack the occurrence of some traversal (where that traversal will map onto a period of time), and so it won't lack some time's coming to pass—nor [does it imply] that it continuously remains the same during any instant of that time.[14]

(8) One might say that *being in place* and not having been there before or afterwards is a universal intelligible and does not actually exist (and the same holds for the relation to it—namely, the thing that they designate an *instant*). The fact is [the objector might continue] that what actually exists is only being in *this* place, not having been there before or afterward (and, likewise, for the relation to *this* [instance of] being [in place]), whereas the universal (as the practitioners of the discipline have agreed) is determined only by its individual instances and is

13. Avicenna's treatment here may be taken as a response to a possible objection to his view that there can be motion at an instant, for it seems that every motion requires a period of time, and yet an instant is not a period of time. Thus, there apparently cannot be motion at an instant. See Aristotle, *Physics* 6.3.234a24–234b9.

14. Avicenna's point here is that although his conception of motion allows for motion at an instant, it also entails that there will be a traversal (for, according to him, the mobile cannot remain in the same state for more than an instant, and so the motion necessarily involves the transition from one state to another), in which case there will be a period of time that corresponds to this traversal.

(٧) والذي يقال من أنَّ كل حركةٍ ففي زمانٍ؛ فإمّا أنْ يعني بالحركة الحالة التي للشيء بين مبدأ ومنتهى وصل إليه فيقف عنده أو لا يقف؛ فتلك الحالة الممتدة هي في زمان، وهذه الحال فوجودها على سبيل وجود الأمور في الماضي وتباينها بوجهٍ آخر؛ لأن الأمور الموجودة في الماضي قد كان لها وجودٌ في آنٍ من الماضي كان حاضراً ولا كذلك هذا، فتكون هذه الحركة يعني بها القطع. وإمّا يعني بالحركة الكمال الأول الذي ذكرناه، فيكون كونه في زمانٍ لا على معنى أنَّه يلزمه مطابقة الزمان، بل على أنّه لا يخلو من حصول قطع؛ ذلك القطع مطابقٌ للزمان، فلا يخلو من حدوث زمانٍ لا أنَّه كان ثابتاً في كل آنٍ من ذلك الزمان مستمراً فيه.

(٨) فإنْ قال قائل: إنَّ الكون في المكان لم يكن قبله ولا بعده فيه وكذلك الإضافة إليه، والأمر الذي يجعلونه آناً هو أمرٌ كلّي معقول وليس بموجودٍ بالفعل، بل إنّما الموجود بالفعل الكون في هذا المكان ولم يكن قبله ولا بعده فيه، وكذلك الإضافة إلى هذا الكون، والأمر الكلّي إنّما يثبت بأشخاصه ولا يكون شيئاً واحداً موجوداً بعينه، كما اتفق عليه

not one and the same existing thing.[15] We say: Insofar as *being in place* is predicated of many placed things, the issue is undoubtedly as has been described. Insofar as it is predicated of a single placed thing, but not simultaneously, the matter, however, is problematic. [That] is because it is not improbable that a certain generic account is predicated of a single subject at two moments while not remaining one and the same—as, for example, when a black body becomes white. In that case, when the body was black, there was blackness in it, and blackness is a color, and color is, for instance, like a part of blackness, and there is blackness through a specification of what is joined to [color].[16] When it is white, however, we cannot say that the very same thing to which the specification had been accidentally joined remains. Now there is a different specification joined to it. [The situation would], for instance, be like a board existing in a house according to one specification—namely, that it is part of a wall—and then the very same [piece] becoming part of the roof and having a different relation and different specification as a part of the roof. The present case is not like that; rather, an example of it would be, for instance, if the wall and the board in it ceased to be, and then there came to be in the house a new wall and, in it, another board like the previous one. That is because blackness's difference does not cease, while its share in the nature of the genus to which it is joined remains the same. Otherwise, it would not be a species-making difference but instead would be some accident that does not make a species. This has been explained elsewhere.[17]

---

15. The objection seems to be this: loosely, the form of motion, according to Avicenna, is a mobile's being at some place for only an instant, where this form will hold of the mobile at every particular instant and at every particular place it happens to be during its motion. The objection, then, is that, for the Aristotelian, such as Avicenna, the form that exists in the world is always of a particular, whereas the single form that is predicated of many exists only in the intellect. Consequently, since Avicenna's form of motion is a single form applying to the mobile at many particular places and many particular instants, it cannot exist in the world but must exist only in the mind; and yet Avicenna claimed that it is this form that, in fact, is found in concrete particulars out in the world.

16. Although not noted in **Y**'s edition, **Z**'s apparatus notes that **T** transposes the text's *qārinuhu mā* so as to read *mā qārinuhu,* which is followed here.

17. Cf. *Kitāb al-madkhal* 1.13 and *Kitāb al-maqūlāt* 1.4.

أهل الصناعة. فنقول: أمّا الكون في المكانّ – من حيث يقال على متمكنات كثيرينّ – فلا شك أنَّ الحال فيه على ما قد وصف، وأمّا من حيث يقال على متمكنٍ واحدٍ ولكن لا مَعاً فالأمر فيه مُشْكل؛ فإنَّه لا يبعد أنْ يكون معنى جنسي يقال على موضوع واحدٍ في وقتين ويكون لم يثبت واحداً بعينه؛ مثل الجسم الأسود إذا ابيضّ، فإنَّ الجسم إذا كان أسود فقد كان فيه سواد، وكان السواد لوناً، وكان اللون كالجزء من السواد مثلاً، وبتخصيصٍ ما قارنه كان سواداً. فلّما ابيضّ فلا يمكننا أنْ نقول إنَّ ذات الشيء الذي كان عرض له مقارنة التخْصيص باقية وقارنه تخصيص آخر كخشبةٍ مثلاً موجودة في بيت على تخصيص أنَّها جزء حائط، ثم صارت هي بعينها جزء سقف ولها إضافة أخرى وتخصيص آخر أنَّه جزء سقفٍ، فإنَّ ذلك ليس كذلك، بل مثله مثل أنْ يُعدم الحائط والخشبة التي فيه، ثم يُحدَث في البيت حائط وفيه خشبة أخرى مثل تلك الخشبة، وذلك لأنَّ السواد لا يبطل فصْله وتبقى حصته من طبيعة الجنس التي كانت مقارنة له بعينها؛ وإلاّ فليس بفصْلٍ منوّع، بل هو عارضٌ لا ينوّع، قد علم هذا من مواضع أخرى.

(9) When the situation is like this, we need to consider [which of the following descriptions best describes it]. On the one hand, it might be that the status of *being in the existing place* with respect to the placed thing is sometimes joined to a specification that it is in *this* existing place and at other times joined to another specification, like color's status. On the other hand, it might not be like that, and instead its status would be like heat that sometimes acts on this and at other times acts on that, or wetness that is sometimes acted upon by this and at other times acted upon by that, while being one and the same; or some other accident that remains one and the same while one specification after another follows upon it.

(10) First, we say that this specification of *this* and *that* in the case of place is not something actually existing in itself, as will become obvious to you later.[18] [That is so] because what is continuous does not have actual parts but is accidentally divided into parts, owing to certain causes that divide the [one] spatial magnitude and so make it [many] spatial magnitudes, according to one of the types of division. Now, what is between the limiting points of that division are also spatial magnitudes, which neither an instant nor a motion encompasses. [At any rate, the motion would not occur] in the manner that we said there is [motion] at an instant; rather, the motion would be like the traversal that maps onto a period of time. Nor would the thing that we call an *instant* actually be many in it. Because of that, it is not actually many except when the spatial magnitude actually is made many. When [the instant] is not actually many and the motion is along the single subject (I mean the spatial magnitude that truly exists and is not numerically many), then [the motion] must be numerically one. It would not be like the state, with respect to color, that exists in the subject during the state when [the subject] is black and the state when it is white, where the state of

18. The reference appears to be 3.2.8–10, where the true nature of continua is given and divisions of the continua are associated with products of the estimative faculty.

(٩) فإذا كان الأمر على هذا، فلننظر هل حكم الكون في المكان الموجود في المتمكن، تارة مقارناً لتخصيص أنّه في هذا المكان الموجود، وتارة مقارناً لتخصيص آخر؛ حكمه حكم اللون أو ليس كذلك، بل حكمه حكم حرارة تارة تفعل في هذا وتارة في هذا، أو رطوبة تارة تنفعل عن هذا وتارة تنفعل عن ذلك وهي واحدة بعينها، أو عرض آخر من الأعراض يبقى واحداً بعينه، ويلحقه تخصيص بعد تخصيص؟

(١٠) فنقول: أولاً إنَّ هذا التخصيص بهذا وبذاك في أمر المكان ليس أمراً موجوداً بالفعل نفسه، كما يظهر لك بعد. إذ المتصل لا أجزاء له بالفعل، بل يعرض أنْ يتجزأ لأسباب تقسم المسافة فتجعلها بالفعل مسافات على أحد أنواع القسمة. وما بين حدود تلك القسمة أيضاً مسافات لا يشتمل عليها آنٌ وحركةٌ - على النحو الذي قلنا إنّها تكون في آنٍ - بل الحركة التي على نحو القطْع يكون الزمان مطابقاً لها، ولا يكون المعنى الذي سميناه آناً هو متكثّر فيها بالفعل، لأنَّ ذلك لا يتكثّر بالفعل إلاّ بتكثّر المسافة بالفعل. وإذا لم يكن متكثّراً بالفعل، وكانت الحركة على الموضوع الواحدّ - أعني المسافة حقاً موجودةً - ولم تكن كثيرة بالعدد؛ كانت بالضرورة واحدة بالعدد، ولم تكن على النمط الذي يكون عليه الحال في اللون ووجوده في الموضوع في حال سواده وفي حال بياضه،

the relation, which specifies both of them, is actually [related] to the subject. [That] is because the motion does not actually require a discontinuity; rather, the continuity persists in such a way that this state accompanying it need not change relative to the subject until some enduring feature individually disappears from it. Indeed, it is the actual relation to an actually differing thing that is different, whereas what is actually one becomes many on account of the relation only when the relation is actually many; but when the spatial magnitude is one by continuity, no difference occurring in it, then a given relation to it does not become different. Because of that, then, the number of something that is one does not differ. When, thereafter, the spatial magnitude accidentally has a certain division and difference—which neither depends upon the motion nor does the motion depend upon it, neither one necessitating nor being necessitated by the other—then the duality that occurs is not essentially but only accidentally many and is by means of the one's relation to the many, where the relation is external, not internal, to the thing itself.

(11) In summary, this state is not the state of color that, in fact, differs not in relation to some external thing, [but] by being joined to the difference of blackness and whiteness. Also, what is undergoing motion is not in a certain place absolutely, becoming many by being many with respect to *this* place and *that* place. [That] is because there is no actual discontinuity with respect to the spatial magnitude of the motion, one place being designated to the exclusion of some other, such that *there* it is possible to be in place absolutely, whether generically or specifically, producing either a species or an individual because of its relation to actually many places.

وحال النسبة التي تخصص كلاً إلى الموضوع بالفعل، لأنَّ الحركة لا توجب بالفعل إنفصالاً بل يستمر الاتصال استمراراً لا يجب معه تغيّر هذه الحال بالقياس إلى الموضوع حتى يعدم منه أمرٌ ثابت بالشخص. فإنَّه إنّما تختلف النسبة بالفعل إلى مختلفٍ بالفعل، وإنّما يكثر الواحد بالفعل تكثّراً من قبل النسبة إذا كانت النسبة متكثّرة بالفعل. وإذا كانت المسافة واحدة بالاتصال لا اختلاف فيها لم تختلف إليها نسبة فلم يختلف بسبب ذلك عدد شيء واحد. ثم بعد ذلك إذا عرض للمسافة قسمة ما واختلاف، ولم يكن ذلك ممّا يتعلق بالحركة ولا الحركة تتعلق به، ولا أحدهما موجب الآخر ولا موجبه، كانت الإثنينية التي تعرض غير متكثّرة بالذات بل بالعرض، ومن طريق نسبة الواحد إلى كثير، وتكون النسبة خارجية غير داخلة في ذات الشيء.

(١١) وبالجملة لا تكون هذه الحال حال اللون الذي هو بالحقيقةّ - لا بالقياس إلى أمرٍ خارجٍ - يختلف بمقارنة فصْل السواد والبياض، ولا كون المتحرّك في مكانٍ مطلقاً؛ يصير كثيراً بكونه كثيراً في هذا المكان وذلك المكان، لأنَّه ليس في مسافة الحركة إنفصالٌ بالفعل ومكانٌ معيّن دون مكان، حتى يجوز أنْ يكون هناك كونٌ في المكان مطلقاً؛ جنسياً أو نوعياً يتنوّع أو يتشخص بسبب نسبته إلى أمكنةٍ كثيرة بالفعل.

(12) Know that motion frequently depends upon six things. These are (1) the mobile, (2) the mover, (3) that with respect to which,[19] (4) a *terminus a quo,* (5) a *terminus ad quem,* and (6) time. Its dependence upon the mobile is obvious. Its dependence upon the mover is because either the mobile has motion of itself insofar as it a natural body, or it proceeds from a cause. Now, were the [mobile] to move owing to itself and to no other cause at all, the motion would never cease as long as there existed the selfsame natural body by which it is a mobile; but there are many cases of bodies where the motion ceases but [those bodies] themselves still exist. Also, were the mobile itself the cause of the motion such that it is [both] mover and mobile, the motion would be necessary of itself; but it is not necessary of itself, since the very same natural body exists while it is not undergoing motion. So, if there is found a natural body that is always undergoing motion, it is because it has some attribute additional to its natural corporeality, whether within it (if the motion is internal) or outside of it[20] (if [the motion] is external). In short, the thing itself cannot be a cause of its motion. [That] is because one [and the same] thing is not [both] mover and what is moved, unless it is a mover through its form and is moved through its subject, or [unless] it is a mover when taken together with something and what is moved when taken together with something else. What will make it plain to you that nothing moves itself is that, when the mover produces motion, it does so inasmuch as it [itself] either is being moved or is not. Now, if the mover produces motion when it [itself] is not being moved, then it is absurd that the mover be what is being moved—it is, in fact, different from it. If it produces motion inasmuch as it is being moved and it produces motion through the motion that is actually in it—where the sense of *to produce motion* is to make an actual motion in something that is potentially

19. While the locution *mā fīhi* (literally, "that in which") might seem to refer to the medium through which the mobile passes, Avicenna more regularly uses this locution to refer to the category *in which* there is motion; see par. 24 below.

20. **Y** apparently omits the phrase *in kānat al-ḥarakah laysat min khārij wa-immā khārijan ʿanhu* (if the motion is internal, or outside of it) by homeoteleuton; it appears in **Z**, **T**, and the Latin.

(١٢) واعلم أنَّ الحركة قد تتعلق بأمور ستة وهي: المتحرّك، والمحرّك، وما فيه، وما منه، وما إليه، والزمان. أمّا تعلّقها بالمتحرّك فأمرٌ لا شبهة فيه، وأمّا تعلّقها بالمحرّك فلأن الحركة إمّا أنْ تكون للمتحرّك عن ذاته من حيث هو جسمٌ طبيعي، أو تكون صادرة عن سبب. ولو كانت الحركة له لذاته لا لسببٍ أصلاً؛ لكانت الحركة لا تعدم البتة ما دام ذات الجسم الطبيعي المتحرّك بها موجود، لكن الحركة تعدم عن كثير من الأجسام وذاته موجودة. ولو كانت ذات المتحرّك سبباً للحركة حتى يكون محرّكاً ومتحرّكاً؛ لكانت الحركة تجب عن ذاته - لكن لا تجب عن ذاته - إذ توجد ذات الجسم الطبيعي وهو غير متحرّك. فإنْ وجد جسمٌ طبيعيٌ يتحرك دائماً فهو لصفةٍ له زائدة على جسميته الطبيعية؛ إمّا فيه - إن كانت الحركة ليست من خارج، وإمّا خارجا عنه إنْ كانت عن خارج. وبالجملة لا يجوز أنْ تكون ذات الشيء سبباً لحركته، فإنّه لا يكون شيءٌ واحدٌ محرّكاً ومتحرّكاً إلاّ أنْ يكون محرّكاً بصورته ومتحرّكاً بموضوعه، أو محرّكاً وهو مأخوذ مع شيء، ومتحرّكاً وهو مأخوذ مع شيء آخر. ومما بيّن لك، أنَّ الشيء لا يحرّك ذاته؛ أنَّ المحرّك إذا حرّك لم يخلُ إمّا أنْ يكون يحرّك لا بأنْ يتحرّك، وإمّا بأنْ يكون يحرّك بأنْ يتحرّك، فإنْ كان المحرّك يحرّك لا بأنْ يتحرّك، فمحالٌ أنْ يكون المحرّك هو المتحرّك، بل يكون غيره. وإنْ كان تحرّك بأنْ يتحرّك، وبالحركة التي فيه بالفعل يُحرّك - ومعنى يُحرّك أنّه يُوجِد في شيءٍ متحرّك بالقوة حركة بالفعل - فيكون حينئذ إنّما يُخرِج شيئاً من

moved—then, in that case, it would make something pass from potency into actuality through something actually in it—namely, the motion. Now, it is absurd that that thing be actually in it while the very same thing is potentially in it. For example, if it is hot, then how can it itself be becoming hot through its heat? In other words, if it is actually hot, then how can it be potentially hot so as acquire from itself a prior heat, so as to be simultaneously in actuality and potentiality?

(13) In general, the nature of corporeality is a certain nature of a substance having length, breadth, and depth, where this standing is something common that does not require that there be motion. Otherwise, the [motion] itself would be common. Now, if, in addition to this status, there is some other account such that motion follows upon the body—and to the extent that there is a substance possessing length, breadth, and depth and together with which there is some other property on account of which it is moved—then there is a certain principle of motion in it in addition to the condition by which, when it exists, there is a body. That is all the more obvious when [the principle] is external.

(14) In establishing that every mobile has a mover, a dialectical account has been given of which the best explanation is ours: namely, that every mobile is divisible (as will become clear later)[21] and has parts whose corporeal nature does not prevent the estimative faculty from imagining them at rest, and, in fact, if it is prevented, it is so because of something in addition to [the nature of corporeality]. Now, everything that the estimative faculty imagines, [provided that the thing's] nature does not prevent it, is possible in relation to the act of the estimative faculty *qua* that nature. Now, it is not impossible (save conditionally) for the estimative faculty to imagine a part of the mobile *qua* body at rest,

21. See 3.6.7–9.

القوة إلى الفعل بشـيءٍ فيه بالفعل؛ وهو الحركة. ومحالٌ أن يكون ذلك الشيء فيه بالفعل وهو بعينه فيه بالقوة، فيحتاج أنْ يكْتسبه مثلاً - إنْ كان حاراً - فكيف يسخِّن نفسه بحرارته؛ أي إنْ كان حارّاً بالفعل فكيف يكون حارّاً بالقوة حتى يكتسب من ذي قبل حرارة عن نفسه، فيكون بالفعل وبالقوة معاً؟!

(١٣) وبالجملة طبيعة الجسمية طبيعة جوهرٍ له طولٌ وعرضٌ وعمقٌ، وهذا القدر مشترك فيه لا يوجب حركة، وإلاّ لاشترك فيها بعينها، فإنْ زيد على هذا القدر معنى آخر حتى يلزم الجسم حركة وحتى يكون جوهراً ذا طول وعرض وعمق وخاصيّة أُخرى مع المذكور؛ يتحرّك بسبب ذلك فيكون فيه مبدأ حركة زائد على الشرط الذي إذا وجد كان به جسماً، وإنْ كان من خارج؛ فذلك فيه أظهر.

(١٤) وقد قيل في إثبات أنّ لكل متحرّك محرّكاً؛ قولٌ جدلي. وأحسن العبارة عنه ما نقول: إنّ كلّ متحرّك - كما يتبيّن من بعدّ - منقسمٌ وله أجزاء لا يمنع عن توهمها ساكنة طبيعة الجسمية التي لها، بل إنْ مُنِعَ مُنِعَ لأمرٍ زائد عليها. وكل توهّم شيء لا تمنعه طبيعته؛ فهو من التوهّم الممكن من حيث تلك الطبيعة. فتوهّم جزء المتحرّك ساكناً - من حيث هو جسمّ - توهّمٌ لا يستحيل إلاّ بشرط، وذلك الجزء ليس هو ذلك الكل. وكل

where that part is not that whole, whereas to posit that whatever undergoes motion essentially is at rest is to posit what is not the case. The fact is that it would not rest, especially when [something's undergoing motion essentially] is neither impossible in itself, nor is its resting required with respect to the estimative faculty. So, to posit that each body is in a state of rest makes it necessary that the whole be resting in the way that the cause makes the effect necessary, because, as should be equally clear to you, the state of rest that belongs to the whole is the collection of the states of rest of the parts, when the posited parts or the like occur. Therefore, then, nothing of the bodies undergoes motion essentially.

(15) One might say: Your claim that what is moved essentially does not rest when something else is posited as resting is true only when it is possible to posit that other thing's coming to rest, not [when it is] impossible. That, then, indicates that a certain state of rest necessarily entails that there be something that can rest with it [and] is not impossible. When it is impossible that it rest, it might necessarily follow from positing that it rests that what essentially undergoes motion is resting, despite the fact that it is an absurdity, just as many absurdities follow upon absurdities. Now the truth is, it is absurd that what essentially undergoes motion should be at rest; however, when some other absurdity is posited, then the absurdity of its resting might necessarily follow. So it is impossible only for it to rest in reality, [not in the estimative faculty]. The impossibility of its necessarily being at rest when an absurdity is posited does not, in fact, contradict the truth that its essentially undergoing motion is eliminated as a result of that [absurdity] when it is posited. [That] is because the one is a categorical statement, while

مـا هو متحرّك لذاتـه ، ففرض ما ليس هو بل هو غير سـاكناً ، وخصوصاً إذا كان غير محالٍ في نفسـه ، لا يوجب في الوهم سـكونه . وكل جسم فإنَّ فرض سكون الجزء منه يوجب سـكون الكلّ ، إيجاب العلّة للمعلول ؛ لأنَّ السكون الذي للكل هو ، كما تبيّن لك ، مجموع سـكونات الأجزاء إذا حصلت أجزاء تفرض ، أو غير ذلك ؛ فاذن ولا شيء من الأجسام متحرّكٌ لذاته .

(١٥) فإنَّ قال قائل : إنَّ قولكم إنَّ المتحرّك لذاته لا يسكن إذا فرض غيره ساكناً ؛ إنّما يصح إذا كان فرض سكون ذلك الغير ممكناً غير مستحيل ، فيدل ذلك على أنَّ سكون ما يلزم أنْ يسكن معه جائز غير مستحيل . وأمّا إذا كان سكونه مستحيلاً ، فيجوز أن يكون فرضه ساكناً يلزم عنه سكون المتحرّك لذاته مع أنَّه محال ، كما أنَّ كثيراً من المحالات يلزمها محالات ، فحقٌ أنَّ سكون المتحرّك لذاته محال ، لكنه إذا فُرض محالٌ آخر جاز أنْ يلزمه سكونه المحال ، فإنَّه إنّما يستحيل سكونه في الوجود ، وأمّا لزوم القول بسكونه عند فرض محالٍ لا يمكن ، بل عند فرض ما يسـقط عنه كونه متحرّكاً لذاته فأمرٌ غير مناقض لذلك

the other is a conditional statement. It is as if we were to assume that a hundred is part of ten, in which case wouldn't ten be a hundred and something, even though that is not the case? Because of that, however, it does not necessarily follow that we are wrong to say that ten is not more than a hundred. In the same way, even if the estimative faculty can imagine a state of rest of a part of what is self-moved *qua* body, it might not be the case *qua* part of what essentially undergoes motion and according to its nature. In other words, even if that is possible for it *qua* the nature of its genus, it is not possible for it *qua* its specific nature, and, in fact, it is impossible to assume it—just as it is not impossible that human *qua* animal should fly, whereas it is impossible *qua* human. So, when the former [assumption] is impossible, then one absurd assumption has necessarily followed from another.

(16) We took for granted only that whatever is self-moved would not come to rest through another's coming to rest either when that other's resting exists in reality or [when] the estimative faculty takes the imagining [of the rest] as one of its proper objects[22]—namely, something possible. Concerning the other reason, we too say that when a certain rest that is [itself] absurd is assumed in another, then that which is self-moved might come to rest. To that, we then say that it is not impossible for part of the body *qua* body to rest, and so the impossibility of resting would be on account of something appearing in it other than the corporeality, in which case the cause of motion in any body is something additional to the corporeality; and this we conceded.

---

22. Literally, "the imagining of the estimative faculty is imagined by the estimative faculty to be an object of the estimative faculty."

الحقّ، لأنَّ ذلك حمليٌ وهذا شـــرطي. كما لو فرضنا المئة جزء للعشـــرة، أليست العشرة تكون حينئذ مئة وشـــيئاً؟ وذلك ما لا يكون، وليس يلزم لذلك أنْ يكون قولنا إنَّ العشرة ليست أكثر من مئة باطلاً. وكذلك؛ فعسى أن المتحرّك بذاته - وإنْ أمكن توهّم سكون جزئه من حيث هو جسمّ - فليس يمكن من حيث هو جزء المتحرّك لذاته وعلى طبيعته؛ أي وإنْ كان يمكن ذلك له من حيث طبيعة جنسه، فليس يمكن ذلك له من حيث طبيعته الخاصّة، بل يستحيل فرضه. كما أنَّ الإنسان من حيث هو حيوان لا يمنع أنْ يكون طائراً؛ ويمتنع من حيث هو إنســـان، فإذا كان ذلك ممتنعاً فقـــد لزم فرض المحال من فرض المحال.

(١٦) ونحن إنّما نســـلّم أنَّ ما هو متحرّك بذاته فلا يســـكن بسكون غيره إذا حصل ســـكون غيره في الوجود؛ أو تُوهّم بوهمِ المتوهّم؛ أي الممكن. وأمّا على وجهٍ آخر فإنّا نقول: إنّه قد يلزم أنْ يســـكن المتحرّك بذاته إذا فُرض ســـكون محالٍ في غيره؛ فنقول في جواب ذلك إنَّ جزء الجسمّ - من حيث هو جسمّ - لا يمتنع عليه السكون، فإنَّ امتنع السكون يكون لمعنى عارض عليه غير الجسمية، فإذا كان كذلك، فيكون علّة الحركة في كل جسم أمراً زائداً على الجسمية وهذا نسلّمه.

(17) Still, one might rightly ask us: What forced you to focus on the *part* (given that this is the key premise of the argument) and not just stipulate that it is not impossible for the estimative faculty to imagine the whole *qua* body at rest, in which case it happens to have some account in addition to the corporeality by which the self-moved necessarily moves such that it is impossible to posit the [state of] rest? If the former argument works for you, this one should all the more so. If the intention of the argument is different from this one, [the objector continues] (as if the original speaker in no way believes or means it—its being merely grandiloquence on your part for the sake of his argument) and he himself neither believes nor considers this assumption possible concerning it *qua* body[23] and, rather, says that anything that must be at rest should the estimative faculty imagine something else's being at rest is not moved essentially, and so this is not granted, but, rather, the case is as we explained when initially setting out the problem, then [given all that] something might very well undergo motion on account of itself. Moreover [the objector continues] the estimative faculty imagines a certain absurdity and then, from the imagining of the estimative faculty, it accidentally becomes something that is not self-moved. That absurdity, however, does not entail that the status of [what essentially undergoes motion] changes as a result of a certain absurdity that necessarily follows upon the former absurdity. In fact, what is moved essentially might not be such that, should the estimative faculty imagine a part of it at rest, it comes to rest, but, rather, in that case, it would necessarily cease to be. If it is said, 'This is absurd!' it is responded, 'Yes, but it is an absurdity that necessarily follows upon a prior absurdity.'

23. Reading *min haythu huwa jism* with **Z**, **T**, and the Latin for **Y**'s (inadvertent) *min haythu haur jism* ("*qua* destruction of a body"?).

(١٧) لكن بالحري أنْ يقول لنا قائل: فما اضْطرّكم إلى أنْ اشتغلتم بالجزء، إنْ كان مأخذ الاحتجاج هو هذا، ولم تنصّوا في أول الأمر على الكلّ أنَّه إذا تُوهم ساكناً - من حيث هو جسمّ - لم يستحل، فقد عرض له معنى أريد من الجسمية به صار متحرّك الذات واجب الحركة فيستحيل فرض السكون؟ وإنْ كان ذلك الاحتجاج يكفيكم؛ فهذا أكفى. وإنْ كان الغرض في الاحتجاج غير هذا الغرض، وكأنْ لم يدهب إليه القائل الأول ولا أراده بوجهٍ، وإنّما هو تحسينٌ منكم لكلامه، وهو نفسه لم يذهب إلى إمكان هذا الفرض فيهّ - من حيث هو جسمّ - ولا اعتبر الإمكان، بل قال: إنَّ كل ما كان يُوهم غيره ساكناً يوجب كونه ساكناً، فليس متحرّكاً لذاته. فليس هذا مسلّماً، بل الأمر على ما أوضحناه في التقرير الأول للشّك. فإنَّه يجوز أنْ يكون الشيء متحرّكاً لذاته، ثم يُتوهم محالٌ فيعرض من توهمه أنْ يصير هو غير متحرّك بذاته، ولا يلزم ذلك المحال أنْ يتغيّر حكمه بمحال يلزم ذلك المحال، بل يجوز أنْ لا يكون المتحرّك لذاته، بحيث إذا تُوهم جزؤه ساكناً سكنَ لكنه حينئذ يجب عدمه. فإنْ قيل إنَّ هذا محال، قيل نعم؛ وقد لزم محال محالاً فُرض قبله.

(18) This is a position for which I have no convincing response, although perhaps someone else might. I suspect that the key premise of the argument is not wholly forced into this, and that is because, if this premise is conceded, then bringing about rest is either absurd or not, and then the argument [follows]. The premise that I mean is "anything whose motion is impossible on account of positing a rest in something else is not something moved essentially," which is different from "Anything whose motion is impossible on account of positing a rest in something else is either absurd or not." Even if we were to say, "Anything whose motion is impossible on account of positing an absurdity in something else is not something essentially undergoing motion" (in which case there would be agreement about that), our argument and inference would [still] be valid. Still, the issue concerns the truth of this premise. So, let someone else who is particularly impressed by this argument try to show that this premise is true, and perhaps he will have better luck with this difficulty than we did.

(19) Another doubt [can be raised] against this argument—namely, that even if what is continuous can be supposed to have parts, the estimative faculty cannot imagine those parts as either resting or undergoing motion except by supposition because, as long as they are parts of what is continuous, they neither possess a *where* nor a position except by supposition, which is something that will be explained later.[24] So, if the estimative faculty's act of imagining a state of rest in the part turns out true only when there is an actual discontinuity, then either this argument does not have a relevant key premise or it requires that the estimative faculty imagine a division and then simultaneously a rest. If your estimative faculty were to imagine a rest at some supposed point, while there

---

24. See 3.2.8–10.

(١٨) فهذا القول ممّا ليس يحضرني له جواب أقنع به. ولا يبعد أنْ يكون عند غيري له جوابٌ، وأظنّ أنَّ مأخذ الاحتجاج لا يلجيء إلى هذا كل الإلجاء، وذلك إنْ كانت هذه المقدمة مسلّمة كان التسكين محالاً أو غير محال. ثمَّ الاحتجاج. أعني بالمقدمة قولنا كلّ ما تمتنع حركته لفرض سكونٍ في غيره فليس متحرّكاً لذاته، وهذا غير قولنا كل ما تمتنع حركته لفرض سكونٍ في غيره محال أو غير محال، حتى لو قلنا كل ما يمتنع أنْ يتحرّك لفرض محالٍ في غيره لم يكن متحرّكاً لذاته؛ فسلّم ذلك، لصحّ لنا القول والقياس، ولكن الشأن في صحة هذه المقدمة؛ فليجتهد غيرنا من المتعصبين لهذا الاحتجاج في تصحيح هذه المقدمة، فربما تيسرت له هذه المتعسرّة علينا!

(١٩) وعلى هذا الاحتجاج شكٌ آخر؛ وهو أنَّ المتصلّ - وإنْ كان يمكن أنْ تُفرض له أجزاءٌ - فلا يمكن أنْ تتوهم تلك الأجزاء ساكنة أو متحرّكة إلاّ بالفرض لأنّها غير ذات أيْن ما دامت أجزاء المتصل إلاّ بالفرض، ولا ذات وضْع؛ وهذا شيء سيبين بعد. فإذا كان توهّم السكون في الجزء ممّا لا يتحقّق توهّماً إلاّ وينفصل بالفعل، لم يكن لهذا الاحتجاج مأخذ سديد، أو يدّعي توهّم فصلٍ ثم إسكانٍ معاً. ولو أنّك توهّمتَ في الجزء المفروض

is a continuous [motion], it would have been taking *rest* as an equivocal term, whereas *rest* in its [proper] definition cannot be imagined in that part by the estimative faculty any more than absurd things can altogether be imagined in the intellect and the imagery faculty. So, let someone else who is interested in confirming this key premise puzzle over it and take over the responsibility from us.

(20) Motion's dependence upon the termini *a quo* and *ad quem* is derived from its definition because [motion] is a first perfection occurring in something that has a second perfection by which it terminates at [that second perfection], as well as having a state of potentiality that precedes the two perfections—namely, the state that the first perfection leaves behind when it is directed toward the second perfection. Sometimes the termini *a quo* and *ad quem* are two contraries. Sometimes they are between two contraries, but one is nearer to one contrary and the other is nearer to the other contrary. Sometimes they are neither contraries nor between contraries, but belong to a class of things that have a relation to contraries or opposites in a certain way such that they do not simultaneously occur together, such as the states that belong to the celestial sphere. [That] is because the point from which the motion begins is not contrary to where it ends, but neither do they simultaneously occur together. Sometimes, that which is at the termini *a quo* and *ad quem* remains there for a time so that there is a state of rest at the two limits. Sometimes ([such as] when [the terminus] is actually by supposition, like a limiting point), it occurs at it only for an instant—as in the case of the celestial sphere, for its motion leaves behind a certain starting point while being directed toward a certain end, and yet it stops at neither one.

سكوناً وهو متصل، فقد توهمتَ معنى مشاركاً للسكون في الإسم، وأمّا السكون بحدّه فلا يمكن أنْ يتوهم في ذلك الجزء، كما لا يمكن أن تتوهم الأمور المحالة في العقل والخيال جميعاً. فليكن هذا المأخذ ممّا يسأل 〈عنه〉 غيرنا ممّن يقف على تحقيقه، أنْ ينوب عنّا فيه.

(٢٠) وأمّا تعلّق الحركة بما منه وبما إليه فيستنبط من حدّها لأنَّها أول كمالٍ يحصل لشيءٍ له كمال ثانٍ ينتهي به إليه وله حالة القوة التي قبل الكمالين؛ وهي الحالة التي الكمال الأول تركها وتوجّه إلى الكمال الثاني. وربّما كان ما منه وما إليه ضدّين؛ وربّما كانا بين الضدّين، لكن الواحد أقرب من ضدّ والآخر أقرب من ضدّ آخر. وربّما لم يكونا ضدّين ولا بين ضدّين؛ ولكن كانا من جملة أمور لها نسبة إلى الأضداد وأُمور متقابلة بوجهٍ مّا فلا تجتمع معاً كالأحوال التي للفلك؛ فإنّه لا يتضادّ مبدأ حركة منه لمنتهاها، لكنها لا تجتمع معاً. وربّما كان ما منه وما إليه ممّا يثبت الحصولان فيهما زماناً حتى يكون عند الطرفين سكونٌ، وربّما لم يكن الحصول فيه إذا فُرض كأنّه حدٌّ بالفعل إلاّ آناً كما للفلك؛ فإنّ في حركته ترك مبدأ وتوجهاً إلى غاية، لكن لا وقوف له عند أحدهما.

(21) Here, someone might say that, according to your school of thought,[25] the limiting points in a continuum do not actually exist but only potentially exist, becoming actual only either by a certain dividing (whether like touching or being parallel) or by some accident or supposition, as we shall explain.[26] So, then, as long as one of these [delimiting] causes is not actual, there is no given starting or ending point; and as long as there is no determinate start or end from which the motion begins and at which [it ends], then neither is there motion. So, as long as the celestial sphere does not have a certain cause of being delimited, it does not undergo motion, which is absurd.

(22) We say in response that motion has the end and starting points through a certain sort of actuality and potentiality, where potentiality is taken in two respects: proximate in actuality and remote in actuality.[27] For example, the mobile, at any given moment that it is being moved, has a certain limiting point in proximate potency (which is up to you to posit) and at which it has arrived at some instant (which you posit). So [what is undergoing the motion] has that [limiting point], [even] while in itself it is truly in potentiality. It becomes an actual limiting point by the occurrence of some actual positing and actual dividing, but with the former it does not stop but continues on. Now, a future limiting point (inasmuch as it is a limiting point of motion) cannot be designated as actually such either by some positing or by some actual delimiting cause; rather, in order to have this description, it requires that the distance up to it be completely covered. I mean that here there is that which you can posit as a starting point or as an ending point and, in general, some limiting point that you posit with respect to the motion. So, any of the celestial sphere's motions to which you can point at some determinate time or come to know has that [starting or ending point] by supposition.

---

25. Cf. Aristotle, *Physics* 8.8.263a23–b9.

26. See 3.2.8–10.

27. For a discussion of Avicenna's analysis of circular motion as presented in this passage, see Jon McGinnis, "Positioning Heaven: The Infidelity of a Faithful Aristotelian," *Phronesis* 51 (2006): 140–161, esp. §4.

(٢١) ثم لقائل أنْ يقول؛ إنَّ الحدود في المتصل على مذهبكم ليست موجودة بالفعل بل بالقوة، وإنّما تصير بالفعل إمّا بقطع وإمّا كمماسّة أو موازاة أو بعرضٍ أو بفرضٍ كما سنذكره. فيكون إذنّ - ما لَمْ يكن أحدٌ هذه الأسباب بالفعلّ - لا يكون مبدأ ولا منتهى. وما لَمْ يكن مبدأ ولا نهاية معينّين عنه تبتدىء الحركة أو تنتهي إليه؛ لا تكون حركة، فالفلك، ما لَمْ يكن له سببٌ محدود، لا يكون متحرّكاً، وهذا محال.

(٢٢) فالذي نقوله في جوابه: إنَّ النهاية والمبدأ تكون للحركة بضرب فعلٍ وبضرب قوةٍ، والقوة تكون على وجهين: وجهٌ قريبٌ من الفعل ووجهٌ بعيدٌ من الفعل، مثال ذلك؛ إنَّ المتحرّك في حال ما يتحرّك له بالقوة القريبة حدٌّ، ولك أنْ تفرضه، وقد وصل إليه في آنٍ تفرضه. فيكون ذلك له وفي نفسه بالحقيقة بالقوة، وإنّما يصير بالفعل حدّاً بحصول الفرض بالفعل والقطع بالفعل؛ ومع ذلك لا يقف بل يستمر. وحدّ مستقبلٍ لا يمكنّ - من حيث هو حدّ حركةّ - أنْ يُجعل بالفعل حدّ حركة بفرضٍ أو بسببٍ محدّدٍ بالفعل، بل يحتاج أنْ يستوفي المسافة إليه حتى يصير بهذه الصفة، أعني أنْ يكون هناك ما يمكنك أنْ تفرضه مبدأ أو يمكنك أنْ تفرضه منتهى، وبالجملة حدّاً تفرضه من الحركة. فكل حركةٍ من حركات الفلك تشير إليها في وقتٍ معيّنٍ وتحصّلها فإنَّها يفترض لها ذلك، فتارة يفرض

So, sometimes the starting and ending points are distinct. In other words, they are two different points, both being limiting points of that which is posited of the motion during that time that you determine.[28] At other times, one and the same point is a starting and ending point—a starting point because the motion is from it and an ending point because the motion is toward it—but that [designation] belongs to it at two [different] times.[29] So, motion's dependence upon a starting or ending point (whether [the motion] is in [the category of] place or position) is that, when you designate a given motion or distance, an independent starting and ending point become designated along with that. The dependence upon the starting and ending point of what undergoes local motion is that it has that either in actuality or in potentiality proximate to actuality, according to whichever of the two works, since we have not stipulated that a particular one of them be assigned to it. In short, [motion] depends upon the starting and ending point according to this form and the aforementioned condition, not inasmuch as both are actual.

(23) Next, it is commonly accepted that motion, moving [something], and being moved are a single thing. When it is taken with respect to itself, it is counted as *motion;* if it is taken in relation to that with respect to which [there is motion], it is called *being moved;* and if it is taken in relation to that from which it results, it is called *moving* [*something*]. We should, however, investigate and consider this position with more precision than is commonly done. Now, we say that there is something apart from this form, and that is because being moved is a state that what is moved has, whereas motion is something related to what is moved inasmuch as motion has a certain state in it that what is moved does not have. [That] is because motion's relation to matter means

28. For example, one might designate the Sun's rising in the east and then setting in the west as beginning and ending points, in which case the Sun's motion would have two distinct termini, but only by supposition.

29. For example, one might designate the Sun's being immediately overhead as a single point by which to mark off solar motion; and, in fact, a sidereal day corresponds to the amount of time between the Sun's being immediately overhead and its subsequent return to that position.

المبدأ والمنتهى متباينينّ – أي نقطتين هما حدّا ذلك المفروض من الحركة في ذلك الوقت الذي تعيّنه، وتارة يكون نقطة واحدة هي بعينها مبدأ ومنتهى. أمّا مبدأ فلأنَّ الحركة عنها، وأمّا منتهى فلأنَّ الحركة إليها؛ ويكون ذلك لها في زمانين. فالحركة المكانية والوضعية تعلّقها بالمبدأ أو المنتهى هو أنّك إذا عيّنت حركة ومسافة تعيّن – ومع ذلك – مبدأ ومنتهى قائم بنفسه. والمتحرّك المكاني تعلّقه بالمبدأ والمنتهى هو أنْ يكون ذلك له بالفعل أو بالقوة القريبة من الفعل؛ ذلك على أي وجهٍ كان منهما جازَ، فإنّا لم نشترط الوجه المعيّن فيه منهما. وبالجملة فإنَّها تتعلّق بالمبدأ والمنتهى على هذه الصورة والشرط المذكور، لا من حيث هما بالفعل.

(٢٣) ثم من المشهور أنَّ الحركة والتحريك والتحرّك ذات واحدة، فإذا أُخذت باعتبار نفسها فحسب كانت حركة، وإنْ أُخذت بالقياس إلى ما فيه سُمّيت تحرّكاً، وإنْ أُخذت بالقياس إلى ما عنه، سُمّيت تحريكاً. ويجب أنْ نتحقّق هذا الموضع ونتأمله تأمّلاً أدقّ من هذا المشهور فنقول: إنَّ الأمر بخلاف هذه الصورة؛ وذلك لأنَّ التحرّك حال للمتحرّك، وكون الحركة منسوبة إلى المتحرّك بأنَّها فيه حالٌ للحركة لا للمتحرّك. فإنَّ نسبة الحركة إلى المادة في المعنى، غير نسبة المادة إلى الحركة وإنْ تلازما في الوجود. وكذلك التحريك حالٌ للمحرّك لا للحركة، ونسبة الحركة إلى المحرّك حال للحرك لا للمحرّك. فإذا كان كذلك؛ كان التحرّك نسبة المادة إلى الحركة لا الحركة منسوبة إلى المادة، ولم يكن

something different from the matter's relation to motion, even if in reality the two mutually entail one another. Likewise, moving [something] is a state that the mover has that motion does not have, and the relation of motion to the mover is a state that motion has that the mover does not. Consequently, being moved is matter's relation to motion, not the motion as something related to matter. Also, being moved and moving [something] are not motion in the subject. There is no question that motion's being related to matter as well as to a mover are intelligible concepts, yet these names do not indicate these concepts.

(24) Motion's dependence upon the categories *with respect to which* there is motion does not refer to the subject of [the motion], but to the thing that is the goal that gives rise to the motion. [That] is because the mobile, while it is being moved, is described as situated in between two things, one that was left behind and another that is the goal (whether a *where,* a quality, or the like), provided that the motion does not change the thing all at once. Therefore, [the mobile] is something in between two limiting points, both belonging to a certain category (whether *where,* quality, or the like) and so motion is said to be with respect to that category. This will become clearer for you after you learn motion's relation to the categories.[30]

30. See 2.3.

التحرّك هو الحركة بالموضوع، وكذلك لم يكن التحريك هو الحركة في الموضوع. ولا يناقش في أنْ يكون كون الحركة منسوبة إلى المادة معنى معقولاً، وكذلك إلى المحرّك، لكن هذين المعنيين لا يدل عليهما بهذين الإسمين.

(٢٤) وأمّا تعلّق الحركة بما فيه الحركة من المقولات فليس يعني به الموضوع لها، بل الأمر الذي هو المقصود حصوله في الحركة؛ فإنَّ المتحرّك عندما يتحرّك موصوف بالتوسّط بين أمرين: أمرٌ متروك وأمرٌ مقصود؛ إمّا أيْنٌ أو كيْفٌ أو غير ذلك، إذ كانت الحركة تغيّر الشيء لا دفعة، فاذن يكون متوسطاً بين حدّين ولهما مقولة؛ إمّا أيْنٌ وإمّا كيْفٌ وإمّا غير ذلك، فيقال إنَّ الحركة في تلك المقولة، وقد تزداد لهذا بياناً بعد أنْ تعرف نسبة الحركة إلى المقولات.

Chapter Two

The relation of motion to the categories

(1) There has been a disagreement about motion's relation to the categories. Some said that motion is the category of passion, while others said that the term *motion* applies purely equivocally to the kinds under which it falls.[1] Still others said that *motion* is an analogical term similar to the term *existence,* which includes many things neither purely univocally nor equivocally, but analogically;[2] however, [they continued,] the kinds primarily included under the terms *existence* and *accident* are the categories [themselves], whereas the kinds included under the term *motion* are certain species or kinds from the categories. So there is a stable *where* and a flowing *where* (namely, motion with respect to place); there is a stable quality and a flowing quality (namely, motion with respect to quality—that is, alteration); there is also a stable quantity and a flowing quantity (namely, motion with respect to quantity—that is, augmentation and diminution). Some of them might even take this position to such an extreme as to say that there is also a stable substance and a flowing substance (namely, motion with respect to substance—that is, generation and corruption). They said that flowing quantity is one of the species of

1. Both Philoponus (*In Phys.* 349.5–6) and Simplicius (*In Aristotelis Physicorum,* ed. H. Diels (Berlin: G. Reimer, 1882), 403.13–23 (henceforth Simplicius, *In Phys.*) mention that Alexander of Aphrodisias argued that *motion* is an equivocal term. While Alexander's commentary on the *Physics* is no longer extant, it was translated into Arabic and may very well have been one of the sources for Avicenna's present discussion. Another likely sources is Plotinus' *Enneads* 6.1.15ff. and 3.20ff.

2. The Arabic *tashkīk* literally means "ambiguous"; however, Avicenna consistently contrasts this term with "equivocal" and "univocal" in a way reminiscent of Aristotle's *pros hen* equivocation and anticipating Aquinas's theory of analogy (see part 6 below). As for the specific group under discussion, I have not been able to identify any source where *motion* is described using the language of *tashkīk,* or even as a "flow" (*sayyāl*), a term that this group also apparently used to describe motion.

<الفصل الثاني>

في نسبة الحركة إلى المقولات

(١) إنَّه قد اُختلف في نسبة الحركة إلى المقولات، فقال بعضهم إنَّ الحركة هي مقولة أنْ ينفعل، وقال بعضهم إنَّ لفظة الحركة تقع على الأصناف التي تحتها بالاشتراك البحْت، وقال بعضهم بل لفظة الحركة لفظة مشكّكة مثل لفظة الوجود تتناول أشياء كثيرة لا بتواطؤ ولا باشـتراك بحْت، بل بالتشكيك. لكن الأصناف الداخلة تحت لفظة الوجود والعرض دخولاً أولياً هي المقولات، وأمّا الأصناف الداخلة تحت لفظة الحركة فهي أنواعٌ أو أصنافٌ من المقولات. فالأين منه قارّ ومنه سـيّال هو الحركة فـي المكان، والكيْف منه قار ومنه سـيّال هو الحركة في الكيف أي الاسـتحالة، والكمّ منه قارّ ومنه سـيّال هو الحركة في الكمّ، أي النمو والذبول، وربّما تمادى بعضهم في مذهبه حتى قال: والجوهر منه قارّ ومنه سيّال وهو الحركة في الجوهر أي الكون والفساد. وقالو إنَّ الكمّ السيّال نوعٌ من أنواع الكمّ

continuous quantity because it is possible to find a common limiting point in it, but [flowing quantity] is distinct from the other in that it has no position, whereas continuous [quantity] has a position and stability. Also, they said that blackening and blackness are a single genus, except that blackness is stable, while blackening is not. In short, motion is the flow in each genus. Some of them said: When [motion], however, is related to the cause in which it is, then it is the category of passion, or, [when related] to the cause from which it results, it is the category of action, whereas one group applied this consideration specifically to flowing quality, deriving from it the categories of action and passion.

(2) The proponents of this school of thought—I mean, the doctrine of the flow—disagreed. Some of them made the distinction between blackness and blackening one involving a species-making difference, while others distinguished it by something other than a species difference, since, [this second group argued,] it is like something that is added to a given line, so that the line becomes larger without thereby departing from its species. The first group, [in contrast,] argued: Blackening *qua* blackening is a flowing blackness, and it does not have this [flowing] as something outside of its essence *qua* blackening; and so, then, it must be distinguished from the blackness that remains the same by a species difference. We can show the falsity of both of these arguments: the first is undermined by number, and the second by whiteness and the fact that it is not something outside of the essence of whiteness *qua* whiteness without there being a species difference.

المتصل لا مكان وجود الحدّ المشترك فيه، إلاّ أنَّه يفارقه بأنّه لا وضْع له وللمتصل وضْعٌ واستقرار. قال <وا> والتسوّد والسواد من جنس واحد إلاّ أنَّ السواد قارّ والتسوّد غير قارّ، وبالجملة فإنَّ السيّال في كلّ جنْس هو الحركة. وقال بعض هؤلاء لكنها إذا نسبت إلى العلّة التي هي فيها كانت مقولة أنْ ينْفعل، أو إلى العلّة التي هي عنها صارت مقولة أنْ يفعل. وقومٌ خصّوا هذا الاعتبار بالكيْف السيّال وأخرجوا منها مقولتي يفعل وينفعل.

(٢) واختلف أصحاب هذا المذهب، أعني القول بالسيّال، فمنهم مَنْ جعل الافتراق الذي بين السواد والتسوّد افتراقاً فصْلياً منوّعاً ومنهم مَنْ جعله افتراقاً بمعنى غير فصْلي، إذ هو كزيادةٍ تعرض على خطٍ فيصير خطّاً أكبر ولا يخرج به عن نوعه. وقال الأولون، بل التسوّد بما هو تسوّدٌ هو سوادٌ سيّال، وليس هذا له أمراً خارجاً عن هويّته بما هو تسوّدٌ، فهو إذن يمايز السواد الثابت بفْصل. ويمكن أنْ نبيّن بطلان الحجتين جميعاً: أمّا الأولى فتنتقض بالعدد، وأمّا الثانية فبالبياض وكونه أمراً غير خارجٍ عن هويّة الأبيض بما هو أبيض من غير أنْ يكون فصلاً.

(3) There is also a third school of thought[3]—namely, of those who say that the kinds under which the term *motion* falls (even if it is an analogical [term], as was said) are not species of categories in the aforementioned way. So, blackness is not a species of quality, and locomotion is not a species of *where*. Indeed, motion does not occur in the [category of] quality in such a way that quantity is its genus or, likewise, its subjects; for all motions are only in the substance *qua* subject, and there is neither difference nor distinction among them in this sense. Still, when [the subject's] substantiality is replaced, then that replacement is called a motion with respect to substance as long as it is in process; and if it is with respect to *where*, then it is called a motion with respect to *where*. In general, if the termini *a quo* and *ad quem* are a quality, the motion is with respect to quality, and if [they are] a quantity, the motion is with respect to quantity. Accordingly, *motion* is not said univocally, for *perfection*, taken as a genus in the description of [motion *qua* first perfection of the potential as such], belongs to the class of terms similar to *existence* and *unity*. Now, you know that quantity, quality, and *where* are not included under a single genus; and when these categories are neither included under a single genus nor does the first perfection's relation to them contain them in the manner of a genus, we have no way to make motion a generic concept. Instead, this description includes a certain concept, something like which only an analogical term will indicate.

(4) Concerning this topic of inquiry, these are the three positions that need to be considered. I do not like the middle position and, in fact, detest its claim that blackening is a quantity and augmentation a quality. It is not right that blackening is a blackness that is undergoing intensification; rather, it is an intensification of its subject with respect to its blackness. That is because, when you assume that some blackness has

3. The position mentioned here has some affinities with that of Abū Bishr Mattá; see the Arabic *Physics*, 179.

(٣) وهـا هنا مذهبٌ ثالـثٌ؛ وهو مذهب مَنْ يقول إنَّ لفظـة الحركة، وإنْ كانت مشـكّكة كما قيل، فإنَّ الأصناف الواقعة تحتها ليسـت أنواعاً من المقولات على السبيل المذكورة، فلا التسوّد نوعٌ من الكيف ولا النقلة نوعٌ من الأين، فإنَّ وقوع الحركة في الكيف ليـس علـى أنَّ الكيف جنسٌ لها ولا أيضاً موضوعٌ لها، فـإنَّ جميع الحركات إنّما هي في الجوهر من حيث هي في موضوع لا غير ولا تمايز بينها في هذا المعنى. ولكن إذا تبدّلتْ جوهريته سـمي ذلك التبدّل - ما دام في السـلوك - حركة في الجوهر، وإنْ كان في الأين سـميّ حركة في الأين. وبالجملة إنْ كان ما عنه وما إليه كيْفاً فالحركة في الكيف، وإنْ كان كمّـاً فالحركة في الكمّ. وتقال الحركة على هذه لا بالتواطؤ، فإنَّ الكمال المأخوذ في رسـمها أخذ الجنْس هو من الألفاظ المجانسـة للوجود والوحدة، وأنت تعلم أنَّ الكمّ والكيف والأيْن ليست داخلة تحت جنس واحد. وإذا لم تكن هذه المقولات داخلة تحت جنسٍ واحد ولا نسـبة الكمال الأول إليها أيضاً أمراً حاصراً إيّاها حصر الجنس؛ لم يكن لنا سبيل إلى أنْ نجعل الحركة معنى جنسياً؛ بل هذا الرسم يتناول معنىً إنّما يدلّ على مثله لفظٌ مشكّكٌ لا غير.

(٤) والمذاهـب الملتفت إليها فـي هذا المطلوب هي هذه الثلاثـة، وليس يعجبني المذهب الأوسط، بل أستكره ما يقال به من أنَّ التسوّد كيْفية وأنَّ النمو كميّة. وبالحري أنْ لا يكون التسوّد سواداً يشتدّ، بل اشتداد سوادٍ، بل اشتداد الموضوع في سواده؛ وذلك لأنّه لا يخلو إذا فرضنا سـواداً اشـتدّ إمّا أنْ يكون ذلك السـواد بعينه موجوداً - وقد

undergone intensification, then either the very same blackness exists and with the intensification there happened to be a certain increase, or [the blackness] does not exist. On the one hand, if it does not exist, then it is absurd to say that what does not exist but has passed away is this thing right now undergoing intensification, for its having an existing description requires that it be something existing that remains the same. Now, if the blackness remains the same, then there is no flow (that is, a flowing quality), as they maintained. Instead, it is something always remaining the same to which there accidentally belongs a certain increase whose amount does not remain the same—and, in fact, at each instant there is some other amount—in which case this continuous increase is the motion, not the blackness. So motion is either the intensification of the blackness and its flow, or the intensification of the subject with respect to blackness and its flow with respect to it; it is not the intensifying blackness. It is obvious from this that the intensification of blackness brings about [the blackness's] departure from its original species, since it is impossible to point to whatever of [the blackness] that exists, which while conjoined to it increases it. The fact is that any limiting point that it reaches is a simple quality; however, people name all the limiting points *black* that resemble a given one and anything resembling white (that is, what is close to it) *white*. Absolute blackness is one [species] (which is an obscure limit), and the same holds for [absolute] whiteness and the rest, such as what is a mix [of the two]. Now, what is a mix [of the two] is not one of the two extreme limits—it shares nothing in common in reality but the name. Only different species arise between the two extremes, but, owing to the proximity to one of the two limits, it is accidentally associated with [one of the extreme limits]. Indeed, sometimes sensation does not distinguish between the two, and so we suppose that they are single species, when that is not the case. This will be confirmed in the universal sciences.[4]

---

4. While there seems to be little question that "universal sciences" here refers to the *Ilāhīyāt*, it is not clear what the exact reference is. For now, see 3.7, where Avicenna does return to the issue of what is involved in the change of a specific color.

عرضتْ له عند الاشتداد زيادةٌ - أو لا يكون موجوداً . فإنْ لم يكن موجوداً فمحالٌ أنْ يقال إنَّ ما قد عدم وبطل هو ذا يشتدّ فإنَّ الموصوف بصفةٍ موجودة يجب أنْ يكون أمراً موجوداً ثابت الذات . وإنْ كان السواد ثابت الذات فليس بسيّالٍ كما زعموا من أنَّه كيفية سيّالة بل هو ثابت على الدّوم ، تعرض عليه زيادة لا يثبت مَبْلَغُها ، بل يكون في كل آنٍ مبلغ آخر ؛ فتكون هذه الزيادة المتصلة هي الحركة لا السواد . فاشتداد السواد وسيلانه أو اشتداد الموضوع في السواد وسيلانه فيه ، هو الحركة لا السواد المشتدّ . ويظهر من هذا أنَّ اشتداد السواد يخرجه عن نوعه الأوّل ، إذ يستحيل أنْ يشير إلى موجودٍ منه وزيادة عليه مضافة إليه ، بل كل ما يبلغه من الحدود فكيْفية بسيطة . لكن الناس يسمون جميع الحدود المشابهة لحدّ واحدٍ سواداً ، وجميع المشابهة للبياض - أي المقاربة له - بياضاً . والسواد المطلق هو واحد وهو طرف خفي ، والبياض كذلك ، وما سوى ذلك كالممتزج ؛ والممتزج ليس هو أحد الطرفين ولا يشاركه في حقيقة المعنى ، بل في الإسم . وإنّما تتكون الأنواع المختلفة في الوسط ، لكنه يعرض لمّا يقرب من أحد الطرفين أنْ ينسب إليه . فالحسّ ربّما لم يميّز بينهما فظنّهما نوعاً واحداً وليس كذلك ، وتحقيق هذا في العلوم الكلّية .

(5) The last position [that we mentioned] shows better judgment than this one and does not follow it except for a common feature that both positions entail. Underlying [that common feature] is the fact that those who assign this number to the number of the categories [i.e., ten] are forced into either one of two situations: either they allow that motion is one of the ultimate genera [i.e., one of Aristotle's ten categories], or they must increase the number of categories, since the kinds of motion are not subsumed under one of their genera—not even the category of passion—whereas [*motion*] is a universal concept generically predicated of many. So, if they are going to be obstinate about the categories' being ten, then they should be indulgent and concede that the category of passion is motion, even if [it means that], with respect to this category, they give up on and do not even try to preserve the pure univocity with which I see them being so particularly impressed. In fact, they were so indulgent in the case of [the category of] possession that it [ought] to convince them all the more so in the case of motion.

(6) Be that as it may, it is quite likely that, even if the expressions *perfection* and *action* apply to substance and the remaining nine [categories] analogically, their application to the kinds of motion is not purely analogical. That is because analogy expresses a single concept, but the things that that concept includes differ with respect to it in priority and posteriority—such as *existence,* since [existence] belongs to substance primarily and to the accidents secondarily. As for the concept of motion—that is, the first perfection belonging to what is in potency insofar as it is in potency—it has nothing to do with [the situation] where one thing called *motion* is derived from another. So locomotion's having this description [namely, being a motion] is not a cause of alteration's having this description. The existence of locomotion might, in fact, be a

(٥) وأمّا المذهب الأخير فهو أحصف من هذا المذهب، ولا يلزمه إلاّ أمرٌ مشتركٌ يلزم المذهبين، ومبناه على أنَّ الواضعين لعدد المقولات هذا العدد يلزمهم أحد أمرين: إمّا أنْ يجوّزوا أنْ تكون الحركة جنساً من الأجناس العالية، وإمّا أنْ يزيدوا في عدد المقولات زيادة ضرورية، إذ كانت أصناف الحركة لا تدخل في جنس منهاّ – ولا في مقولة ينفعلّ – وهي معانٍ كلّية مقولة على كثيرين قول الأجناس. فإنْ تشدّدوا في عشرية المقولات، فواجبٌ أنْ يسامحوا ويجعلوا مقولة أنْ ينفعل هي الحركة، وأنْ لا يطلبوا في مقولة أنْ ينفعل من صريح التواطؤ ما أراهم يتعصبون فيه ولا يحفظونه. فإنَّهم قد فعلوا في أمر مقولة الجدة من المسامحة ما يحملهم على أكثر من ذلك في الحركة.

(٦) على أنَّه لا يبعد أنْ تكون لفظة الكمال والفعلّ – وإنْ كان وقوعهما على الجوهر والتسعة الباقية وقوعاً بالتشكيكّ – فإنَّ وقوعهما على أصناف الحركة لا يكون بالتشكيك الصريح. وذلك لأنَّ التشكيك هو أنْ يكون اللفظ واحد المفهوم؛ ولكن الأمور التي يتناولها ذلك المفهوم تختلف بالتقدّم والتأخّر فيه كالوجود؛ فإنّه للجوهر أولاً وللأعراض ثانياً. وأمّا مفهوم الحركةّ – وهو الكمال الأول لما بالقوة من حيث هو بالقوةّ – فليس ممّا يستفيده بعض ما يسمّى باسم الحركة من بعض. فليس كون النقلة بهذه الصفة علّة لكون الاستحالة بهذه

cause for the existence of alteration, in which case the priority and posteriority would concern the concept expressed by *existence,* but not the concept expressed by *motion.* It is just as the couplet precedes the triplet with respect to the concept of existence, while not preceding it with respect to the concept of being a number, for both have a number simultaneously. The triplet does not have a number because the couplet has a number in the way that the triplet's existence is dependent upon the existence of the couplet: the concept of existence is different from the concept of number, the sense of which you have learned elsewhere.[5] So it is quite likely that, even if *perfection* is analogical in relation to other things, it is univocal in relation to these [that is, the kinds of motion], just as it quite likely that it is equivocal in relation to certain things while univocal in relation to what falls under some of them.

(7) Returning to where we were, we ask of both groups what they will say about the category of passion: Is it motion itself, or is it, as they say, one of motion's [various] relations to the subject? If it is motion itself, then is it motion itself absolutely or a certain motion? If it is motion itself absolutely, then motion is one of the [ultimate] genera. If it is a certain motion—as, for example, locomotion or alteration—then the number of the categories must be increased. [That] is because, if locomotion is a genus, then alteration and motion with respect to quantity are equally genera, since each one of these is just as deserving as any other. If locomotion is not a genus but an analogical term, something that is a genus [namely, the category of passion] will exist under it, even though [locomotion] is more specific than [the category of passion] taken as a whole. Now, if the category of passion is not motion absolutely, but motion's relation to matter, then [this relation] must belong either

5. See, for instance, *Physics* 1.8.3.

الصفة، بل يجوز أنْ يكون وجود النقلة سبباً لوجود الاستحالة، فيكون التقدّم والتأخّر في المفهوم من لفظة الوجود لا في المفهوم من لفظة الحركة كما أنَّ الإثنينية قبل الثلاثية في مفهوم الوجود وليس قبله في مفهوم العددية، فإنَّ العددية لهما معاً، ليست العددية للثلاثية من جهة العددية للثنائية، كما أنَّ الوجود للثلاثية يتعلق بالوجود في الثنائية، ومفهوم الوجود غير المفهوم من العدد، وأنت قد عرفتَ هذا المعنى في مواضع أُخرى. فلا يبعد أنْ يكون الكمالّ - وإنْ كان مشكّكاً بالقياس إلى أشياء أُخرىّ - هو متواطيء بالقياس إلى هذه، كما لا يبعد أنْ يكون مشتركاً بالقياس إلى أشياء ومتواطئاً بالقياس إلى ما تحت بعضها.

(٧) ونرجع إلى ما كنا فيه فنقول للطائفتين جميعاً: ما قولكم في مقولة أنْ ينفعل؛ أهي نفس الحركة أمْ نسبة للحركة إلى الموضوع كما يقولون؟ فإنْ كانت نفس الحركة، أفهي نفس الحركة المطلقة أمْ نفس حركةٍ مّا؟. فإنْ كانت نفس الحركة المطلقة، فالحركة أحد الأجناس، وإنْ كانت نفس حركةٍ مّا مثلاً نفس النقلة أو نفس الاستحالة، فيجب أنْ يزاد في عدد المقولات، فإنّه إنْ كانت النقلة جنساً فالاستحالة أيضاً جنس، والحركة في الكمّ جنس، فإنَّ كلّ واحدٍ من هذه يستحق ما يستحقه الآخر. وإنْ كانت النقلة ليست جنْساً بل إسماً مشكّكاً فيوجد تحته معنى هو جنس، وإنْ كان أخصّ من عمومه. وإنْ لم تكن مقولة أنْ ينفعل هي الحركة مطلقة؛ بل كانت نسبة الحركة إلى المادة؛ فلا يخلو إمّا أنْ تكون

[(1)] to absolute motion or [(2)] to a certain motion. If, on the one hand, [the relation] belongs to absolute motion, then absolute motion must be predicated either [(1a)] univocally or [(1b)] analogically of its kinds. If [(1a)] it is predicated univocally, then motion considered in itself is a genus, and so the genera become greater than ten! (The fact is that it is better suited to be a genus through itself than through its relation to its subject, and even if not better suited, at least no less suited). If, on the other hand, [(1b)] [absolute motion] is predicated analogically [of its kinds]—and likewise for the category of passion, which [on the present supposition] is the relation of this thing that is analogical in name to its subject—then there is no genus. If [(2)] the category [of passion] is the relation of a certain kind of motion [to matter], then all the other kinds [of motions] are equally entitled to be the same as it [that is, a genus]. Moreover, [each kind of motion] would be one genus in itself and another in relation to the subject, increasing the genera greatly. Likewise, they [namely, the proponents of either (1) or (2)] must ask themselves why they made quality itself a genus while not making its relation to the subject a genus, whereas here they make the relation of either absolute motion or a certain motion a genus while not making motion itself a genus. If what they are considering is the natures of things considered in themselves as abstract essences without their accidents of relations and the like, then they should make the category of passion the very state of passivity, not its relation to something. The whole of this discussion will be confirmed once you understand what we said earlier about action and being acted upon by motion and being moved.[6] So it is most fitting that they make the category of passion and motion belong to a single class.

---

6. Avicenna discusses the association of the categories of action and passion to motion at *Kitāb al-maqūlāt* 6.6, although the reference may also be to *Physics* 2.1.23–24.

للحركة المطلقة أو لحركةٍ مّا . فإنْ كانت للحركة المطلقة فلا يخلو إمّا أنْ تكون الحركة المطلقة مقولة على أصنافها بالتواطؤ أو بالتشكيك، فإنْ كانت مقولة بالتواطؤ فالحركة باعتبار ذاتها جنس؛ فصارت الأجناس أكثر من عشرة! . ولأنْ تكون بذاتها جنْساً أولى من أنْ تكون بنسبتها إلى موضوعها جنساً ، وإنْ لم تكن أولى فليس دونه في الاستحقاق؛ وإنْ كانت مقولة بالتشكيك. وكذلك مقولة أنْ ينفعل التي هي نسبة هذا المشكّك إسمه إلى موضوعه مقولة بالتشكيك فليس بجنْس. وإنْ كانت المقولة هي النسبة لصنْفٍ من الحركة فيستحق مثله سائر الأصناف، ومع ذلك فيكون بنفسه جنساً وبالقياس إلى الموضوع جنساً آخر، وتتزيّد الأجناس تزيّداً كثيراً. وكذلك يلزم أنْ يطالبوا بالسبب الذي جعلوا له نفس الكيفية جنساً ولم يجعلوا نسبتها إلى الموضوع جنساً، وهناك أخذوا نسبة الحركة المطلقة، أو حركة مّا فجعلوها جنساً ولم يجعلوا الحركة نفسها جنساً. وإنّ كان مأخوذهم طبائع الأمورّ - وذواتها مجردة الماهيات لامع عوارض لها من نسبٍ وغير ذلكّ - فيجب أنْ يجعلوا مقولة أنْ ينفعل هي نفس حالة الانفعال لا ما هو نسبةٌ لها إلى شيء. وهذا الكلام إنّما يتحقّق كلّه بعد أنْ تعرف ما قلناه قديماً من حال الفعل والانفعال بالتحريك والتحرّك، فالأولى بهم أنْ يجعلوا مقولة أنْ ينفعل والحركة من بابةٍ واحدة.

(8) We ourselves are not that obstinate about preserving the received canon—namely, that the genera are ten and that each one of them is truly generic and that there is nothing outside of them. You can also give this same explanation to whoever makes *motion* an absolutely equivocal term. So, when the positions for which we have given evidence but [have] not accepted are repudiated, the truth alone remains: namely, the first position. Since we have explained the manner of motion's relation to the categories and made clear the sense of our saying that motion is in the category, what is there, then, but to let us now explain in how many categories motion occurs?

(٨) وأمّا نحن فإنّا لا نتشدّد كلّ التشدّد في حفظ القانون المشهور من أنَّ الأجناس عشرة، وأنّ كلّ واحدٍ منها حقيقي الجنسية ولا شيء خارج منها، ويمكنك أنْ تبيّن هذا البيان بعينه لمن جعل الحركة إسماً مشتركاً على الإطلاق. فإذا انفسخت المذاهب التي أثبتناها ولم نقبلها بقي الحق واحداً وهو المذهب الأول. فإذ قد بيّنا وجه نسبة الحركة إلى المقولات، وأوضحنا معنى قولنا إنَّ الحركة في المقولة ما هو؛ فلنبيّن الآن أنَّ الحركة في كم مقولةٍ تقع.

Chapter Three

Concerning the list of those categories alone in which motion occurs

(1) Let us lay a foundation, even if might include a repetition of some of what was said. So, we say that the statement "Motion is in such-and-such a category" might possibly be understood in four ways, the first of which is that the category is a certain real subject of [motion] subsisting in itself. The second is that, even if the category is not [motion's] substantial subject, it is by means of [the category] that [the motion] really does belong to the substance, since it exists in it primarily, just as smoothness belongs to the substance only by means of the surface. The third is that the category is [motion's] genus, and [motion] is a species of it. The fourth is that the substance is moved from a certain species of that category to another and from one kind to another. Now, the sense that we adopt is this last one.

(2) We say: Motion is said to be in [the category of] substance [only] metaphorically. Indeed, motion does not occur in this category, because when the substantial nature corrupts and comes to be, it does so all at once, and so there is no intermediate perfection between its absolute potentiality and absolute actuality. That is because the substantial form is not susceptible to increase and decrease, which, in turn, is because if it

<الفصل الثالث>

في بيان المقولات التي تقع الحركة فيها وحدها لا غيره

(١) إنّا لنضع أصلاً، وإنْ كان ربّما اشتمل على تكرار بعض ما قيل، فنقول: إنَّ قولنا إنَّ مقولة كذا فيها حركة؛ قد يمكن أنْ يُفهم منه أربعة معانٍ؛ أحدها أنَّ المقولة موضوع حقيقي لها قائم بذاته، والثاني أنَّ المقولة، وإنْ لم يكن الموضوع الجوهري لها، فبتوسطها تحصل للجوهر، إذ هي موجودة فيه أولاً، كما أنَّ الملاسة إنّما هي للجوهر بتوسط السطح. والثالث أنَّ المقولة جنس لها وهي نوعٌ لها. والرابع أنَّ الجوهر يتحرّك من نوع لتلك المقولة إلى نوع آخر، ومن صنْفٍ إلى صنْفٍ، والمعنى الذي نذهب إليه هو هذا الأخير؛

(٢) فنقول: أمّا الجوهر؛ فإنَّ قولنا إنَّ فيه حركة هو قولٌ مجازيٌ؛ فإنَّ هذه المقولة لا تعرض فيها الحركة، وذلك لأنَّ الطبيعة الجوهرية إذا فسدت تفسد دفعة، وإذا حدثت تحدث دفعة، فلا يوجد بين قوتها الصرفة وفعلها الصرف كمال متوسط، وذلك لأنَّ الصورة الجوهرية لا تقبل الاشتداد والتنقّص، وذلك لأنّها إذا قبلت الاشتداد والتنقّص لم يخل إمّا

is so susceptible, then, when it is in the middle of increasing and decreasing, its species must either remain or not. Now, on the one hand, if its species remains, then the substantial form has not changed at all, but only some accident belonging to the form has changed, in which case that which is decreasing or increasing has ceased to exist while the substance has not; and so this is a case of alteration or the like, not generation. On the other hand, if the [same] substance does not remain with the increase, then the increase would have brought about another substance. Likewise, at every moment assumed during the increase, another substance would come to be once the first has passed away, and it would be possible for there to be a potential infinity of substantial species (as in the case of qualities); but it is a known fact that this is not the case. Therefore, the substantial form passes away and comes to be all at once, and whatever has this description does not have an intermediary between its potentiality and actuality, which is motion.

(3) We also say that the subject of the substantial forms does not actually subsist except by receiving the form (as you have learned),[1] like the material that does not, in itself, exist except as something potential. Now, it is impossible that something that does not actually exist should be moved from one thing to another. So, if there is substantial motion, it involves some existing moving thing, where that moving thing will have a form by which it is actual and is an actually subsisting substance. So, if it is the substance that was before and so it is found to exist up to the moment that the second substance exists, then it has neither corrupted nor changed with respect to its substantiality, but [merely] with respect to its states. Now, if it is some substance other than the species from which and to which [there is purportedly motion], then the substance has first

1. See 1.2.5.

أنْ يكونّ – وهو في وســط الاشتداد والتنقصّ – يبقى نوعه أوْ لا يبقى، فإنْ كان يبقى نوعــه فما تغيّرت الصورة الجوهرية البتة، بل تغيّر عارض للصورة فقط؛ فيكون الذي كان ناقصاً واشــتدّ قد عدم، والجوهر لم يعدم، فيكون هذا إستحالة أو غيرها، لا كوناً. وإنْ كان الجوهر لا يبقى مع الاشــتداد، فيكون الاشــتداد قد جلب جوهراً آخر، وكذلك في كل آنٍ يُفرض للاشــتداد يحدث جوهراً آخر، ويكــون الأول قد بطل، ويكون بين جوهرٍ وجوهــرٍ إمكان أنواعٍ جوهرية غيــر متناهية بالقوة، كما في الكيفيات. وقد علم أنَّ الأمر بخــلاف هذا. فالصورة الجوهرية إذن تبطل وتحدث دفعة، وما كان هذا وصفه فلا يكون بين قوته وفعله واسطة هي الحركة.

(٣) ونقــول أيضاً إنَّ موضوع الصور الجوهرية لا يقــوم بالفعل إلاّ بقبول الصورة كما علمتّ – كالهيولىّ – وهي في نفســها لا توجد إلاّ شــيئاً بالقوة، والذات غير المحصّلة بالفعل يســتحيل أنْ تتحرّك من شيءٍ إلى شيء. فإنْ كانت الحركة الجوهرية موجودة فلها متحــرّكٍ موجود، وذلك المتحــرّك تكون له صورة هو بها بالفعــل، ويكون جوهراً قائماً بالفعل. فإنْ كان هو الجوهر الذي كان قبل؛ فهو حاصل موجود إلى وقت حصول الجوهر الثانــي لم يفســد ولم يتغير في جوهريته، بل في أحوالــه. وإنْ كان جوهراً، غير الجوهر الذي عنه والذي إليه، فيكون قد فسد الجوهر أولاً إلى الجوهر الوسط، وتميّز إذن جوهران

corrupted into the intermediary substance, and thus two substances are actually distinguished. The discussion about [this intermediary substance] is just like the discussion about the substance from which the motion was assumed [to begin]. [That] is because either it possesses the nature into which it first changed during that entire period of time, and then all at once changes into the second, or it preserves its original species during part of that time, while at another part it becomes that other species without some intermediate state, which comes down to our earlier claim about transitioning from one species to another all at once. That period of time, then, corresponds with some motion other than ones that would produce [a new] species of substance, since transitions with respect to substantiality do not [occur] over a period of time. Now, it cannot be said that this argument equally holds with respect to the motion of alteration. That is because the material, in our view, requires the existence of certain actual forms for its subsistence, whereas when the form exists, a certain species actually exists; and so the substance that is between two substances must exist in actuality, not merely by supposition. That is not the case with respect to accidents that the estimative faculty imagines between, for instance, two qualities. So [accidents] are dispensable with respect to the actual subsistence of the subject.

(4) Sometimes they establish that there is no motion with respect to substance because [substance's] nature has no contrary.[2] Now, when its nature has no contrary, it cannot increasingly and decreasingly go[3] from one nature to another such that the state in which it is when there is motion is in between two extreme limits that are not together and between which there is the maximum degree of separation—namely,

2. Compare Aristotle, *Physics* 5.2.225b10–11.

3. Reading *yantaqilu* with **Z**, **T**, and the Latin (*permutetor*) for **Y**'s *yanfaṣilu* (to separate oneself).

بالفعــل. والكلام فيه كالكلام فــي الجوهر الذي فرضت الحركة منه؛ فإنّه إمّا أنْ يكون في تلــك المدة كلّها على طبيعة الجوهر المتغيّر إليه أولاً فيكون التغيّر إلى الثاني دفعة، وإمّا أنْ يكــون في بعض تلك المدة حافظاً لنوعــه الأول، وفي بعضها الآخر واقعاً في النوع الآخر بلا توســط. فيلزم منه ما قيل من الانتقال من نــوع دفعةً، فتكون تلك المدة مطابقة لحركةٍ غيــر حركات نوعية الجوهــر؛ إذ كانت الانتقالاتُ في الجوهرية لا فــي مدةٍ وزمان، ولا يمكن أنْ يقال إنَّ هذا القول يلزم أيضاً على حركة الاســتحالة، وذلك لأنَّ الهيولىّ - فيما نحن فيــهّ - محتاجة في قوامها إلى وجود صور بالفعــل، والصورة إذا وجدت بالفعل حصلــت نوعاً بالفعل. فوجَبَ أن يكون الجوهر الذي بــين الجوهرين أمراً محصّلاً بالفعل ليس بالفرض، ولا كذلك في الأعراض التي تتوهم بين كيفيتين مثلاً فإنّها مُستغنى عنها في قوام الموضوع بالفعل.

(٤) وقد يثبتون أنَّ الجوهر لا حركة فيه لأنَّ طبيعته لا ضدّ لها، وإذا لم يكن لطبيعته ضدٌّ اســتحال أنْ ينتقل عن طبيعةٍ إلى طبيعةٍ أخرى على سبيل التنقّص والاشتداد حتى تكون الحالة التي هو فيها عند الحركة حالة متوسطة بين طرفين لا يجتمعان، وبينهما غاية

two contraries. We should consider this proposition in some detail. We say that, in the definition of contrariety, either the matter or the subject must be taken. Now, if by *subject* one means the real subject actually subsisting as a species that receives those accidents that belong to that species, then substantial forms are not contraries, because they are in a material, not a subject. If by that subject one means any substrate whatsoever, then it seems that the form of fire is contrary to the form of water—and not merely their quality (for there is no doubt about [the contrariety of their qualities]), but, rather, the forms from which their qualities proceed. That is because the two forms share a substrate upon which they successively follow, and there is a maximal degree of difference between them. On account of this, there has been a tendency[4] to try to show that the celestial sphere is not generated because its form has no contrary, as if it were taken as an axiom that the form of whatever is generated has a contrary toward which it goes. In that case [namely, on the assumption that whatever is generated has a contrary toward which it goes], fire, air, water, and earth would represent contraries of form. So why was the substantial forms' having a contrary denied absolutely? It seems that between the contrary that we mentioned here and some other thing there is a maximal degree of difference, where there is *a maximal degree of difference* between it and that one only when some third thing together with it has less than a [maximal degree of] difference (namely, what is intermediary), such that its being borne toward it involves an extension like the extension, in an interval between two things. Now, the substantial forms with respect to which there is primary alteration are not intermediaries having this description, just as there is no intermediary between fire and air. The idea seems to be that the

4. Following **Z** and **T**, which reads *min al-sha'n* for **Y** *fī al-shubbān* (among the young men). The Latin *et hoc etiam amplius* also strongly suggests that that translator read *sha'n*, which can also mean *importance* (≈*amplius*).

البُعْـد ؛ وهما الضدّان . ويجب أنْ نتأمل نحن هذه القضية فنقول : إنَّه لا بدّ من أخذ المادة أو الموضوع في حدّ التضادّ ، فإنْ عنى بالموضوع الموضوع الحقيقي القائم بالفعل نوعاً القابل للأعـراض الذي لذلـك النوع ؛ فلا تكون الصور الجوهرية متضـادّة لأنَّها في هيولى لا في موضوع . وإنْ عنى بذلك الموضوع أي محلٍ كان ، فيشبه أنْ تكون الصورة النارية مضادّة للصورة المائية لا كيّفيتاهما فقط . فذلك لا شكّ فيه ، بل الصور التي عنها تصدر الكيْفيات التي لهما ، وذلك لأنَّ الصورتين مشتركتان في محلٍ وتتعاقبان عليه وبينهما غاية الخلاف . ولهذا من الشـأن ما اشتغل من بين أنَّ الفلك لا يتكون لأنَّه لا ضدّ لصورته ، كأنَّه وضع أنَّ كلّ متكونٍ فلصورته ضدٌّ وإليه يكون انتقاله ، فيجعل النار والهواء والماء والأرض متضادّة الصورة ، فلِمَ أنكر أن يكون للصور الجوهرية ضدّ البتة؟ فيشبه أنْ يكون الضدّ الذي نذكره هاهنـا هو الذي بينه وبين شـيء آخر غاية الخلافّ - وإنّما يكـون بينه وبين ذلك غاية الخلافّ - إذا كان لشيءٍ ثالث معه خلاف دونه وهو الواسطة ، بحيث يحتمل استمراراً فيه كالاستمرار في بُعْدٍ بين شيئين . وليس بين الصور الجوهرية التي فيها الاستحالة الأولية واسـطة بهذه الصفة ، كما ليس بين النار والهواء واسـطة . ويشبه أنْ يكون يرى التعاقب

succession taken in the definition of *contrary* is a succession between two things between which there is a maximal degree of difference. As we said, however, this can happen without an intermediary, and so this contrary can be eliminated and another succeed it without some other successor intervening between the two; but again, if, as is frequently the case, the intermediary (if there is an intermediary) successively follows, then the transition is something extending continuously between the two extreme limits.

(5) Moreover, one does not see the substrate's receiving the form of fire successively upon the [form] of water without its first receiving the intermediary form of air (never mind a continuous extension!); rather, it must inevitably come to rest possessing the form of air. So the form of water is not contrary to the form of fire,[5] since the transition does not extend from one to the other, but from fire to air; whereas the form of fire is not contrary to the form of air, since there is not a maximal degree of difference between the two. If this is the intention, then the interpretation of it comes down to the first explanation we tried out—namely, that the nature of substantiality is not cast off gradually, since it does not undergo an increase or weakening such that its increasing and weakening have two extreme limits, which, in this inquiry, are specified by the name *contraries*. In First Philosophy, we will also provide you with a more detailed explanation that the substantial form does not undergo increase and weakening.[6]

5. With **Z** and the Latin, secluding the subsequent phrase, *nor is the form of fire the contrary of the form of air,* which appears in MSS S and M but appears to be either a transposition or a duplication of the phrase in the next line. While the phrase appears in **T**, someone has gone back and marked through it. Also see *Kitāb fī al-samāʾ wa-l-ʿālam* 6, where Avicenna argues specifically against the suggestion that substances can undergo intensification and weakening with respect to their substance.

6. The reference may be to *Ilāhīyāt* 2.3, where the vocabulary of "increasing" and "weakening" is used in speaking about the species form belonging to matter; or it may be to *Ilāhīyāt* 2.4, where much of the argumentation—though not the vocabulary—of the present discussion is repeated.

المأخوذ في حدّ الضدّ هو تعاقبٌ بين شيئين بينهما غاية الخلاف. وهذا، على ما قلنا، يصحّ أنْ يكون بلا واسطة، فيصحّ أنْ يرتفع هذا الضدّ ويعقبه الآخر من غير أنّ يتخلّل بينهما عاقبٌ آخر، وإنْ كان قد يصحّ أيضاً أنْ يكون يتعقّب المتوسّط - إنْ كان هناك متوسّط - فيكون الانتقال مستمراً من الطرفين على الاتصال.

(٥) ثم لا يرى أنَّ المحل يقبل الصورة النارية عقيب المائية، من غير أنْ يقبل أولاً صورة الهواء المتوسط لا على استمرارٍ متصل، بل وَجَبَ أنْ يسكن لا محالة على الصورة الهوائية. فلا تكون الصورة المائية مضادّةً للصورة النارية، إذ لا يستمر الانتقال من إحداهما إلى الأخرى إلاّ من النارية إلى الهوائية، إذ ليس بينهما غاية الخلاف. فإنْ كان القصد هذا القصد، كان التعبير عنه يرده إلى البيان الأول الذي حاولناه نحن؛ وهو أنَّ طبيعة الجوهرية لا تنسلخ يسيراً يسيراً؛ إذ لا يقبل الشدّة والضعف قبولاً يكون لاشتداده ولضعفه طرفان يخصّان في هذا النظر باسم الضدّية. وسنبيّن لك أيضاً في الفلسفة الأولى أنَّ الصورة الجوهرية لا تقبل الاشتداد والضعف، ببيانٍ أشرح.

(6) Still, on the basis of observing semen gradually developing into an animal and the seed gradually into a plant, it is imagined that there is a motion here [namely, with respect to substance]. What should be known is that, up to the point that the semen develops into an animal, it happens to undergo a number of other developments between which there are continuous qualitative and quantitative alterations; and so, all the while, the semen is gradually undergoing alteration. In other words, it is still semen until it reaches the point where it is divested of its seminal form and becomes an embryo. Its condition [remains] like that until it is altered [into] a fetus, after which there are bones, a nervous system, veins, and other things that we do not perceive, [remaining] like that until it receives the form of life. Then, in like fashion, it alters and changes until it is viable and there is parturition. Someone superficially observing the transformation imagines that this is a single process from one substantial form to another and therefore supposes that there is a motion with respect to the substance, when that is not the case and, instead, there are numerous motions and rests.[7]

(7) That there is motion with respect to quality is obvious. Still, among the people[8] there are those who do not believe that there is motion in all the species of quality, but [only] in the kind related to the senses. They say that state and habit are the sort that depends upon the soul and that their subject is not the natural body. As for power and impotence, hardness and softness, and their like, they follow upon certain accidents that the subject just happens to have; and the subject, together with some of those accidents, becomes their subject. In that case, then, the subject for power is the same subject for lack of power, and the same holds in the case of hardness and softness. Shapes and what are like them come to exist all at once in the matter that received them only because they are not susceptible to strengthening and weakening. I do not know what they

7. For a brief discussion of this objection and Avicenna's response, especially with reference to his biological works, see Jon McGinnis, "On the Moment of Substantial Change: A Vexed Question in the History of Ideas," in *Interpreting Avicenna; Science and Philosophy in Medieval Islam,* ed. J. McGinnis (Leiden: E. J. Brill, 2004), 42–61, esp. §3.

8. It is not clear who the author of this and the subsequent views is. In neither Aristotle nor the Graeco-Arabic commentary tradition treating the topic of those categories in which motion occurs have I been able to find a discussion paralleling Avicenna's discussion here. The closest discussion I have been able to find is at *Enneads* 6.110–12, where Plotinus distinguishes between sensible qualities and qualities of the soul.

(٦) لكنه لمّا رأى أنَّ المنيّ يتكون حيواناً يسيراً يسيراً، والبذر نباتاً يسيراً يسيراً؛ توهّم من ذلك أنَّ هناك حركة، والذي يجب أنْ يعلم هو أنَّ المنيّ إلى أنْ يتكون حيواناً تعرض له تكونات أخرى تصل ما بينهما استحالات في الكيف والكم، فيكون المنيّ لا يزال يستحيل يسيراً يسيراً - وهو بعْد منيّ - إلى أنْ يبلغ حدّاً تنْخلع عنه صورة المنويّة وتصير عَلَقة، وكذلك حالها إلى أنْ تستحيل مُضْغة، وبعدها عظاماً وعصباً وعروقاً، أو أموراً أُخر لا ندركها، وكذلك إلى أنْ يقبل صورة الحياة. ثم كذلك يستحيل ويتغيّر إلى أنْ يشتدّ فينفصل. لكن ظاهر الحال يُوهم أنَّ هذا سلوكٌ واحدٌ من صورةٍ جوهريةٍ إلى صورة جوهرية أخرى، ويُظنّ لذلك أنَّ في الجوهر حركة وليس كذلك، بل هناك حركات وسكونات كثيرة.

(٧) وأمّا كون الحركة في الكيف فذلك ظاهر، لكن في الناس مَنْ لم يرَ الحركة في أنواع الكيف كلّها؛ إلاّ في الصنف المنسوب إلى الحواس؛ فقال: أمّا نوع الحال والملكة فهو يتعلق بالنفس وليس موضوعه الجسم الطبيعي. وأمّا القوة واللاقوة والصلابة واللين وما أشبه ذلك؛ فإنَّها تتبع أعراضاً تعرض للموضوع ويصير الموضوع مع بعض تلك الأعراض موضوعاً لها، فلا يكون حينئذ الموضوع للقوة هو بعينه الموضوع لعدم القوة، وكذلك الحال في الصلابة واللين. وأمّا الأشكال وما يشبهها فإنَّها إنّما توجد في المادة التي تقبلها دفعة، إذ لا تقبل التشدّد والتضعّف. ولا أدري ماذا يقولون في الانحناء والاستقامة وغير ذلك؟

would say about being curved or rectilinear and the like. My opinion is that the situation is not as they say. The fact is that, in the subject of the state and habit—whether it is a soul, a body, or the two together in a common state—there exists a certain potential perfection *qua* potential belonging to a given substance. Those who said that the subject for hardness and softness and for power and weakness is not one and the same are undone by augmentation and diminution, which, according to their position, could not be motions. The fact is, however, that with respect to these things, we mean by *subject* only the nature of the species that bears the accidents; and, so as long as that nature remains, the species does not change and the substantial form does not corrupt. The subject is something that endures regardless of whether we consider it [as the subject] of some accident it happens to have; or [consider it] as something additional that is added to it, becoming a proximate subject for the state in which there is motion; or [consider it] in itself. We concede that the status of shapes does not appear to be like that of other qualities with respect to their alteration, since [shapes] occur all at once.

(8) There is also motion with respect to quantity, and that in two ways, one of which is through either a certain increase being superadded, owing to which the subject is augmented, or a certain decrease that takes a part away through separation, owing to which the subject is decreased. In both cases, however, its form remains. This is called *augmentation* and *diminution*. [The second] is not by either a certain increase being added to it or a certain decrease decreasing it, but in that the subject itself receives a certain greater or lesser magnitude, whether by rarefying or condensing, without a separation [or addition] occurring in its parts. Now, when this entails an alteration of some underlying thing (namely, with respect to quality), then that is different from its increasing or decreasing in quantity;[9] but because this state is a gradual process from potency to act, it is a perfection of what is in potency and so is a motion.

9. The difference between the two ways that there might be motion in the category of quantity might be understood better if we anachronistically consider *quantity* here as mass. In the first case, the mass of some object has been either increased or decreased (and, presumably, the volume it occupies as well). In the second case, the mass is neither increased nor decreased, but the volume that is occupied is either increased or decreased. Avicenna, like al-Fārābī before him, would appeal to this second kind of quantitative change to explain away certain phenomena frequently explained by appealing to a void; see al-Fārābī, *Fârâbî's Article on Vacuum,* ed. and trans. N. Lugal and A. Sayili (Ankara: Türk Tarih Kurumu Basımevi, 1951); and Avicenna, *Physics* 2.9.17 and 20–21.

وعنـدي إنَّ الأمر ليس على ما يقولون، وأنَّ موضوع الحال والملكة - كان نفساً أو بدناً أو هما معاً - بحال الشركة فإنّه يوجد فيه كمالٌ مّا بالقوة من جهة ما هو بالقوة لجوهر مّا. والذين قالوا إنَّ الموضوع ليس واحداً للصلابة واللين، أو القوة والضعف، فينتقض عليهم في النمو والذبول. وكان يجبّ - على قولهمّ - أنْ لا تكونا حركتين، بل إنّما نعني بالموضوع في هذه الأشياء طبيعة النوع الحاملة للأعراض. فما دامت تلك الطبيعة باقية لم يتغيّر النوع ولم تفسد الصورة الجوهرية؛ فإنَّ الموضوع ثابت من غير أنْ نبالي أنّه لعارضٍ يعرض له أو زيادة تنضاف إليه، يصير موضوعاً قريباً للحالة التي فيها الحركة أو لذاته. نعم، الأشكال يشبه أنْ لا يكون حكمها حكم سائر الكيفيات في وقوع الاستحالة فيها؛ لأنَّها تكون دفعة.

(٨) والكمّ ففيه أيضاً حركة، وذلك على وجهين: أحدهما بزيادةٍ مضافة فينمو لها الموضوع، أو نقصانٍ يقتطع بالتحلّل فينقص لها الموضوع، وصورته في الأمرين باقية، وهذا يسمّى ذبولاً ونمواً. وقد يكون لا بزيادةٍ تزاد عليه أو نقطانٍ ينقص منه، بل بأنْ يقبل الموضوع نفسه مقداراً أكبر أو أصغر بتخلّخلٍ أو تكاثف؛ من غير انفصالٍ في أجزائه. وهذاّ - وإنْ كان تلزمه استحالة قوامٍ وهي من الكيفّ - فتلك غير إزدياده في الكمّ أو نقصانه فيه، ولأنَّ هذه الحالة سلوك من قوةٍ إلى فعلٍ يسيراً يسيراً، فهو كمالٌ مّا بالقوة فهو حركة.

(9) One, however, may have doubts and say that small and big are not contraries, whereas all motions are between contraries. We say, first, that we ourselves are not all that strict in requiring that every motion be only between contraries; rather, we say that something is undergoing motion when there are certain opposing things that are not simultaneously together and the thing gradually proceeds from one of them to the other, even if there is no contrariety there. Additionally, the big and the small between which the augmented and diminishing things are moved are not some absolute, relative to big and small; rather, it is as if nature has assigned to the animal and plant species certain limiting points with respect to big and small that they cannot exceed but between which they are moved. So, here there is an absolute huge with respect to the species that does not become small relative to some other huge thing, and the same holds for an absolute small. Consequently, it is not at all unlikely that they are, in a way, like contraries and, in fact, there is a certain contrariety.

(10) One might also object that augmentation is a certain motion with respect to place [rather than quantity] because the place changes during [the augmentation]. The answer is that when we say that augmentation is a certain motion with respect to quality, it is not that there cannot thereby be a motion with respect to place accompanying it. Nothing prevents two changes—a change of quantity and a change of *where*—being in the subject of augmentation. In that case, there would be two motions in it simultaneously.

(٩) لكنه قد يتشكك فيقال إنَّ الصغير والكبير ليسا بمتضادّين؛ والحركات كلّها بين المتضادّات؛ فنقول: أمّا أولاً، فلسنا نحن ممّن يتشدّد كل التشدّد في إيجاب كون الحركات كلّها بين المتضادّات لا غير، بل إذا كانت أشياء متقابلة لا تجتمع معاً، وسلك الشيء من أحدهما إلى الآخر يسيراً يسيراً سمينا الشيء متحرّكاً، وإنْ كان لا تضادّ هناك. على أنَّ الصغير والكبير اللّذين يتحرك فيما بينهما النامي والذابل ليسا الصغير والكبير الإضافي المطلق، بل كأنَّ الطبيعة جعلت للأنواع الحيوانية والنباتية حدوداً في الصغر وحدوداً في الكبر لا تتعداهما وتتحرّك فيما بينهما، فيكون العظيم هناك عظيماً على الإطلاق، ولا يصير صغيراً بالقياس إلى عظيم آخر في ذلك النوع، وكذلك الصغير يكون صغيراً بالإطلاق - وإذا كان كذلك - لم يبعد أنْ تتشاكل المتضادات؛ بل تكون متضادّة.

(١٠) فإنْ قال قائل إنَّ النمو حركة في المكان لأنَّ المكان يتبدل فيه، فالجواب أنّه ليس إذا قلنا إنَّ النمو حركة في الكم فإنَّ ذلك نمنع به أنْ يكون معه حركة في المكان، فإنّه لا يمتنع أنْ يكون في موضوع النمو تبدلان؛ تبدّل كمّ وتبدّل أينٍ، فتكون فيه حركتان معاً.

(11) As for the category of relation, it seems that the lion's share of transitions in it are from one state to another, [occurring] only all at once. Even if there is variation[10] in some cases, the real and primary change is in another category to which the relation just happens to belong, since the relation is characteristically concomitant with other categories and does not, in itself, really exist. So, when the category is something susceptible to increase and weakening, then the relation happens to be like that as well, for, since *heat* is susceptible to increase and weakening, so is *hotter.* So it is the subject of the relation that is primarily susceptible and upon which that necessarily follows, in which case the motion essentially and primarily is in the thing that accidentally has the relation, while belonging to the relation accidentally and secondarily.

(12) That motion exists in the category of *where* is perfectly clear, whereas for the category of *when,* it would seem that the transition from one *when* to another occurs all at once, like the transition from one year to the next or one month to the next. Alternatively, the situation concerning *when* might be like that of relation, in that there is no transition from one thing to another with respect to the *when* itself; but, rather, the primary transition is with respect to quality and quantity, where time necessarily follows on account of that change, and so, because of it, there is accidentally change with respect to [*when*]. As for what is unchanging, you will learn[11] that it is not in time. So how can it have a motion in it?

---

10. Reading *ikhtalafa* (m.) with **Z** and **T** (*ikhtalafat* [f.]) and the Latin (*diversificantur*) for **Y**'s *akhlafa* ("to not hold true" or, literally, "to break a promise").

11. See 2.13.6.

(١١) وأمّا مقولة المضاف ، فيشـــبه أنْ يكون جُلّ الانتقال فيها إنّما هو من حالٍ إلى حالٍ دفعةً ، وإنْ اختلف في بعض المواضع فيكون التغيّر بالحقيقة ، وأولاً ، في مقولةٍ أخرى عرضت لها الإضافة ، إذ الإضافة من شأنها أنْ تلحق مقولات أخرى ، ولا تتحقق بذاتها . فإذا كانت المقولة ممّا تقبل الأشـدّ والأضعف ، عــرض للإضافة مثل ذلك . فإنَّه لمّا كانت السـخونة تقبل الأشدّ والأضعف ، كان الأسخن يقبل الأشدّ والأضعف ، فيكون موضوع الإضافة يقبل ، ويلزم ذلك قبولاً أولياً ، فتكون الحركة في الأمر العارض له الإضافة بالذات وأولاً ، وللإضافة بالعرض وثانياً .

(١٢) وأمّــا مقولة الأين فإنَّ وجود الحركة فيها بيّنٌ واضح . وأمّا مقولة متى فيشــبه أنْ يكون الانتقال من متى إلى متى أمراً واقعاً دفعةً ، كالانتقال من ســنةٍ إلى ســنةٍ أو من شــهر إلى شهر . أو يشــبه أنْ يكون حال متى كحال الإضافة ، في أنَّ نفس متى لا تنتقل فيه عن شــيءٍ إلى شــيءٍ ، بل يكون الانتقال الأول في كيفٍ أو كمّ ، ويكون الزمان لازماً لذلك التغيّر ؛ فيعرض بسببه فيه التبدّل . وأمّا ما لا تغيّر فيه ؛ فستعلم أنَّه ليس في الزمان ، فكيف تكون له حركة فيه؟

(13) Now, it has been said that there is no motion whatsoever in the category of position, since there is no contrariety with respect to position.[12] Also, [it has been said] that when someone goes from standing to sitting, he is still judged to be standing until, all at once, he is seated. The truth requires that there be motion with respect to position, whereas there is no great need for real contrariness at motion's two extreme limits, which should be obvious to you by considering the motion of the celestial sphere. Additionally, it is not out of the question that there is a contrariety with respect to [position] to the extent that lying face up is contrary to lying face down. As for the claim that the transition toward sitting occurs all at once, if one means by it that the sitting, which is the extreme limit, is attained all at once, then it is true; but the blackness and *where* that are extreme limits are likewise attained all at once. If one means by it that that transition, which involves every position from which the sitting results, [occurs] all at once, then it is false, because one gradually goes from standing to sitting until one comes to the end, which is sitting, exactly like the case during the transition from down to up.

(14) The way that motion exists with respect to position is for the whole of something to change its position without leaving its place at all, and, instead, the relation of its part to either its place's parts or sides undergoes change. Inevitably, then, it is something moved with respect to position because it has not changed its place, but only its position in its place has changed, where the place itself is the initial one. Now, when there is change with respect to position and, moreover, it proceeds gradually by degrees, that change is motion with respect to position, since every motion is a change of state having this description and vice versa, being related to the state that is changing, not to something else that has not changed. By this I do not mean that everything undergoing

12. The reference may be to Philoponus, who says that, though it might seem that there is motion in the category of position, the motion is in fact in the category of place (Arabic *Physics,* 512–13), although nothing like the reason Avicenna mentions here is put forth there. For an extended discussion of Avicenna's account of motion with respect to the category of position, see Jon McGinnis, "Positioning Heaven" §4.

(١٣) وأمّا مقولة الوضْع؛ فقد قيل إنّها لا حركة فيها البتة، إذ لا تضادّ في الوضْع، وأنه إذا انتقل شيءٌ من قيام إلى قعودٍ فإنَّه لا يزال في حكم القائم إلى أنْ يصير قاعداً دفعةً، وكذلك إذا انتقل من قعودٍ إلى قيامٍ فإنّه لا يزال في حكم القاعد إلى أنْ يصير قائماً دفعةً. والحقّ يوجب أنْ يكون في الوضْع حركة، وأنَّه لا كثير حاجةٍ إلى التضادّ الحقيقي في طرفي الحركة، يتبيّن ذلك لك؛ بتأمل حركة الفلك. على أنَّ الوضْع لا يبعد أنْ يكون فيه تضادّ؛ حتى يكون المستلقي مضادّاً للمنبطح. والذي قيل من أنَّ الانتقال إلى القعود يكون دفعةً، إنْ عنى به أنَّ القعود الذي هو الطرف يحصل دفعةً؛ فهو صادق. وكذلك السواد الذي هو الطرف يحصل دفعةً، وكذلك الأين الذي هو الطرف الذي يحصل دفعةً. وإنْ عنى به أنَّ كل وضْعٍ ينتقل عنه إلى القعود يكون ذلك الانتقال دفعةً فهو كذب؛ لأنَّ الانتقال عن القيام إلى القعود يكون قليلاً قليلاً حتى يوافي النهاية التي هي القعود؛ كالحال في الانتقال من السفل إلى العلو بعينه.

(١٤) أمّا كيْفية وجود الحركة في الوضع؛ فهو أنَّ كل مستبدل وضْعٍ من غير أنْ يفارق بكلّيته المكان، بل بأنْ تتبدل نسبة أجزائه إلى أجزاء مكانه أو إلى جهاته؛ فهو متحرّك في الوضْع لا محالة؛ لأنَّ مكانه لم يتبدل بل تبدّل وضعه في مكانه، والمكان هو الأول بعينه. وإذا كان التبدّل في الوضْع، وكان مع ذلك متدرجاً يسيراً يسيراً، كان ذلك التبدّل حركة في الوضْع، إذ كانت كلّ حركة هي تبدّل حالٍ بهذه الصفة وبالعكس، وتكون منسوبة إلى الحالة التي تبدلت لا إلى شيءٍ آخر لم يتبدّل. ولست أعني بهذا أنَّ

motion with respect to position remains in its place. So it is not necessary from my account—namely, that everything remaining in its place that gradually changes its position is something moved with respect to position—that everything moved with respect to position is like that [namely, not changing its place at all]. The fact is that nothing prevents something from changing its position only after having changed its place, just as nothing prevents something from changing its quantity only after having changed its place.[13] Instead, the intention is to show that motion exists in the [category of] position by showing that there is something that is moved with respect to position. As for whether something can change its position alone without changing its place, let its possibility be recognized from the motion of the celestial sphere; for, on the one hand, it might be like the outermost celestial sphere, which is not in a place in the sense of the containing limit that exactly encompasses [what it contains], which is what we mean by *place.*[14] On the other hand, it might be in a place, but it would absolutely not leave its place; and, instead, what changes is only the relation of its parts to the parts of its place with which it is in contact. When there is only this [type of] change—where there is no change of the place—and *this change* is change of position and there is nothing but this change, then there is only this motion, which is with respect to position.

(15) That the outermost celestial sphere does not move from [its] location they[15] take to be patently obvious. Moreover, it does not undergo motion with respect to quality, quantity, substance, or some category other than position. So, when you go through each one of the categories, you do not find this motion fitting well with [any] of them, except position or *where;* but it is not *where,* so position remains. Someone might say that every part of the celestial sphere undergoes motion with respect to place, and [for] everything of which every[16] part undergoes motion with

13. Cf. the objection considered in par. 10.

14. Cf. Aristotle, *Physics* 4.4.212a2–6 and Avicenna, *Physics* 2.9.1.

15. The "they" here is probably a reference to certain Aristotelians who denied motion with respect to position, mentioned at the beginning of par. 13.

16. Reading *kull* with MSS A and M, **T**, and the Latin (*omnis*) for **Y** and **Z**'s *kāna,* which if retained, could be translated, "whatever a part of which is undergoing motion with respect to place."

كل متحرّكٍ في وضْعٍ فهو ثابت في مكانه؛ فليس يجب من قولي: إنَّ كل ثابت في مكانه يستبدل وضعه بالتدريج فهو متحرّك في الوضع؛ أنّ كل متحرّك في الوضْع كذلكّ - بل لا أمنع أنْ يكون الشيء لا يتغيّر وضعه إلاّ وقد تغيّر مكانه، كما لا أمنع أنْ يكون شيءٌ لا يتغيّر كمّه إلاّ وقد تغيّر مكانه، بل الغرض هو أنْ يثبت وجود المتحرّك في الوضْع بإثبات متحرّكٍ مّا في الوضْع. وأمّا أنّه هل يمكن أنْ يكون الشيء يتبدل وضعه وحدّه ولا يتبدل مكانه؛ فليعلم إمكانه من حركة الفلك. فإنَّه أنْ يكون كالفلك الأعلى الذي ليس في مكان؛ بمعنى نهاية الحاوي الشامل المساوي الذي إياه نعني بالمكان، وإمّا أنْ يكون في مكانه لكنه لا يفارق كلّية مكانه، بل إنّما تتغيّر عليه نسبة أجزائه إلى أجزاء مكانه التي يلقاها. وإذا لم يكن هناك إلاّ هذا التغيّر والمكان ثابت، وهذا التغيّر تغيّر الوضْع، وليس هناك غير هذا التغيّر، فليس هناك غير هذه الحركة التي في الوضّع.

(١٥) وأمّا كون حركة الفلك الأعلى غير مكانية؛ فواضحٌ عندهم بيّنٌ. ثم ليس تحرّكه في كيْفيةٍ ولا كميّةٍ ولا جوهريةٍ ولا في مقولةٍ غير الوضْع. فإنك إذا تعقّبت مقولة مقولة لم تجد هذه الحركة تلائمها ما خلا الوضْع أو الأين، ولا أيْن؛ فبقي الوضْع. فإنْ قال قائل إنَّ الفلك كلّ جزء منه متحرّك في المكان، وكلّ ما كلّ جزء منه متحرّك في المكان؛

respect to place, the whole of it undergoes motion with respect to place. The response is that this is not the way things stand. On the one hand, the celestial sphere does not have some actual part such that it undergoes motion. Even if we were to posit parts for it, they would not leave their places; rather, each part thereof would leave as a part of the place of the whole, if the whole of it is in a place. Now, the place of the part is not part of the place of the whole (although part of the place of the whole can, in fact, be part of the place of the part). That is because part of the place of the whole is not contained by the part, whereas, as you know, the *place* is what contains. The fact is that the parts of something continuous might be in place only potentially, and indeed this has been clearly explained to them in their books.[17] On the other hand, it is not the case, when every part leaves its own place, that the whole leaves its own place, since there is a distinction between *each part* and *the whole of the parts*. In other words, each part has a certain description, while the whole does not have that description because the whole has a certain proper reality distinct from a certain reality of each one of the parts. For starters, don't you see that each part is a part, while the whole is not a part? So each part of ten is one, but ten is not one. The fact is, returning to the issue at hand, that we say that some place might well enclose something possessing actual parts, like the sand [of a desert] and the like, and then every part of it leaves its place, whereas the whole does not leave its place. Indeed, by admitting [this] we should have no doubt that, even if we concede that each part of it leaves its proper place, the whole does not leave its proper place; and so the doubt about the whole not undergoing motion with respect to place would not have arisen, even if each part is moved. It seems to me that whoever considers what we have said and weighs the evidence will come to believe with certainty that there is a motion with respect to position.

---

17. Cf. Aristotle, *Metaphysics* 5.26.

فالـكل منه متحرّك في المكانّ – فالجواب عن هذا أنَّ الأمر بخلاف ذلك. أمّا الفلك فلا جزء له بالفعل حتى يتحرّك، ولو فرضنا له أجزاء فليســت تفارق أمكنتها، بل يفارق كل جــزء منها جــزءاً من مكان الكل؛ إنْ كان كله في مكان. وليس مكان الجزء جزء مكان الكلّ، بل عسى أنْ يكون جزء مكان الكل جزء مكان الجزء؛ وذلك لأنَّ جزء مكان الكل لا يحيط بالجزء، والمكان كما تعلم محيط، بل عسى أنَّ المتصل ليست أجزاؤه في مكانٍ إلاّ بالقــوة، بــل قد صُرِّح لهم بهذا في كتبهم. وبعد هــذا؛ فليس إذا كان كل جزء يفارق مكان نفسه فالكلّ يفارق مكان نفسه؛ لأنَّه فرقٌ بين قولنا كلّ جزء وبين قولنا كل الأجزاء. وذلك أنَّ كل جزء قد يكون بصفةٍ، والكلّ لا يكون بتلك الصفة لأنَّ للكلّية حقيقة خاصّة مباينة لحقيقة كل واحدٍ من الأجزاء. ألا ترى أول شيء إنَّ كل جزء هو جزء، والكلّ ليس بجزء، فكلّ جزء من العشرة واحد والعشرة ليست بواحد. بل نرجع إلى مسألتنا فنقول: إنَّه يجوز أنْ يكون مكان يشــتمل على شيءٍ ذي أجزاء بالفعل كالرمل وغير ذلك؛ ثم كل جزءٍ منه يفارق مكانه والكلّ لا يفارق مكانه. بل ما نحن بسبيله لانشكّ أنّاّ – وإنْ سلّمنا فيهّ – أنَّ كل جزء منه يفارق مكانه الخاصّ، فالكلّ لا يفارق مكانه الخاص، فلم يقع الشك في أنَّ الكلّ غير متحرّكٍ في المكان، وإنْ كان كل جزء متحرّكاً. وعندي أنَّ كل مَنْ يتأمل ما قلناه ثم ينصف، سيعتقد يقيناً أنَّ الوضْع فيه حركة.

(16) Perhaps someone would say that the sense of motion with respect to place is not that the mobile leaves its place but that it is something moved that is in a place, even if it does not depart it. In that case, the response is that its being moved and changed must have some sense. Now, on the one hand, if its being moved and changed is not dependent on something that leaves, but belongs to it, then, in fact, there is no motion or change, and both the terms *motion* and *change* have been taken equivocally. On the other hand, if it depends upon something other than the place that changes, then there is a certain state that changes, with respect to which the motion is proper. Even if something is in a place, undergoing alteration while being in a place, that fact does not make the alteration necessarily an alteration of location, even though it is in a place. Nor is it our intent that the meaning of "motion with respect to such-and-such" is "to be moved in such-and-such," as you would have known.

(17) As for the category of possession, I have not as of yet undertaken an independent investigation of it. Now, it is said that this category indicates a body's relation to what it contains and is inseparable from it during transition.[18] So the change of this relation would primarily be only with respect to the containing surface and place, in which case, as I suspect, there would be no motion essentially and primarily with respect to it.

(18) As for the categories of action and passion, one might suppose that there is motion with respect to them for a number of reasons. One of them is that something [initially] is either not acting or not being

18. Cf. Aristotle, *Categories* 15.

(١٦) ولعل قائلاً يقول إنَّ معنى الحركة في المكان ليس هو أنْ يكون المتحرّك يفارق المكان، بل أنْ يكون متحرّكاً وهو في مكانٍ وإنْ لم يفارقه، فيقال له حينئذ يجب أنْ يكون لكونه متحرّكاً متغيّراً معنى. فإنْ كان كونه متحرّكاً ومتغيراً غير متعلقٍ بأمرٍ يفارقه وأمر يوجد له، فلا حركة في الحقيقة ولا تغيّر، بل الحركة والتغيّر المذكوران هما باشتراك الإسم إنْ كان يتعلق بأمر يتغيّر وهو غير المكان فهناك حالة تتبدل فيها الحركة الخاصّة، وإنْ كان الشيء في مكان كون الشيء مستحيلاً وهو في مكان، وذلك لا يوجب أنْ تكون الاستحالة استحالة مكانية وإنْ كانت في مكان، ولا غرضنا في أنَّ الحركة في كذا معناه والمتحرّك في كذا؛ على ما علمت.

(١٧) وأمّا مقولة الجدة فإني إلى هذه الغاية لم أتحقّقها - والذي يقال إنَّ هذه المقولة تدل على نسبة الجسم إلى ما يشمله ويلزمه في الانتقال، فيكون تبدل هذه النسبة على الوجه الأول إنّما هو في السطح الحاوي وفي المكان، فلا يكون فيها، على ما أظنّ، لذاتها وأولاً حركة.

(١٨) وأمّا مقولة يفعل وينفعل فربما ظنَّ أنَّ فيهما حركة من وجوه. من ذلك أنَّ الشيء يكون لا يفعل، أو لا ينفعل، ثم يتدرج يسيراً يسيراً إلى أنْ يصير يفعل أو ينفعل،

acted upon, and thereafter there is a gradual progression until it is acting or being acted upon, in which case its acting and being acted upon are a certain end for that progression—as, for example, blackness is a certain end for blackening—and so it is supposed that there is a motion with respect to these two categories. Also, something might change from not being acted on by part[19] (or acting on it) to being acted upon by part (or acting upon it), where that occurs gradually, and so it is supposed that that is a motion. Again, being acted upon might be slow and then gradually progress until it is increasingly faster, and vice versa, so it is thought that that is moving toward fastness. As for the first reason, I say that the motion is not with respect to action and passion, but is with respect to acquiring the disposition and form by which the action and passion are able to arise. What we'll explain below[20] will resolve the second reason—namely, that it is impossible to proceed continuously from becoming cold to becoming hot, or from heating to cooling, except through a pause and intervening stop. As for the third reason, I know of no one who makes the gradual alteration from potentially fast to actually fast a motion (that is, a perfection of what is potency *qua* potency). The fact is that that is with respect to fastness and slowness, which are neither two motions nor actions nor passions, but two accidents, qualities, or dispositions belonging to either [motion], action, or passion.

---

19. **T** has *ḥarr* (heat), which corresponds with the Latin *calore* (by heat), which in Arabic script could be confused with *juzʾ* (part). If *heat* were accepted, the text would make more immediate sense, reading: "Also, something might change from not being acted on by heat (or acting on it) to being acted upon by heat (or acting upon it), where that occurs gradually, and so it is supposed that that is a motion." This reading has evidence in its behalf in Avicenna's response to this argument (at the end of this paragraph and again in the next), which involves becoming hot and heating (albeit the terms there are not derived from Ḥ-R-R, but S-Kh-N). So the response would seem to be immediately relevant to this position, if understood in terms of heat, whereas it is not clear that it is when *part* is read. Still, all the older Arabic manuscripts agree in reading *juzʾ*, and the principle of *lectio difficilior* suggests that it should be retained. If *juzʾ* is retained, then perhaps the case Avicenna has in mind is diffusion—as, for example, when one part of a quantity is heated and then, so affected, that part acts upon another part and heats it. This is a best conjecture.

20. See par. 19; also, the general issue of whether two contrary motions, such as heating and cooling, could be continuously joined is discussed at length at 4.8.

فيكون أنْ يفعل وأنْ ينفعل غاية لذلك التدرّج؛ مثل السواد فإنّه غاية للتسوّد، فظنَّ أنَّ في هاتين المقولتين حركة. وأيضاً فإنّه قد يتغيّر الشيء من أنْ لا يكون ينفعل بالجزء أو يفعله إلى أنْ ينفعل بالجزء أو يفعله، ويكون ذلك قليلاً قليلاً فيظنّ أنَّ ذلك حركة. وأيضاً فإنَّ الإنفعال قد يكون بطيئاً فيتدرج يسيراً يسيراً إلى أنْ يسرع ويشتدّ وبالعكس، فيظنّ أنَّ ذلك حركة إلى السرعة. فأقول: أمّا الوجه الأول فلا تكون الحركة فيه في الفعل والانفعال، بل في اكتساب الهيئة والصورة التي بها يصحّ أنْ يصدر الفعل أو الانفعال. وأمّا الوجه الثاني فيحلّه ما سنبيّن بعد من أنّه لا سبيل إلى أنْ يتصل السبيل من تبردٍ إلى تسخّن، أو تبريدٍ إلى تسخينٍ إلاّ بانقطاع وتخلّل وقفة. وأمّا الوجه الثالث فلا أعرف مَنْ يجعل الاستحالة من السرعة بالقوة إلى سرعةٍ بالفعل يسيراً يسيراً حركة، وهو استكمال لما بالقوّة – من حيث هو بالقوّة – لكن ذلك في السرعة والبطء، وليسا بحركتين ولا فعلين ولا انفعالين، بل عارضين وكيفيتين وهيئتين لها ولفعلٍ أو انفعال.

(19) In general, it is not admitted that, with respect to the nature of passion and action, there is a motion in the way that motion is said to be in a category; for if it is admitted that there is a gradual transition from becoming cold[21] to becoming hot, then [that transition] must occur either while that cold itself is being produced or when the production of cold ends. On the one hand, if [the heating] occurs while the process of becoming cold[22] is still occurring—and the transition to becoming hot undoubtedly involves the nature of becoming hot, which in its turn involves the nature of heat—[the transition] would have becoming cold as its goal at the very same time that it has becoming hot as its goal, which is absurd. If, on the other hand, [the transition from becoming cold to becoming hot] occurs when the production of cold ends, it will be after coming to rest at cold and ending (as you will learn).[23] Additionally, in that case, the transition [from becoming cold to becoming hot] must be either the very state of becoming hot or a transition to becoming hot. If, on the one hand, [the transition from becoming cold to becoming hot] is the becoming hot itself, then, as you know, unless there is some period of time of resting or a certain instant during which there is neither a motion nor a rest, it won't be between becoming cold and becoming hot. On the other hand, if [that transition] is the progression to becoming hot, then that progression, in its turn, must either involve the nature of becoming hot or not. Now, if it does not, then that is not an alteration at all! If it does, then it inevitably involves the nature of heat. To involve the nature of heat, however, is to become hot, in which case the transition and advancement toward becoming hot would be an existing state of becoming hot [rather than, as was assumed, a transition to that state]—that is, unless it is assumed that *becoming hot* is to become hot at the extreme degree, while there is the transition toward it inasmuch as

21. The Arabic *tabarrud* has the sense of "to be or become cold"; however, since Avicenna has earlier contrasted it with *tabrīd* (to make or produce cold) and the present context is clearly about the category of passion and so *being acted upon*, I have in some cases overtranslated it in terms of "to be made cold" or "the production of cold" in order to bring out the passive nature under consideration. Similar comments hold for *tasakhkhun* (to be or become hot).

22. Rejecting **Y**'s proposed addition of *yantahī* (to end), which does not occur in **Z**, **T**, or the corresponding Latin. If retained, the sense of the text would be "when the state of cold terminates after being cold."

23. See 4.8.

(١٩) وبالجملة لا يجوز أنْ يكون في طبيعة أنْ ينفعل وأنْ يفعل حركة على سبيل ما تقال الحركة في المقولة، فإنَّه إنْ جاز أنْ يكون انتقالٌ من التبرّد إلى التسخّن يسيراً يسيراً، فلا يخلو إمّا أنْ يكون ذلك التبرّد تبرّداً، أو عندما ينتهي التبرّد. فإنْ كان عندما التبرّد بعد تبرّدٌ – ومعلوم أنَّ الانتقال إلى التسخّن أخْذٌ من طبيعة التسخن، وفي طبيعة التسخّن أخْذٌ من طبيعة السخونةّ – فيكون عندما يقصد الحر يقصد البرد وهذا محال. وإنْ كان عند منتهى التبرّد؛ فهو بعد الوقوف على البرد وبعد الانتهاء؛ كما ستعلم. ومع ذلك، فحينئذ لا يخلو إمّا أنْ يكون ذلك الانتقال نفس التسخّن أو انتقالاً إلى التسخّن؛ فإنْ كان نفس التسخّن فليس بين التبرّد والتسخّن إلاّ زمان سكون أو آن لا حركة فيه ولا سكون كما تعلمه. وإنْ كان المصير إلى التسخّن فلا يخلو إما أنْ يكون في المصير إلى التسخّن آخذ من طبيعة التسخن أو لايكونّ – فإن لم يكن فليس ذلك إسحالة البتة. وإن كان، فهناك أخذ لا محالة من طبيعة السخونة؛ والأخْذ من طبيعة السخونة هو تسخّن، فيكون عند الانتقال إلى التسخّن والتوجّه إليه تسخّنٌ موجودٌ، اللّهم إلاّ أنْ يفرض التسخّن ما هو في الغاية تسخّنٌ ويكون الانتقال إليه بما هو أضعف منه. ثم التسخّن نفسه وكل حركة فإنّه

it is weaker than it. In that case, [the response is that,] becoming hot (and every motion in fact) is itself divisible by time, as you will learn.[24] Now in the case where the heat is perfected at a certain instant, it is not *becoming* hot. [That] is because if it is becoming hot, it will be divisible into parts, where each part of becoming hot is assumed to be an instance of becoming hot, while the part proceeding it will be weaker and so will not be in the extreme degree. In that case, it is not an instance of becoming hot on the present assumption; but it was posited as an instance of becoming hot. This is a contradiction. Either becoming hot is not divisible, in which case there is no motion but a state of being hot, or it is divisible, in which case its becoming hot is not an extreme degree. Therefore, being at the extreme degree is not a condition of becoming hot but, rather, involves the state of being hot, not the process of being heated so as to be at that extreme. Now that you know the account about *being made hot* [that is, a passion], so have you learned the account about *making hot* [that is, an action], and this much should be enough.

(20) Now that we have finished going over every position concerning this topic, it will have become clear to you from this summary, once you have applied yourself to motion's relation to the categories, that motion occurs in only four of them: quality, quantity, *where,* and position. Now that we have explained the nature of motion, we should explain the nature of rest.

---

24. See 2.11.3 and 2.12.7.

منقسمٌ بالزمان على ما ستعرف. وحينئذ تستكمل السخونة في آنٍ فلا يكون تسخّن، فإنْ كان تســـخّناً فهو منقسم إلى أجزاء، ويكون كل جزء من التسخّن يفرض تسخّناً، ويكون الجزء المتقدّم فيه أضعف فلا يكون بالغاية، فلا يكون تسخّناً بهذا المعنى، وفُرض تسخّناً، هذا خُلْف. وأمّا أنْ يكون التسخّن غير منقسم البتة؛ فلا تكون حركة بل سخونة، وأمّا أنْ يكون منقســـماً فلا يكون من التســـخّن ما هو غاية، فليس إذن من شرط التسخّن أن يكون في الغاية، بل أنْ يكون آخذاً في الســـخونة ولا يتســـخّن في الغاية. وإذ قد عرفت الكلام في التســـخّن عرفتَ <الكلام> في التســـخين، ويجب أنْ يكون هذا القدر كافياً.

(٢٠) ونرفـــض جميـــع ما يذهب به <في> هـــذا الموضع. فقد ظهـــر لك من هذه الجملـــة أنَّ الحركة إنّما تقع في المقولات الأربع التي هي الكيف والكم والأين والوضْع، فقد وقفتَ على نسبة الحركة إلى المقولات. وإذ قد عرفنا طبيعة الحركة، فحريٌّ بنا أنْ نعرف طبيعة السكون.

Chapter Four

Establishing the opposition of motion and rest

(1) There is some difficulty concerning the topic of rest. That is because it is generally accepted among the school of natural philosophers that rest is the opposite of motion in the way that a privation, not a contrary, is the opposite of possession.[1] Moreover, it is obvious that the only opposition that can be assumed between [motion and rest] is one of these two—I mean, being a privation or being a contrary—but we have already made the term *motion* something applying in the sense of a form, not a privation, since we said that it is a *first perfection.* So, if the opposition occurs in the way that privation is opposite of possession, then, of the two, motion cannot be the privation. On the other side, however, we do maintain that the body is said to be at rest when [(1)] it is [experiencing] a privation of motion but is of the character to be moved, where what we mean by *of the character to be moved* is that it exists as something with which motion is associated—namely, for instance, it occurs in a certain place during a certain time. It is equally said to be resting, however, when [(2)] it exists at a single place for a certain time. So here there are found two senses of *resting*, one of which is the privation of motion while being of the character to be moved, and the other [of which] is to exist at some *where* for a time. So, if rest is the first of the two (and the latter is just a necessary accident), then rest is a privation, whereas if rest is the second of the two (and the first is just a necessary accident), then rest is not in the sense of a privation.

1. Cf. Aristotle, *Physics* 5.2.226b15–16.

<الفصل الرابع>

في تحقيق تقابل الحركة والسكون

(١) إنَّ أمر السكون فيه إشكال أيضاً ؛ وذلك لأنَّ المشهور من مذهب الطبيعيين أنَّ السكون مقابلته للحركة هي مقابلة العدم للقنية لا مقابلة الضد . ثم من البيِّن أنَّه لا يصلح أنْ تُفرض بينهما مقابلة إلاّ إحدى هاتين المقابلتين ؛ أعني العدمية والضدّية وقد جعلنا لفظ الحركة واقعاً على معنى صوري ليس عدمياً ؛ إذ قلنا إنّها كمالٌ أول . فإن كانت المقابلة مقابلة العدم للملكة لم يمكن أنْ تكون الحركة منهما هي العدم ، بل نقول إنَّ الجسم إذا كان عادماً للحركة وكان من شأنه أنْ يتحرّك ؛ قيل له ساكن . ومعنى قولنا من شأنه أنْ يتحرك ؛ أن يكون ما تتعلّق به الحركة موجوداً ، وهو أنْ يكون مثلاً في مكانٍ وزمانٍ ، وأيضاً إذا كان له حصول في مكانٍ واحدٍ زماناً ؛ فيقال له إنّه ساكن . فها هنا معنيان موجودان في الساكن ؛ أحدهما عدم الحركة ومن شأنه أنْ يتحرّك ، والآخر أيْنٌ له موجوداً زماناً . فإنْ كان السكون منهما هو الأولّ - وهذا لازمٌ لهّ - كان السكون عدماً ، وإنْ كان السكون هو الثاني منهماّ - والأول لازمٌ لهّ - لم يكن السكون معنى عدمياً .

(2) Of the two, let us assume [for now] that the rest opposing motion is some formal factor and that its definition indicates a form. So, when we intend to compare this definition and *motion*'s definition, we must derive either *motion*'s definition from this definition or this definition from *motion*'s definition, as is required by the general rule for testing the adequacy of the definition of a contrary derived from the definition of its contrary. (Now, I am not saying that the way to define the contrary is that we derive [it] from the definition of its contrary, for this is something that we prohibited with respect to demonstrative teaching, although allowing it in a certain way with respect to dialectical teaching.[2] Again, I am not saying that [this] is the way to hunt down the definition; rather, I am saying that that contrary, even if it is not necessary, will be something possible. I mean that the definition of the contrary thereby will parallel the definition of its contrary, and [so] it is a way to test for adequacy.) So, if the two definitions are contraries and opposites, then *rest*'s being a possession is possible. If the two definitions are not opposites, then this account will not belong to rest (because rest is the opposite of motion), and instead it will be an account that necessarily follows upon the account of rest, while rest will be the account that the privative definition indicates.

(3) We say, firstly, that this description [namely, to exist at some *where* for a time] is not the one opposite of what motion is said to be, when the expression *motion* is understood in our technical vocabulary. So, when we intend [for example] *a first perfection belonging to what is in potency inasmuch as it is in potency* to pick out local motion specifically, it becomes the following: a first perfection with respect to *where* belonging to what potentially has a certain *where* inasmuch as it is in potency. Now, this definition is not the opposite of the definition of rest that we had

2. The reference appears to be to *Kitāb al-burhān* 4.3.

(٢) فلنضع أنَّ السكون المقابل للحركة هو المعنى الصوري منهما، وأنَّ حدّه هو الدالّ على كونه صورياً، فإذا أردنا أنْ نقايس بين هذا الحدّ الحركة وجب أنْ يكون لنا أن نقتضب إمّا حدّ الحركة من هذا الحدّ أو نقتضب هذا الحدّ من حدّ الحركة على ما يوجبه القانون الامتحاني في اقتضاب حدّ الضدّ من حدّ ضدّه. لستُ أقول إن سبيل التحديد للضدّ أنْ نقتضب من حدّ ضدّه، فهذا شيء منعنا عنه في تعليم البرهان، ورخّصنا فيه بوجهٍ مّا في تعليم الجدل. بل نقول إنَّ ذلك الضدّ - وإنْ لم يكن واجباً ولم يكن طريقاً لاقتناص الحدّ - فهو ممكنٌ؛ أعني أنْ يكون حدّ الضدّ يوازي به حدّ ضدّه، ويكون للامتحان سبيل إليه، فإنْ كان الحدّان متضادين ويتقابلان جاز حينئذ أنْ يكون السكون قنْية. وإنْ كان الحدّان لا يتقابلان لم يكن حينئذ هذا المعنى هو للسكون لأنَّ السكون مقابل للحركة، بل معنى يلزم معنى السكون، والسكون هو المعنى الذي يدل عليه الحدّ العدّمي.

(٣) فنقول: أمّا أولاً، فإنَّ هذا الرسم لا يقابل الرسم المقول للحركة الذي هو باصطلاحنا مفهوم لفظة الحركة؛ فإنَّ قولنا كمالٌ أولٌ لما بالقوة من حيث هو بالقوة، إذا أردنا أنْ نخصّصه بالحركة المكانية صار هكذا؛ وهو أنّه كمالٌ أولٌ في الأين لما هو بالقوة، ذو أيْن من حيث هو بالقوة. وهذا الحدّ ليس بمقابلٍ لحدّ السكون الذي حدّدناه، بل عسى

defined [again, as existing at some *where* for a time]. It might, in fact, necessarily follow upon what is opposite of that, but this is not something that we precluded; for we concede that the account of each of the two assumed descriptions of rest entails the other, while not itself being [that account]. If we wish to derive the definition of *rest* from the definition of *motion* (assuming that rest is a formal factor), we find ourselves at a loss but to say that either it is a first perfection belonging to what is actually a *where* inasmuch as it is actually a *where,* or it is a second perfection belonging to what is potentially a *where* inasmuch as it is in potentiality.[3] On the one hand, the first of the two definitions does not necessarily entail rest. [That] is because rest as rest does not need to be some first perfection such that the thing has some second perfection, since the intellect can conceive of the rest, as a rest, where there is no perfection in the thing other than what is in it.[4] On the other hand, the second definition stipulates a certain condition of the essence of rest as rest—namely, that motion has preceded it—which is not something necessary.[5] If, however, we omit the expression *first* and *second,* then we have not preserved the condition of opposition in the definition. If we make some other change, it would not have the right meaning at all—namely, if we mean to take the opposite of *perfection* to be *potentiality.* In this case, then, rest would be a member of the class of privatives, since we clearly cannot derive from the definition of *motion* a definition[6] that corresponds with the definition of *rest,* where rest is an opposite of [motion] and nonetheless is also a possession.

---

3. Secluding the phrase *bi-l-fiʿl ayna aw naqūlu innahu kamāl thānin li-mā huwa bi-l-qūwah aynu min haythu huwa* (. . . actually a *where,* or it is a second perfection belonging to what is potentially a *where* inasmuch as it is . . .) which does not appear in all the manuscripts and is omitted in **Z**, **T**, and the Latin translation. The phrase is almost certainly a result of dittography, since Avicenna will only explicitly address two, not three, possible definitions in his response.

4. For instance, according to ancient and medieval elemental theory, the natural place of the element earth was at the center of the universe, understood as the center of the planet Earth. Hence any of the element earth resting at the center of the planet Earth would have no second perfection toward which it would be naturally directed.

5. Again, appealing to ancient and medieval elemental theory and the belief that the cosmos was eternal, the elemental earth at the center of the cosmos could have been at rest there from all eternity without its having previously been in motion.

6. **Y** (inadvertently) repeats a line, which does not appear in **Z**, **T**, or the Latin.

أنْ يلزمه ما يقابل ذلك، وهذا ممّا لا نمنعه. فإنّا نسلّم أنَّ معنى كل واحدٍ من الرسمين المفروضين للسكون يلزم الآخر، وليس هو هو. فإنْ شئنا أنْ نقتضب من حدّ الحركة حدّ السكونّ - على أنَّ السكون معنى صوريّ - لم نجد إلاّ أنْ نقول بأنَّه كمالٌ أولٌ لما هو بالفعل أيْنٌ من حيث هو بالفعل أيْنٌ، أو نقول إنَّه كمالٌ ثانٍ لما هو بالقوة أينٌ من حيث هو بالقوة فيكون الأول من هذين ليس حدّاً لازماً للسكون، فإنَّ السكون من حيث هو سكونٌ ليس يحتاج أنْ يكون كمالاً أولاً حتى يكون للشيء كمالٌ ثانٍ، فإنَّه يجوز أنْ يعقل السكون سكوناً والشيء لا كمال فيه غير ما فيه. وأمّا الحدّ الثاني فإنّه يجعل من شروط ماهيّة كون السكون سكوناً أنْ يكون قد تقدمته الحركة؛ وهذا ليس بواجب، فإنْ حذفنا لفظ الأول والثاني لم نكن قد حفظنا شرط التقابل في الحدّ؛ وإنْ غيّرنا تغييراً آخر، لم يكن له مفهوم صادق أُصلاً. وإنْ أردنا أنْ نأتي بمقابل الكمال كان القوة فالتحق حينئذ السكون بالعدميات. فقد بانَ أنَّه ليس يمكن أنْ نقتضب من حدّ الحركة حدّاً يطابق حدّ السكون ويكون السكون مقابلاً لها، ويكون السكون مع ذلك قنية.

(4) If we make the definition of *rest* the original one that we mentioned [namely, to exist at some *where* for a time], then either time or something associated with time is immediately included in [the definition]. Time, however, is defined in terms of motion, and so rest would be defined in terms of motion; but one contrary is not part of the description of the other. Likewise, time would enter into the definition of motion, because it is something entering into what enters into its [own] definition. Now, motion precedes time conceptually, in which case motion cannot be a privation (assuming that rest is a possession) because *privation* does not enter into the concept of *possession.* In fact, just the reverse is the case, since the motion entering into the definition of time, which is [itself] entering into the aforementioned definition of rest, is a formal factor. Obviously, then, in this derivation we cannot say that motion is for the body not to have a single *where* for a period of time.

(5) So it is up to us to consider whether this derivation can occur in some other way. The best that can be said here is that rest is to be at a single *where* for a moment and to being at it both before and after [that moment], whereas motion is to be at a single *where* without being at it before or after. In so understanding [motion and rest], however, we have appealed to a temporal *before* and *after,* both of which are defined in terms of time; and again, time is defined in terms of motion, and so motion itself would have been taken in what is understood by it. So motion is apparently not understood in this way, and so this is not a description. Even weaker than this is to take a [temporal] expanse in [the definition] and so say that rest is to be at a single *where* for a period of time, whereas motion is to be at a single *where* for no period of time, for this entails the objection just given. Also, [this definition of motion and rest] would be shared in common with the mobile's state at the first and last moment of motion (for that is to be at a single place for no period of time), but [that state] is not a motion or a rest. So it has become evidently clear that there is no way to confirm the opposition between the definitions of motion and rest when the definition of *rest* [in the sense of] is a possession,[7] and so it remains that the definition of *rest* is in the sense of a privation.

---

7. Following **Z** and **T** that have the adjectival form of *qunya* (possession), which also corresponds with the Latin *habitus,* for **Y**'s *yaqīnī* (known with certainty).

(٤) فإنْ جعلنا الأصل حدّ السكون الذي ذكرناه؛ دخل فيه أول شيء الزمان، أو ما يتعلق بالزمان؛ والزمان يتحدّد بالحركة، فيكون السكون يتحدّد بالحركة – والأضداد ليس بعضها جزء رسم البعض – ويكون الزمان يدخل أيضاً في حدّ الحركة لأنَّه داخلٌ فيما يدخل في حدّه. والحركة قبل الزمان في التصوّر، فلا يجوز أنْ تكون الحركة حينئذ عدماً إنْ كان السكون قنية، لأنَّ العدم لا يدخل في مفهوم القنية، بل الأمر بالعكس؛ فإنَّ الحركة داخلة في حدّ الزمان الداخل في حدّ السكون المذكور بالمعنى الصوري. فبيّنٌ إذن أنّه لا يجوز أنْ نقول في هذا الاقتضاب إنَّ الحركة هي أنْ لا يكون للجسم أينٌ واحدٌ زماناً،

(٥) فننظر هل يمكن أنْ يكون هذا الاقتضاب على وجه آخر، فنقول: إنَّ أحسن ما يمكن أنْ يقال حينئذ هو أنَّ السكون كونٌ في أينٍ واحدٍ وقتاً، والشيء قبله وبعده فيه، والحركة كونٌ في أينٍ واحدٍ من غير أنْ تكون قبله أو بعده فيه. فيكون قد استعملنا في تفهيمهما القَبْل الزماني والبَعْد الزماني وهما متحدّدان بالزمان؛ والزمان متحدّدٌ بالحركة، فيكون قد صارت الحركة مأخوذة في مفهوم نفسها. فظاهرٌ أنَّ الحركة لا تفهم من هذه الجهة؛ فليس هذا رسماً. وأضعف من هذا أنْ يؤخذ متوسعاً فيه؛ فيقال: إنَّ السكون كون في أينٍ واحدٍ زماناً، والحركة كونٌ في أينٍ واحدٍ لا زماناً، فإنَّ هذا يلزمه ما قيل هناك ويشركه حال المتحرّك في ابتداء الحركة وانتهائها، فذلك كونٌ في مكانٍ واحدٍ لا زماناً وليس بحركة ولا سكون. فقد تبيّن واتضح أنّه لا وجه لتصحيح تقابل حدّ الحركة بحدّ السكون. والسكون حدّه المعنى القني، فبقي أن يكون السكون حدّه المعنى العدمي.

(6) Know that, with respect to every kind of motion, there is some opposing rest. So augmentation has some rest opposing it, and likewise alteration. Also, just as the rest opposing alteration is not the quality existing for a period of time, so likewise the rest opposing locomotion is not a single *where* existing for a period of time, but being at rest in that *where*. So resting is a privation of motion.

(7) Since we have now discussed motion and rest, we should provide the true and real definitions of *place* and *time*, since these are topics closely related to motion.

(٦) واعلم أنَّ في كل صنْفٍ من أصناف الحركة سكوناً يقابله؛ فللنمو سكونٌ يقابله، وللاستحالة كذلك، وكما أنَّ السكون المقابل للاستحالة ليس هو الكيف الموجود زماناً، بل سكونٌ في الكيف، وكذلك السكون المقابل للنقلة ليس هو الأين الواحد الموجود زماناً، بل هو سكونٌ في ذلك الأين، فالسكون عدم الحركة.

(٧) وإذ قد تكلمنا في الحركة والسكون؛ فحريٌ بنا أنْ نعرف حقيقة المعنى المسمّى مكاناً، والمعنى المسمّى زماناً، إذ هما من الأمور الشديدة المناسبة للحركة.

Chapter Five

Beginning the account of place and reviewing the arguments of those who deny and those who affirm it

(1) The first thing that we must investigate about place is its existence and whether or not there is such a thing as place at all; nevertheless, in the following we shall not come to understand place itself, but only its relation to body (in that [the body] rests in it and is moved away and toward it). Certainly, one frequently investigates a thing's existence after identifying its essence. At other times, however, it is before identifying it, as when one knows a certain accident it has—as, for example, knowing that a certain thing has the aforementioned relation [that is, place's relation to body] while not knowing what that thing is. In the case where you understand that essence, you need to explain the existence of [that essence]; and thereafter, if the existence of that relation to which [that essence] belongs is not clear, then we need to explain that it is the essence itself that the relation specifies. This has been explained to you elsewhere.[1]

(2) So we say: There are some who refuse to accept that place has any existence whatsoever, whereas others make its existence necessary. As for the "refuse" of the one group, they could avail themselves of arguments close to what we present here, namely, if place exists, then it must

1. It is not clear what Avicenna's reference is here. Two possibilities are his discussion about the essential (*dhātī*) and accidental at *Kitāb al-madkhal* 1.6, or his discussion of the difference between demonstrations *quia* and *propter quid* at *Kitāb al-burhān* 1.7, although neither reference explicitly addresses the point made here.

<الفصل الخامس>

في ابتداء القول في المكان وإيراد حجج مبطليه ومثبتيه

(١) أول ما يجب أنْ نفحص عنه من أمر المكان وجوده وأنَّه هل ها هنا مكان أم لا مكان البتة على أنّا نحن إنّما نفهم بعد من اسم المكان لا ذاته، بل نسبة له إلى الجسم بأنَّه يسكن فيه وينتقل عنه وإليه بالحركة. فإنَّ الفحص عن وجود الشيء قد يكون بعد تحقّق ماهيّته، وقد يكون قبل تحقّق ماهيّته، إذا كان قد وقف على عارضٍ له؛ مثلاً قد وقف على أنَّ ها هنا شيئاً له النسبة المذكورة ولم يعلم ما ذلك الشيء، وحينئذ تحتاج - إذا فهمتَ تلك الماهيّة - أنْ نبيّن وجودها. ثم إنْ لم يكن وجود تلك النسبة بيّناً لها احتجنا إلى أنْ نبيّن أنّها هي الماهيّة التي تخصّها النسبة، وهذا شيءٌ قد بانَ لك في موضع آخر.

(٢) فنقول: إنَّ من الناس مَنْ نفى أنْ يكون للمكان وجود أصلاً، ومنهم مَنْ أوجب وجوده. فأمّا النفاة منهم فلهم أنْ يحتجوا بحجج منها ما تقرب منه عبارتنا هذه: وهو أنَّ

be either a substance or an accident. On the one hand, if it is a substance, then it is either a sensible or intelligible substance. Now, if it is a sensible substance, and every sensible substance has a place, then place has a place *ad infinitum.* If it is an intelligible substance, then it simply cannot be said that the sensible substance is joined with it and departs from it, because intelligibles cannot be pointed to and do not have a position, whereas whatever the sensible substance is joined to or departs from can be pointed to and has a position.[2] If, on the other hand, it is an accident, then that in which this accident inheres is like that in which whiteness inheres, and that in which whiteness inheres derives its name from it and so is said to be *whitened* and *white.* So the substance in which place inheres should derive its name from it and so be *placed,* in which case the place of the placed would be an accident in [the placed], and it would necessarily follow, then, that it remains permanently in it during local motion and occurs with it wherever it occurs. If that is the case, then nothing can move locally from it, but instead moves with it, whereas place (as you [Aristotelians] allege) is not that *together with* which something is moved locally, but that *in* which something is moved locally.

(3) Again, place must either be a body or not. On the one hand, if it is a body and the placed thing is in it, then the placed thing interpenetrates it; but [the idea] that some bodies interpenetrate others is absurd. Moreover, how could it be a body when it is neither among the simple bodies[3] nor a composite of them? If, on the other hand, it is not a body, then how can they say that it coincides with the body and is coextensive with it, when what is coextensive with body is a body?

---

2. **Y** repeats the phrase *kull mā yuqārinuhu al-jawhar al-maḥsūs aw yufāriquhu fa-huwa dhū ishārah ilayhi wa-lā waḍ*ᶜ *lahā* (whereas whatever the sensible substance is joined to or departs from can be pointed to and has a position), which is not found in **Z**, **T**, or the Latin. That repetition has been omitted here.

3. That is, the elements earth, air, fire, and water.

المكان إذا كان موجوداً فلا يخلو إمّا أنْ يكون جوهراً أو عرضاً ، فإنْ كان جوهراً فإمّا أنْ يكون جوهراً محسوساً أو معقولاً ، فإنْ كان جوهراً محسوساً - وكل جوهر محسوس فله مكانٌ - فللمكان مكان إلى غير نهاية ، وإنْ كان جوهراً معقولاً فيستحيل أنْ يقال إنَّ الجوهر المحسوس يقارنه ويفارقه ، لأنَّ المعقولات لا إشارة إليها ولا وضع لها ، وكل ما يقارنه الجوهر المحسوس أو يفارقه فهو ذو إشارة إليه ووضع له. وإنْ كان عرضاً فالذي يحله هذا العرض هو كالذي يحله البياض ، والذي يحله البياض يشتق له منه الاسم فيقال : مبيّض وابيضّ. فالجوهر الذي يحلّه المكان يجب أنْ يشتق له منه الاسم فيكون هو المتمكن ، فيكون مكان المتمكن عرضاً فيه ، فيلزم أنْ يلزمه في النقلة ويصير معه حيث صار ، وإذا كان كذلك ، لم يكن منتقلاً عنه بل منتقلاً معه ، والمكانّ - كما تزعمونّ - ليس هو المنتقل معه بل المنتقل فيه.

(٣) وأيضاً فإنَّ المكان لا يخلو إمّا أنْ يكون جسماً ، وإمّا أنْ يكون غير جسم ؛ فإنْ كان جسماً ، والمتمكّن يكون فيه ، فالمتمكّن مداخلٌ له ، ومداخلة الأجسام بعضها بعضاً محال. ثم كيف يكون جسماً ولا هو بسيط من الأجسام ولا مركّب منها ؟ وإنْ كان غير جسم ، فكيف يقولون إنَّه يطابق الجسم ويساويه ، ومساوي الجسم جسمٌ؟

(4) Furthermore, locomotion is nothing but change of proximity and remoteness and, just as it might apply to a body, so likewise [locomotion] might apply to a surface, line, and point. In that case, if locomotion requires a place for what undergoes local motion, the surface, line, and, in fact, [even] the point must have a place. Now, it is known that the place of the point must exactly equal it, since you all make place something exactly equaling the placed thing such that nothing else contains it; but what exactly equals the point is a point, and so the place of the point is a point. In that case, why does one of the two points become a place and the other the placed thing? Perhaps, on the contrary, each one of them is [both] a place and a placed thing, and so [the point] is a *placed thing* in the relation going from it to the other, whereas it is a *place* in the relation going from the other to it. This is something you blocked yourself off from when you denied[4] that *place* is something placed in the placed thing in which it is.[5]

(5) They have additionally said: If the point has a place, then they ought to make it have a certain heaviness or lightness. That [argument] is one that the group denying motion specifically pressed, saying that it is senseless to require that the body have a place and motion without equally requiring that the point have a place and motion—in which case, if you permit motion in the point, you have given it a certain inclination, making it have a certain lightness and heaviness. Now, this is commonly accepted as false, given that the point is nothing but the termination of a line, and the termination of a line is a privative notion, and how can a privative notion have a place or motion? The point is a termination of a line because it is an end point, where the end point terminates something such that nothing of it remains. Now, when the point has no place, the body will have no place, since whatever requires the body to have a place would require the point to have a place.

---

4. Reading *abaytum* with **Z** and **T** for **Y**'s *ithbattum* and the Latin *vultis* (affirmed).
5. Cf. Aristotle, *Physics* 4.3.210b8–31.

(٤) وأيضاً فإنَّ الانتقال ليس إلاّ الاستبدال لقربٍ وبُعْدٍ، وكما أنَّ هذا الاستبدال قد يقعُ للجسم؛ فكذلك قد يقع للسطح وللخط وللنقطة. فإنْ كان الإنتقال يوجب للمنتقل مكاناً؛ فيجب أنْ يكون للسطح مكان وللخط مكان بلْ للنقطة مكان! ومعلومٌ أنَّ مكان النقطة يجب أنْ يكون مساوياً لها؛ إذْ جعلتم المكان مساوياً للمتمكن حتى لا يسعه غيره، وما يساوي النقطة نقطة، فمكان النقطة نقطة؛ فلِمَ صارت إحدى النقطتين مكاناً والأخرى متمكنة؟ بل عسى أن يكون كل واحدة منهما مكاناً ومتمكناً؛ فتكون بالقياس الآخذ منها إلى الأخرى متمكنة، وبالقياس الآخذ من الأخرى إليها مكاناً. وهذا ممّا حظرْتموه حين أبيتم أنْ يكون المكان متمكناً في المتمكن فيه.

(٥) وزادوا فقالوا إنْ كان للنقطة مكانٌ، فبالحري أنْ تجعلوا لها ثقلاً أو خفّةً - قال ذلك خصوصاً القوم الذين نفوا الحركةّ - فقالوا لا معنى يوجب للجسم مكاناً وحركةً إلاّ ومثله يوجب للنقطة مكاناً وحركةً - فإنْ جوّزتم في النقطة حركة فقد أعطيتموها ميلاً، وجعلتم لها خفةً وثقلاً وهذا مشهور البطلان. على أنَّ النقطة ليست إلاّ فناء الخط؛ وفناء الخط معنى عدمي، وكيف يكون للمعنى العدمي مكانٌ أو حركةٌ؟ فأمّا أنَّ النقطة فناء الخط فلأنها نهاية، والنهاية هي أن يفنى الشيء فلا يبقى منه شيءٌ، وإذا لم يكن للنقطة مكان لم يكن للجسم مكان، إذ كان ما يوجب للجسم مكاناً يوجب للنقطة مكاناً.

(6) Moreover, in your opinion, place is something indispensable for motion, since you make motion need it and so it is one of the causes of motion. It is not an agent of motion, however. How could it be when you make every motion have an efficient principle that is known to be different from place? Similarly, it cannot be a material principle, since motion subsists only in what is moved, not in the place. Again, it cannot be a formal principle, because place is not motion's form. Moreover, it cannot be a final principle, and that is because [place] is something that, in your opinion, is [just as much] needed before arriving at the end and completion as it is needed upon arriving. So, if the place is an end, it is not because it is place [absolutely], but because it is a certain place with a present actuality for some motion with a present actuality, whereas our discussion concerns place inasmuch as it is absolute place. Were place a perfection so that the mobile desired it, whether by nature or will, then being in the places that one desires would also be one of the human perfections. Also, [one could deny the existence of place, arguing] on the basis that there fall under *perfection* both proper and common [perfections]. Now, the proper [perfection] is a thing's form, but place is neither the form of the mobile nor the form of the motion. As for the common [perfection], it belongs to one thing as well as something else, whereas place, according to your view, is something proper [to the thing].

(7) Also, if body were in a place, then growing bodies would be in a place; but, if they were in a place, then their place would grow with them; and, if their place should grow with them, then their place would move with them and their place would have a place, all of which you deem impossible.

(٦) وأيضاً فإنَّ المكان عندكم أمر لا بُدّ منه للحركة، إذ تجعلون الحركة محتاجة إليه، فهو إحدى عِلل الحركة لكنه ليس بفاعلٍ للحركة؛ وكيف، ولكل حركةٍ تجعلونها في المكان مبدأ فاعلي معلوم غير المكان؟ ولا أيضاً هو مبدأ عنصري له، إذ الحركة إنّما قوامها في المتحرّك لا في المكان، ولا أيضاً مبدأ صوري له؛ لأنَّ المكان ليس هو صورة الحركة، ولا أيضاً مبدأ غائي له؛ وذلك لأنَّه ممّا يحتاج عندكم إليه قبل الوصول إلى الغاية والتمامّ - كما يحتاج إليه عند الوصول. فإنْ كان المكان غاية؛ فليس لأنّه مكان، بل لأنَّه مكان بحالٍ لحركةٍ بحال، وكلامنا في المكان من حيث هو مكانٌ مطلقاً. ولو كان المكان كمالاً - لأنّه يشتاق إليه المتحرّك إمّا طبعاً وإمّا إرادةً - لكان من كمالات الإنسان أيضاً أن يحصل في أمكنةٍ يشتاق إليها. على أنَّ التمام منه خاصٌ ومنه مشتركٌ، والخاص هو صورة الشيء، والمكان ليس هو صورة المتحرّك ولا صورة الحركة، وأمّا المشترك فإنّه يكون للشيء ولغيره، والمكان عندكم خاص.

(٧) ولو كان الجسم في مكان لكانت الأجسام النامية في مكان، ولو كانت في مكان لكان مكانها ينمو معها ولو كان مكانها ينمو معها؛ لكان مكانها يتحرك معها، ولكان لمكانها مكان، وأنتم تمنعون هذا كلّه.

(8) Those who affirm [the existence of] place argued from the existence of locomotion, noting that locomotion certainly involves departing from one thing toward another. Now, that thing that is departed from is not some substance, quality, quantity, or some other thing within the [thing] itself, since all of these remain despite the locomotion. Instead, that [thing] departs from something that the body was in, thereupon replacing it. This [thing] is what we call *place*. Similarly, they argued from the existence of replacement. So [for instance] we observe that [some] body is present, and then we see that it is absent, seeing that some other body has become present where it was—as, for example, water was in a jar, and then afterwards air or oil came to be in it. Now the light of reason requires that what replaces and succeeds this thing do so in something that initially belonged uniquely to that first thing but that it has now left behind. That is neither a quality nor a quantity in the very being of either one of them, nor a substance; but, rather, it is the space in which the first was, and then something else came to be in it. Also, [they affirm the existence of place] because everybody intellectually recognizes that there is an up and down, whereas something does not come to be up owing to its substance or quality or quantity, but owing to that thing that is called *place*. Also, the estimative faculty does not even imagine mathematical figures unless they are distinguished by a certain position and space.[6] Also, were it not that place exists and has with its existence a certain specification, species differences, and properties unique to the species, then some bodies would not be moved naturally up and others down. They said that the power of place has reached such a degree that popular imagination refuses to believe that anything exists

6. In the Arabic paraphrase of John Philoponus's commentary on the *Physics* (see Arabic *Physics*, 276), Philoponus gives the example of a triangle, arguing that one cannot imagine a triangle without imagining a base and apex that are separated by space. Thus, even mathematical reasoning requires some idea of space.

(٨) وأمّا مثبتو المكان فقد احتجوا بوجود النقلة، وذكروا أنَّ النقلة لا محالة مفارقة شيءٍ لشيءٍ إلى شيءٍ، وليس ذلك مفارقة جوهر ولا كيف ولا كمّ في ذاته ولا غير ذلك من المعاني، إذْ جميع هذه تبقى مع النقلة، بل إنّما ذلك مفارقة شيءٍ كان الجسم فيه ثم استبدل به؛ وهذا هو الذي نسميه مكاناً. واحتجوا أيضاً بوجود التعاقب، فإنّا نشاهد الجسم حاضراً ثم نراه غابَ، ونرى جسماً آخر حضر حيث هو؛ مثلاً قد كانت جرّة فيها ماء، ثم حصل بعده فيها هواء أو دهن، والبديهة توجب أنَّ هذا المعاقب عاقب هذا الشيء وخلف في أمرٍ كان لذلك الشيء أولاً، وكان الأول مختصاً به؛ والآن فقد فاته، وذلك لا كيفٌ ولا كمٌّ في ذات أحدهما ولا جوهرٌ، بل الحيّز الذي كان الأول فيه ثم صار الآخر فيه. ولأنَّ الناس كلهم يعقلون أنَّ ها هنا فوق وأنَّ ها هنا أسفل، وليس يصير الشيء فوق بجوهرٍ له أو كيف أو كمّ فيه أو غير ذلك، بل بالمعنى الذي يسمى مكاناً. وحتى أنَّ الأشكال التعليمية لا تتوهم إلاّ أنْ تتخصّص بوضع وحيّز، ولولا أنَّ المكان موجودٌ ومع وجوده له تنوّعٌ وفصولٌ وخواصٌ لما كان بعض الأجسام يتحرّك طبعاً إلى فوق وبعضها إلى أسفل. قالوا: وقد بلغ من قوة أمر المكان أنَّ التخيّل العامّي يمنع وجود شيء لا في مكان،

that is not in place and demands that place be something self-subsistent, requiring that it be a certain preparatory [cause] to the extent that bodies come to exist in it. Also, when Hesiod desired to compose a poem in which he related the order of creation, he did not think that anything preceded the existence of place, and so said: "Place is what God created first, then the broad expanse of Earth."[7]

(9) As for solving the puzzles that were mentioned rejecting place, we shall put them off until the time that we fully grasp the essence of place. So let us first find out the essence of place.

---

7. The reference is to Hesiod's *Theogony,* 116, where Hesiod states that the first of the gods was chaos, or infinite space. Cf. Aristotle, *Physics* 4.1.208b29–31.

ويوجب أنَّ المكان أمرٌ قائم بنفسه، يحتاج أن يكون معدّاً حتى توجد فيه الأجسام. ولمّا أراد 〈هزيود〉 الشاعر أنْ يقول شعراً تحدّث فيه عن ترتيب الخليقة، لم يرَ أنْ يقدّم على وجود المكان شيئاً فقال: إنَّ أول ما خلق اللّه المكان ثم الأرضَ الواسعة.

(٩) فأمّا حل الشكوك التي أوردها نُفاة المكان فسنؤخرها إلى وقت إحاطتنا بماهيّة المكان، فلنعرف أولاً ماهية المكان.

Chapter Six

The various schools of thought about place and a review of their arguments

(1) The common man frequently uses the term *place* in two ways. On the one hand, he sometimes means by *place* whatever something rests on. He does not go on, however, and distinguish whether it is the lower body or the outermost surface of the lower body unless he is one of those who has broken with common opinion a little, in which case some imagine that it is the outermost surface of the lower body, to the exclusion of the rest of it. On the other hand, sometimes they mean by *place* the thing that contains another, like the cask for wine and the house for the human and, in general, whatever something is in, even if it is not resting on it. This is the majority opinion even if they are not aware of it, since the general public thinks that the arrow passes through a place, and those who have a sense of the universe's form believe that Heaven and Earth are at rest in a place even if not supported by something.

(2) Philosophers, however, found certain attributes belonging to the thing to which the name *place* applies in the second sense.[1] Examples [of these attributes of place] include [the following]: something [call it *x*] is in it; [*x*] departs from it during motion; it encompasses [*x*] and nothing

1. For discussions of the earlier Greek philosophical positions surrounding place upon which Avicenna is drawing, see Richard Sorabji, *Matter, Space, and Motion: Theories in Antiquity and Their Sequel* (Ithaca, NY: Cornell University Press, 1988); Keimpe Algra, *Concepts of Space in Greek Thought* (Leiden: E. J. Brill, 1995); and Helen S. Lang, *The Order of Nature in Aristotle's* Physics: *Place and the Elements* (Cambridge, UK: Cambridge University Press, 1998).

<الفصل السادس>

في ذكر اختلاف مذاهب الناس في المكان وإيراد حججهم

(١) إنَّ لفظة المكان قد تستعملها العامة على وجهين: فربما عنوا بالمكان ما يكون الشيء مستقراً عليه؛ ثم لا يتميّز لهم أنَّه هو الجسم الأسفل أو السطح الأعلى من الجسم الأسفل إلاّ أن يتزعزعوا يسيراً عن العامية فيتخيّل بعضهم أنَّه هو السطح الأعلى من الجسم الأسفل دون سائره. وربما عنوا بالمكان الشيء الحاوي للشيء؛ كالدّن للشراب والبيت للناس، وبالجملة ما يكون فيه الشيء وإنْ لم يستقر عليه - وهذا هو الأغلب عندهم - وإنْ لم يشعروا به، إذ الجمهور منهم يجعلون السهم ينفذ في مكانٍ، وأن السماء والأرض عند مَنْ فهم صورة العالم منهم مستقرة في مكانٍ وإنْ لم تعتمد على شيء.

(٢) لكن الحكماء وجدوا للشيء الذي يقع عليه اسم المكان بالمعنى الثاني أوصافاً؛ مثل أنْ يكون فيه الشيء ويفارقه في الحركة ولا يسعه معه غيره، ويقبل المنتقلات إليه؛ ثم

else; and things undergoing local motion are received into it. In gradual degrees, then, their estimative faculties imagined it as a container, since the placed thing is described by its being *in* it. Once they were of a mind to find out the essence and substance of this thing, however, it was as if they became divided among themselves, saying: Whatever is proper to something and nothing else must either be inherent in or extrinsic to the thing itself. If it is inherent in the thing itself, then it is its material or its form. If it is extrinsic to the thing itself and yet exactly equals it and is proper to it, then either it is an extremity of some surface that meets [the thing] and is occupied by contacting it and nothing else and, whether it is what surrounds or [what] is surrounded, it remains at rest, whichever of the two it happens to be; or it is a certain interval exactly equaling the dimensions of [the thing] and so occupies it by permeating it.[2]

(3) Those who maintained that place is the material [asked], "How could it not be, when the material is what is susceptible to replacement?"[3] Those who maintained that place is the form [asked], "How could it not be, when it is the first defining container?"

(4) Those who maintained that place is intervals[4] said that between the extremes of the container holding water are certain naturally disposed[5] intervals that remain fixed, and that the bodies held in the container successively replace them. The situation brought them to the point of saying that this was commonly accepted and that, in fact, the light of reason is naturally disposed to it, for everyone judges that water

2. Cf. Aristotle, *Physics* 4.4.211b6–9.

3. Cf. the argument for place from *replacement* in 2.5.8.

4. The Arabic *buʿd,* which can also be translated "dimension," is almost certainly translating the Greek *diastēma,* which Aristotle mentions as a (rejected) candidate for place (see *Physics* 4.4.211b5ff), but which the Neoplatonist John Philoponus subsequently defended (see his corollaries on place from his commentary on the *Physics*).

5. Reading *mafṭūrah* with **T** for **Y**'s and **Z**'s *maqṭūrah* (dripped). The former not only makes better sense but also brings the text in line with the same expression that appears at the beginning of par. 8 of 2.7, where Avicenna discusses the independent interval. Still, *maqṭūrah* is clearly the *lectio difficilior;* and so, if retained, "dripped" might be understood as "wet," thus referring to the "watery" interval of the contained water. Alternatively, one of the terms Avicenna used for "dimension" (*quṭr*) is derived from the same root as *maqṭūrah,* and so, although it is unlikely, the phrase might mean "certain dimensional intervals." The Latin has *infinita* (infinite) both here and at 2.7.8, where the Arabic is clearly *mafṭūrah;* however, *infinita* is quite possibly a corruption of *insita* (inborn or innate), which would correspond nicely with the Arabic *mafṭūrah.*

تدرجوا قليلاً قليلاً إلى أنْ توهموا أنَّه حاوٍ إذ كان المتمكن موصوفاً بأنَّه فيه. فلما أرادوا أنْ يعرفوا ماهيّة هذا الشـــيء وجوهره، فكأنهم قسموا في أنفسهم، فقالوا إنَّ كل ما يكون خاصّاً بالشيء ولا يكون لغيره فلا يخلو إمّا أنْ يكون داخلاً في ذاته أو خارجاً عن ذاته. فإنْ كان داخلاً في ذاته؛ فإمّا أنْ يكون هيولاه وإمّا أنْ يكون صورته، وإنْ كان خارجاً عن ذاتّه - ويكون مع ذلك يساويه ويخصّه - فهو إمّا نهاية سطح يلاقيه ويشتغل لماسته ولا يماسّــه غيره، إمّا محيط وإمّا محاط مستقر عليه أيّهما اتفقٍ، وإمّا أنْ يكون بُعْداً يساوي أقطاره، فهو يشغله بالاندساس فيه.

(٣) فمنهـــم مَنْ زعم أنَّ المكان هو الهيولى؛ وكيـــف لا ، والهيولى قابل للتعاقب! ومنهم مَنْ زعم أنَّ المكان هو الصورة، وكيف لا؛ وهو أول حاوٍ محدّد!

(٤) ومنهـــم مَنْ زعم أنَّ المكان هـــو الأبعاد فقال: إنَّ بين غايات الإناء الحاوي للماء أبعـــاداً مفطورة ثابتة، وإنها تتعاقب عليها الأجســـام المحصورة فـــي الإناء، وبلغ بهم الأمر إلى أنْ قالوا إنَّ هذا مشـــهور؛ بل مفطور ‹ة› عليه البدنهة، فإنَّ الناس كلّهم يحكمون أنَّ

is in what is between the limits of the container and that the water disappears and departs and air comes be in that very same interval. They also argue by attacking [others'] arguments. So, addressing specifically the advocates of surface,[6] they said that, if place is a surface that meets the surface of another, then motion would be to depart from one surface while advancing toward another. In that case, the bird standing still in a [stream of] air and the rock standing still in [flowing] water, where both [the water and air] are changing (that is, one surface is being departed from for another), must be something undergoing motion.[7] That is because what they have made the place of [the mobile] is changing around it. If it is resting, then its resting is in which place? [The question arises] since a condition for something's being at rest is that it stay put in its place for a time; and this account is frequently what attests to something's being at rest. So, since the surface does not stay put, what stays put is only the interval that it occupies, which neither is snatched away nor changes, but is always one and the same.

(5) Again, they said that [conceptual] analysis results only in simple things, where the estimative faculty removes one thing after another from the things together in the composite until what remains in the estimative faculty after everything else is removed is something simple existing in itself, even if, taken alone, it cannot subsist. It is in this way that we recognize material, form, and simple [elements], which are certain units, in composite things. Once more, then, when our estimative faculty imagines the water and the other bodies as removed and eliminated in the container, it necessarily follows from that that the interval remaining between its limits exist and that that [interval] exist also at the same time [that] the former exists [namely, the water or other bodies].

---

6. These would be those who advocate the Aristotelian account of place as Avicenna understands it. Whether Aristotle himself would have identified place with a surface is, however, an open question; see H. Lang, *The Order of Nature in Aristotle's* Physics: *Place and the Elements,* 104–111.

7. The "bird example," although employed in a different context, can be found in Galen, *On Muscular Movement,* 4.402, 12–403, 10; see *The Hellenistic Philosophers,* ed. A. A. Long and D. N. Sedley, 2 vols. (Cambridge, UK: Cambridge University Press, 1987), vol. 1, 283.

المـاء فيما بين أطراف الأناء، وأنَّ الماء يزول ويفارق ويحصل الهواء في ذلك البُعْد بعينه.

واحتجّوا أيضاً بضروب من الحجج فقالواّ - وهم يخاطبون خاصّة أصحاب السطوحّ - إنَّه إنْ كان المكان سـطحاً يلقى سطح الشيء؛ فتكون الحركة هي مفارقة سطح متوجهاً إلى سـطح آخـر، فالطائر الواقف في الهواء والحجر الواقف فـي الماءّ - وهما يتبدلان عليهّ - وهو يفارق سـطحاً إلى سطح، يجب أنْ يكون متحركاً. وذلك لأنَّ ما يجعلونه مكانه يتبدل عليه، فإنْ كان سـاكناً فسـكونه في أيّ مكان. إذ من شرط الساكن أنْ يلزم مكانه زماناً؟ إذ السـاكن قد يصدق عليه هذا القول. فإذ ليس يلزمه السـطح، فما الذي يلزمه سـوى البُعْد الذي شـغله، الذي لا ينزعج ولا يتبدل؛ بل يكون دائماً واحداً بعينه؟

(٥) وقالوا أيضاً إنَّ الأمور البسـيطة إنّما يؤدي إليها التحليل وتُوهم رَفْع شيء شيء من الأشياء المجتمعة معاً وَهْماً، فالذي يبقى بعد رفْع غيره في الوهم هو البسيط الوجود في نفسـه، وإنْ كان لا ينفرد له قوام، وبهذا السبب عرفنا الهيولى والصورة والبسائطّ - التي هي آحادّ - في أشياء مجتمعة. ثم إنّا إذا توهمنا الماء وغيره من الأجسام مرفوعاً غيـر موجود في الإناء؛ لزم من ذلك أنْ يكون البُعْـد الثابت بين أطرافه موجوداً، وذلك أيضاً موجود عندما تكون هذه موجودة معه.

(6) They further said that body is in place not by means of its surface, but through its volume and quantity. So what it is in by means of its corporeality must be what exactly equals it. In that case, [place must] be an interval, because place exactly equals the placed thing, and the placed thing is a body possessing three dimensions, and so place possesses three dimensions.

(7) Again, they said that place must in no way undergo motion nor disappear; but the extremities of what contains frequently do undergo motion in some way and do disappear.

(8) Moreover, they said that people say that place is sometimes empty and sometimes full, but they do not say that the simple [surface] is empty and full.

(9) They argued that, while the doctrine of the interval sees to it that every body is in a place, the school of thought associated with the advocates of the simple containing [surface] makes it impossible for certain bodies to have a place.[8]

(10) Furthermore, they claim that, in fire's motion upward and earth's motion downward, both seek a place for the whole of themselves. It is absurd, however, that they seek the extremity of a body that is above or below, for it is absurd that the whole of the body should meet with the limit. Hence it is the ordered position in the interval that is sought.

(11) These, then, are the arguments of the advocates of the interval [considered] absolutely. They are, however, of two schools of thought. Some of them deem it absurd that this interval should remain empty without something filling it, requiring that it never be absolutely stripped of what fills it save with the entry of something else that fills it.[9] Others do not deem that absurd but embrace the possibility that this interval

---

8. The body in question almost certainly is that of the cosmos itself, which, according to Aristotelian cosmology, is finite, and "outside" of which there is absolutely nothing, not even void space. Consequently, if *place* is understood as the innermost limit of a containing body, (Aristotle's preferred definition) and yet nothing is outside of the cosmos that could contain it, then the cosmos as a whole cannot have a place.

9. The proponent of this view is John Philoponus; cf. *In Phys*. 568.14ff.

(٦) وقالوا أيضاً إنَّ كون الجسم في مكانٍ ليس بسطحه؛ بل بحجمه وكميّته، فيجب أنْ يكون فيه بجسميته مساوياً له فيكون بُعْداً، ولأنَّ المكان مساوٍ للمتمكن؛ والمتمكن جسم ذو ثلاثة أقطار، فالمكان ذو ثلاثة أقطار.

(٧) وقالوا أيضاً إنَّ المكان يجب أنْ يكون شيئاً لا يتحرك بوجه ولا يزول، ونهايات المحيط قد تتحرّك بوجهٍ مّا وتزول.

(٨) وقالوا أيضاً إنَّ الناس يقولون إنَّ المكان قد يكون فارغاً وقد يكون ممتلئاً، ولا يقولون إنَّ البسيط يكون فارغاً ويكون ممتلئاً؛

(٩) قالوا؛ والقول بالأبعاد يجعل كل جسم في مكان؛ ومذهب أصحاب البسيط الحاوي أوجب أنْ يكون من الأجسام ما لا مكانَ له.

(١٠) وقالوا أيضاً إنَّ النار في حركتها إلى فوق، والأرض في حركتها إلى أسفل يطلبان مكاناً لكُلِّيَّتهما، ومحال أنْ يطلبا نهاية الجسم الذي فوقه أو تحته؛ فإن النهاية محالٌ أنْ تلاقيها كلّية جسم، فإذن يُطلب الترتيب في البُعْد.

(١١) فهذه حجج أصحاب البُعْد مطلقاً. لكن أصحاب البُعد على مذهبين: منهم مَنْ يحيل أنْ يكون هذا البُعد يبقى فارغاً لا مالىء له؛ بل يوجب أنّه لا يتخلّى عن مالىءٍ البتّة؛ إلاّ عند لحوق مالىء. ومنهم مَنْ لا يحيل ذلك، بل يجوّز أنْ يكون هذا البُعد خالياً

is sometimes void and sometimes full—namely, those who advocate the void ([albeit] some of those who defend the void suppose that the void is not an interval, but is nothing, as if to be *something* is to be body).[10] Now, the first thing that incited the imagination to believe in the void was air. That is because the initial common opinion was that whatever is neither a body nor in a body does not exist. Moreover, their initial opinion about existing bodies was that they are perceptible by sight, whereas whatever is not perceptible by sight was supposed not to be a body and so, then, must be nothing. Therefore, it was imagined, in the case of air, not that it was something that fills but that it was nothing. On their initial view, then, the air in the container was[11] imagined to be void[12] intervals, not some thing. The first to rouse them from their [intellectual] slumber did so by showing them that inflated wineskins resist prodding, and so, by prodding, it became obvious to them that air is a body, just like the rest of the bodies *qua* body. Some of those who saw that backtracked and no longer considered void an existing thing, since the thing that they had supposed was a void—that is, air—turned out to be full. Others, not voiding themselves of the void, conceded that air is not a pure void but, rather, a mixture of void and something that fills, since they found certain arguments and reasoning that conclude that the void exists.

(12) One of the arguments for that is their claim that we see bodies rarefy and condense without anything entering or leaving. Thus, rarefaction involves the parts being separated in such a way that what is left between them is a void, whereas condensation is the return of the parts to fill the void produced by rarefaction.

---

10. For Avicenna's sources of the earliest accounts of the void, see Aristotle, *Physics* 4.6.

11. Reading *kāna* with **Z**, **T**, and the Latin (*est*) for **Y**'s *ka-anna* (as if).

12. Reading *khālīyah* with **Z**, **T**, and the Latin (*vacua*) for **Y**'s *ʿālīyah* (high), which simply makes no philosophical sense in the present context and may simply be a typographical error in **Y**'s edition.

تـارة ومملوءاً تارة؛ وهم أصحاب الخـلاء . وبعض القائلين بالخلاء يظن أنَّ الخلاء ليس هو بُعداً ، بل هو لا شيء ؛ كأن الشيء هو الجسم . وأول شيء خيّل اعتقاد الخلاء هو الهواء ، وذلك لأنَّ الظنّ العامي الأول هو أنَّ ما ليس بجسمٍ ولا في جسمٍ فليس بموجودٍ ! ثم ظنّهم الأول في أمر الأجسـام الموجودة؛ هو أن تكون محسوسة بالبصر ، وما لا يحسّ بالبصر يُظن أنه ليس بجسـم ، ثم يوجب أنّه ليس بشـيء . فلذلك يتخيل من أمر الهواء أنّه ليس بملاء ، بل لا شيء . فكان الإناء الذي فيه هواء لا يتخيّل عندهم من أمره ، في أول الأمر ؛ أنَّ فيه شيئاً ؛ بل يتخيّل أنَّ هناك أبعاداً خالية . فأول مَنْ نبههم بأَنْ أراهم الأزقاق المنفوخة تقاوم المسّ ؛ فأظهر لهم بالمسّ أنّ الهواء جسـمٌ كسـائر الأجسام في أنّه جسم . فمن الذين أراهم ذلك مَنْ رجع فلم يرَ ها هنا خلاء موجوداً ، إذ صار الشيء الذي كان يظّنه خلاء وهو الهواء ملاء . ومنهم من سـلّم أنَّ الهواء ليس بخلاءٍ صرف ، بل ملاء ويخالطه خلاء ولم يخلُ عن الخلاء ، إذ قد وجد حججاً وقياسات انتجت أن الخلاء موجود .

(١٢) فمـن الحجج على ذلك قولهم : إنّا نرى أنَّ الأجسـام تتخلخل وتتكاثف من غير دخول شيء أو خروجه ؛ فالتخلخل إذن تباعد الأجزاء تباعداً يترك ما بينها خالياً ، والتكاثف رجوع من غير دخول شيء أو خروجه ؛ فالتخلخل إذن تباعد الأجزاء تباعداً يترك ما بينها خالياً ، والتكاثف رجوع من الأجزاء إلى ملء الخلاء المتخلخل .

(13) They also said that we see something completely filled with ashes being completely filled with water; but were there not a void, it would be impossible that it should be filled with water. Similarly, they said: There is also the wine cask that is completely filled with wine, and then that same wine is placed into a wineskin, and then both [the wine and the wineskin] are placed back into that same wine cask such that the cask simultaneously contains both wine and wineskin.[13] Were there no void in the wine into which the magnitude measuring the wineskin is contracted, it would be impossible that what had been completely filled by the wine alone should now contain both the wineskin and wine together.

(14) They further said that what grows does so only by extending into something that it is in but that, undoubtedly, that thing cannot extend into what is already filled, but only into what is void. Also, some of them generalized this argument, saying that the mobile must be moved into either a void or something already filled. If it moved into something already filled, however, then what is full would move into what is full! Thus, it remains that it is moved into a void. Also falling under that form of their argument is the phial that is sucked on and then inverted into water such that water enters into it, but if [the phial] were already full, then it could not contain some other thing that enters into it.

(15) Again, they said that when the mobile is moved, it must either repel what is full, and so move it, or it must interpenetrate it. Interpenetration, however, is absurd. So it remains that it repels it so as to move it, but the situation concerning what is repelled will be just like what is moved into it, and so, when it is moved, it necessarily follows that the universe is moved. Also, when any given thing is moved violently, there would be a ripple effect such that the universe [itself] would undulate violently imitating the undulation of [that thing].

---

13. Aristotle mentions this example at *Physics* 4.6.213b15–18.

(١٣) قالوا : ونحن نرى إنه مملوءاً من رماد يسع مَلْؤه ماء ، فلولا أنَّ هناك خلاء لا ستحال أن يسع ملؤه ماء . وقالوا أيضاً : والدنّ يُمْلأُ شراباً ، ثم يجعل ذلك الشراب بعينه في زقّ ، ثم يجعلان في ذلك الدنّ بعينه ، فيسع الدنّ الزقّ والشراب معاً . فلولا أنَّ في الشراب خلاء قد انحصر فيه مقدار مساحة الزقّ ، لا ستحال أن يسع الزقَّ والشراب معاً ما كان يملؤه الشراب وحده .

(١٤) وقالوا : إنَّ النامي أيضاً إنما ينمو بنفوذ شيءٍ فيه ، فلا شك أنَّ ذلك الشيء ينفذ لا في الملاء ، بل في الخلاء . وبعضهم جعل هذا الاحتجاج كلّياً فقال : إنَّ المتحرك لا يخلو إما أنْ يتحرك في خلاء أو يتحرك في ملاء ، لكنه إن تحرّك في ملاء دخل ملاء في ملاء ، فبقي أن يتحرك في خلاء . ومن ذلك احتجاجهم بالقارورة التي تُمَصّ ثم تُكبّ على الماء فيدخلها الماء . ولو كانت مملوءة لما وسعتْ شيئاً آخر يدخل فيها .

(١٥) وقالوا أيضاً : إنَّ المتحرك إذا تحرّك فلا يخلو إما أنْ يدفع الملاء فيحرّكه ، وإما أنْ يداخله ؛ لكن المداخلة محالة ، فبقي أنْ يدفعه فيحركه . وكذلك حال المدفوع فيما يتحرك فيه ، فيلزم إذا تحرّك أن يتحرك العالم ، وأن يكون إذا تحرّك متحرّك بعنفٍ أن يتموج العالم تموجاً بعنفٍ ومضاهياً لتموّجه .

(16) Those claiming that whatever the thing is on is *place* draw on [the belief] of the common man, since he calls wherever he sits his *place.* Our concern is not with the one who call this a *place,* but with identifying this place that the placed thing is on, which, in fact, is what contains and is exactly equal [to it] and is necessary for everything that undergoes locomotion, wherever it might be, even if it is not resting on anything supporting it.

(17) As for those who argue that place is the simple [surface], however it might be, they say that, just as the surface of the jar is a place for the water, so likewise is the surface of the water a place for the jar, since a touching surface belongs entirely to any given simple [surface] continuous with it. Also, they say that the outermost celestial sphere is moved. Now, whatever is moved has a place, and so the celestial sphere has a place; however, it does not have a containing limit of some surrounding thing. So, not every place is a containing limit of that which surrounds, but, rather, the place of [the outermost celestial sphere] is the upper surface of the sphere below it.[14] As for those who defend place's being a containing surface, we shall relate the facts about their school of thought and confirm it later;[15] but first we must refute those [other] schools of thought and then follow it up by revealing the errors in their reasoning.

---

14. Cf. Themistius *In Phys.* 121.1–4.
15. See 2.9.1.

(١٦) أما القائلون بأنَّ المكان ما يكون الشيء عليه؛ فيأخذون ذلك من العامة، إذ يسمون مجالسهم أمكنة لهم. ونحن لا نبالي أنْ يسمّي مسمّ هذا مكاناً، لكنا لا نشتغل بتحقيق هذا المكان الذي يكون المتمكن عليه؛ بل الذي قيل إنّه حاوٍ مساوٍ، ولا بُدّ منه لكل منتقلٍ حيث كان، وإنْ لم يكن مستقراً على مستند.

(١٧) وأما القائلون بأنَّ المكان هو البسيط كيف كان؛ فهم يقولون إنّه كما أنَّ سطح الجرّة مكان للماء كذلك سطح الماء مكانٌ للجرّة لأنّه سطحٌ مماس لجملة بسيطٍ متصل به. ويقولون إنَّ الفلك الأعلى متحركٌ، وكل متحرّك فله مكان؛ فالفلك الأعلى له مكان، لكن ليس له نهاية حاوية من محيط. فليس كل مكانٍ هو النهاية الحاوية من المحيط، بل مكانه هو السطح الظاهر من الفلك الذي تحته. وأما القائلون بأنَّ المكان هو السطح الحاوي، فسنذكر مذهبهم ونحقّقه. فيجب أنْ نبدأ أول شيءٍ بإبطال هذه المذاهب، ثم نتبعها بكشف المغالطات في قياساتهم.

Chapter Seven

Refuting the view of those who say that place is matter or form or any indiscriminate contacting surface or an interval

(1) The claim of those who think that the material or form is place is shown to be false in that one knows that the place is left behind when there is motion, whereas the material and form are not left behind. Also, motion is *in* place, whereas motion is not *in* the material and form, but *with* them. Also, motion is *toward* place, while motion is in no way *toward* the material and form. Also, when something is generated, as when water becomes air, its natural place changes, whereas its natural material does not change; and, at the start of the generation, [the generated thing] is *in* the initial place, whereas it is not *in* its form. Also, it is said that the wood *was* a bed, and that vapor was *from*[1] water, and [that] a human was *from* semen, while it is not said that place *was* such-and-such a body or that such-and-such a body was *from* place.

(2) On the view of those who say that place is any simple [surface] that contacts a complete simple [surface], whether as what surrounds or as what is surrounded, it necessarily follows that one body would have two places. In other words, according to their view, the jar must have two places: one place that is the surface of the water that is in it, and another that is the surface of the air that surrounds it. Now, it is known that one

1. For Avicenna's use of the preposition *ʿan* in relation to the material cause, see 1.2.18–19.

<الفصل السابع>

في نقض مذهب من قال إنَّ المكان هيولى أو صورة أو أي سطحٍ ملاق كان أو بُعْداً

(١) أمّا بيان فساد قول مَنْ يرى أنَّ الهيولى أو الصورة مكان ، فبأن يعلم أنَّ المكان يفارق عند الحركة ، والهيولى والصورة لا يفارقان . والمكان تكون الحركة فيه ، والهيولى والصورة لا تكون الحركة فيهما بل معهما . والمكان تكون إليه الحركة ، والهيولى والصورة لا تكون إليهما الحركة البتة . والمتكون إذا تكون استبدل مكانه الطبيعي كالماء إذا صار هواء ولا يستبدل هيولاه الطبيعية ؛ وفي ابتداء الكون يكون في المكان الأول ، ولا يكون في صورته . ويقال إنَّ الخشب كان سريراً ، ويقال عن الماء كان بخاراً ، وعن النطفة كان إنساناً ، ولا يقال إنَّ المكان كان جسم كذا ، ولا عن المكان كان جسم كذا .

(٢) وأمّا القائلون بأنَّ المكان كل بسيطٍ ملاقٍ لبسيطٍ تام كان محيطاً أو كان محاطاً ؛ فيلزمهم أنْ يجعلوا للجسم الواحد مكانين ، وأنه يلزمّ – على مذهبهمّ – أن يكون للجرّة مكانان ، مكان هو سطح الماء الذي فيها ، ومكان هو سطح الهواء المحيط بها . وقد علم

body does not have two places, but that a single placed thing has a single place. They were forced to this position simply because they did not understand the motion of the celestial sphere and supposed that it was local—that is, it is something moved with respect to place—while finding that the outermost body is not in a place that contains from outside. If our view concerning positional motion[2] is recognized, we are spared this inconvenience and saved from this exigency.

(3) As for those who claim that place is the fixed interval between the limits of what contains, we shall single out those of them who deem it absurd that this interval [ever] be devoid of what is placed [in it].[3] This interval must either exist together with the interval belonging to the contained body or not. On the one hand, if it does not exist [together with the interval belonging to the contained body], then a place would not exist together with the thing placed in place, because the placed thing is this contained body and the place is this interval that [is being assumed] to not exist together with the interval of the body. If, on the other hand, it does exist together with it, then one or the other of the following must be the case: It might have an existence that is numerically different from the existence of the interval of the contained body and so is distinct from it, receiving certain properties and accidents that numerically belong to it to the exclusion of those belonging to the interval of the contained body. Alternatively, it might not be different from it and instead might be united with it so as to become identical with it. If, on the one hand, it is different from it, then there is an interval between the limits of what contains, which is place, and another interval in the placed thing that is likewise between the limits of what contains and is numerically

2. For Avicenna's account of motion with respect to position, see 2.3.13–16.
3. The position mentioned is that of John Philoponus; see the note to 2.6.11.

أنَّ الجسم الواحد لا يكون في مكانين، وأنَّ للمتمكن الواحد مكاناً واحداً. وإنّما اضطروا إلى هذا القول بسبب جهلهم بحركة الفلك وظنّهم أنّها مكانية، ووجودهم الجرم الأقصى لا في مكانٍ حاوٍ من خارجٍ وهو متحرّك حركة مكانية. وإذا عُلم مذهبنا في الحركة الوضعية، اُستغني عن هذه الكُلْفة وتخلّص عن هذه الضرورة.

(٣) وأما القائلون بأنَّ المكان هو البُعد الثابت بين أطراف الحاوي؛ فنخصّ الذين يُحيلون منهم خلو هذا البُعد عن المتمكن، أنَّ هذا البُعد لا يخلو إما أنْ يكون موجوداً مع البُعد الذي للجسم المحوي عليه، أو لا يكون موجوداً؛ فإنْ لم يكن موجوداً فليس مع وجود المتمكن في المكان مكانٌ، لأنَّ المتمكن هو هذا الجسم المحوي، والمكان هو هذا البُعد الذي لا يوجد مع بُعْد الجسم. وإنْ كان موجوداً معه؛ فلا يخلو إمّا أنْ يكون له وجود هو غير وجود بُعْد الجسم المحوي بالعدد، فهو ممايز له، يقبل خواص وأعراضاً هي بالعدد أعراضاً له من دون التي لبُعد الجسم المحوي. وإمّا أنْ لا يكون غيره، بل يتحد بل فيصير هو هو وإنْ كان غيره. فهناك بُعْد بين أطراف الحاوي هو مكانٌ وبُعْدٌ آخر في المتمكن أيضاً هو بين أطراف الحاوي، غير ذلك بالعدد. ولكن معنى قولنا البُعْد الشخصي الذي بين

different from the former. Our saying "the individual interval that is between these two things" means that this [interval] is the continuous thing that is susceptible to division between the two [limits] in a particular way so as to be pointed to, and so whatever is between *this* limit and *that* limit is this interval that is between the two. Now, whatever is this interval that is between the two definite limits is necessarily a single individual, not something different, and so whatever is between *this* limit and *that* limit is a single individual interval, not one interval and another. Consequently, between *this* limit and *that* limit there does not exist one interval belonging to the body and also some other interval; however, the interval that belongs to the body between the two limits does exist, and so the "other" interval, the latter, does not exist. If, on the other hand, [the purported interval identified with *place*] is identical with [the interval of the contained body], then there is no interval but this one; and similarly, when some other body replaces it, there would be no interval but that which belongs to the other body. So there simply does not exist between the limits of what contains any other interval than the interval of what is contained, but they do not believe that it is possible for it to be wholly devoid of what is placed [in it]. So, then, the separate interval exists only in the estimative faculty's imagining certain absurdities, similar to its imagining that that containing body remains [something that contains] without some of the internal limits being covered by others nor having any body in it. This is just like one who says that when our estimative faculty imagines that five has been divided into two equal parts, the odd in that case has been increased by one unit.[4] So, when this necessarily follows on the estimative faculty's imagining some absurdity, then it need have no reality in existence.

4. It was a standard conception among ancient and medieval mathematicians that the *even* is what can be divided into two equal parts without remainder, whereas the *odd* is that which cannot be equally divided, but always has a unit left over; see, for instance, Euclid, *Elements* 7, defns. 6–7. Thus, to imagine that five has been divided into two equal parts is tantamount to imagining that what is odd is simultaneously even—an obvious absurdity.

هذين الشيئين هو أنّه هذا الأمر المتصل الذي يقبل بينهما القسمة الواحدة المشار إليها ، فكل ما بين هذا الطرف وهذا الطرف هو هذا البُعْد الذي بين الطرفين . وكل ما هو هذا البُعْد الذي بين الطرفين المحدودين ، فهو لا محالة واحد شخصي لا غير ، فيكون كل ما بين هذا الطرف وهذا الطرف بُعْداً شخصياً واحداً ليس بُعْداً وبُعْداً آخر . وإذا كان كذلك ؛ لم يكن بين هذا الطرف وهذا الطرف بُعْدٌ للجسم وبُعْد آخر . لكن البُعْد الذي للجسم بين الطرفين موجودٌ ، فالبُعْد الآخر ليس بموجودٍ . هذا فأمّا إنْ كان هو هو فليس هناك بُعْدٌ إلاّ هذا . وكذلك إذا تعقّبه جسم آخر لم يكن هناك بُعْدٌ إلاّ الذي للجسم الآخر ، فلا يوجد البتة بين أطراف الحاوي بُعْدٌ هو غير بُعْد المحوي ، ولا يجوز عندهم خلوّه البتة عن المتمكن . فإذن لا يوجد البُعْد المفرد إلاّ في توهم محالات ؛ مثل أنْ يتوهم أنْ يبقى ذلك الجسم الحاوي غير منطبق النهايات الداخلة بعضها على بعضٍ ولا جسم فيه . وهذا كمَنْ يقول إذا توهمنا الخمسة منقسمة بمتساويين فيكون حينئذ زائداً على الفرد بواحدٍ ، فليس يجب ، إذا لزم هذا ، عن توهم محالٍ هناك أنْ تكون له حقيقة في الوجود .

(4) Also, how could it even be possible for the two intervals to be together when it is obvious that two intervals are greater than one interval, because (and for no other reason than) they are *two* and a composite? Now, the whole of any composite interval is greater than one interval, and is thus greater than it because the *greater* is that which is quantitatively more than some other by a number that exceeds the other; and the greater with respect to magnitudes is like the many with respect to number, where whatever is quantitatively more with respect to magnitudes is greater. So, when one interval enters into another, either the interval that is entered into ceases to exist, such that an existing interval has entered into a nonexisting one, or it remains and the one interpenetrates it, forming a composite that is greater than either one of them, such that the two intervals would be greater than a single one. The situation is not like that, however, since the composite of the two would be that which is between the limits, and that is the exact amount of each one of them; and so the composite would not be greater than either one [taken individually].

(5) Here, one might ask about the state of the line when it is folded such that half of it is superimposed upon the other half, in which case there would be two lines, the composite of which does not exceed in length the length of either one of them. This, however, is absurd, because either one of the following cases must hold: On the one hand, each half might have a distinct position from the other, in which case the composite of the two lines would make an interval different from and bigger than the interval of either one of them, whereas if it is not [folded] rectilinearly, then it would not have been folded [as required], nor would there be one interval derived from the composite of both, but, instead, one interval would be distinct from another.[5] On the other hand, the two might unite so as to form a single line (if that is possible), in which case there would not be two lines, but only one.

---

5. The two scenarios Avicenna seems to have in mind are (1) one line is lying on top of another similar to an equal sign (=), in which case the composite is greater, i.e., thicker, than any single line in the composite, or (2) rather than being folded, the line is bent like a horseshoe (⊃), in which case not only is it not folded as instructed, but also there is no actual composite of the two.

(٤) كيف يمكن أنْ يكون بُعْدان معاً؟ ومن البيّن أنَّ كل بُعْدين اثنين أكثر من بُعْدٍ واحدٍ؛ لأنهما إثنان ومجموع لا لأجل شيء آخر. وكل مجموع بُعْدٍ أكبر من بُعْدٍ؛ فهو أعظم منه، لأنَّ العظيم هو الذي يزيد على القدر بعددٍ خارج عن الشيء. والعظيم في المقادير كالكثير في العدد، وكل ما هو أكبر في المقادير قدراً فهو أعظم. فإذا كان بُعْدٌ يدخل في بُعْدٍ فإمّا أنْ يعدم البُعْد المدخول فيه؛ فيكون قد دخل بُعْدٌ موجودٌ في معدوم، وإمّا أنْ يبقى هو والداخل فيه مجموعين أعظم من واحدٍ منهما، فيكون البُعْدان أعظم من واحدٍ، وليس الأمر كذلك؛ لأنَّ مجموعهما هو الذي بين النهايات، وذلك بعينه قدر كل واحدٍ منهما؛ فليس المجموع أعظم من الواحد.

(٥) ولسائلٍ أنْ يسأل ها هنا حال الخط إذا عطف حتى لزم نصفُه نصفَه فيكون خطّان، ومجموعهما في الطول لا يزيد على طول واحد منهما، لكنْ هذا محالٌ؛ لأنّه لا يخلو إمّا أنْ يتميّز كل نصفٍ عن الآخر في الوضع، فيكون مجموع الخطّين يفعل بُعْداً غير بُعْد واحدٍ منهما وأكبر منه – وإنْ كان ليس على الاستقامة لم يكن الإنعطاف، ولا يكون البُعْد الواحد متناولاً لمجموعهما بل يتميّز بُعْد وبُعْد – وإمّا أنْ يتحدا خطّاً واحداً إنْ أمكن ذلك، فحينئذ لا يكون خطّان، بل خطٌ واحد.

(6) [In contrast with lines, however,] bodies are precluded from interpenetrating.[6] Now, what precludes the bodies from doing that is not that there is included in that body some set of forms, qualities, or the like comprising the body as such; for whichever forms or qualities you care to take, were they not to exist, while the body is assumed to exist, the interpenetration would still be impossible. Moreover, the material does not preclude something's interpenetrating some numerically different material.[7] That is because when we say that the material is precluded from interpenetrating some other material, this might be taken in either one of two ways: On the one hand, this might be by way of negation, like saying that sound is not seen and that the soul does not interpenetrate motion, since neither one of these is characterized as being with the other such that the estimative faculty imagines that there is interpenetration. On the other hand, [the material's precluding interpenetration] might not be like this, but in the sense that opposes interpenetration as a proper opposite. For, just as the meaning of *interpenetration* is that anything you take from one of two [interpenetrating] things, you find locally with it something of the other (since one is not locally separate from the other), so that which opposes it is that *this* very thing is locally distinguished from *that* very thing, and so its parts are taken to be distinct from the parts of that one.

(7) If the claim is that the material precludes interpenetration in the negative sense (that is, the first [sense]), then our discussion is not about it, and it is conceded that, in itself, the material does have this description. Our discussion instead concerns the second option. Now, that second option is inconceivable with respect to the material unless [the material] itself is assigned a location; but that happens only accidentally on account of the interval it happens to have, in which case it is [the interval] that resists being partitioned and divided. So the material is so disposed as to bear this opposite (that is, interpenetration), whereas

6. For a detailed discussion of Avicenna's argument against the interpenetration of bodies, see Jon McGinnis, "A Penetrating Question in the History of Ideas: Space, Dimensionality and Interpenetration in the Thought of Avicenna," *Arabic Sciences and Philosophy* 16 (2006): 47–69.

7. Cf. John Philoponus, *In Phys.* 559.9–18, who argues that it is the materiality that precludes two bodies from interpenetrating.

(٦) والأجسام التي تمتنع عن التداخل، ليس الذي منع ذلك من هذا الجسم أنْ يدخل في ذلك الجسم جملة ما يشتمل عليه الجسم من الصورة والكيفيات وغير ذلك. فإنَّ الصورة والكيفيات أيّها فُرضتْ، لو لم تكن وفُرض الجسم موجوداً، كان التداخل ممتنعاً أيضاً، وليس الهيولى هي التي تمتنع عن مداخلة هيولى أخرى بالعدد. وذلك أنّا إذا قلنا إنَّ الهيولى تمتنع عن مداخلة هيولى أخرى، إمّا أنْ يكون هذا على سبيل السلب كقولنا: إنَّ الصوت لا يرى، بل كما نقول إنَّ النفس لا تداخل الحركة؛ إذ ليس من شأن كل واحد منهما أنْ يكون مع الآخر بحيث يتوهم عليه المداخلة. وإمّا أنْ لا يكون على هذا، بل على المعنى الذي يقابل المداخلة مقابلة خاصيّة. فإنّه، كما أنَّ معنى المداخلة هو أنْ يكون أيّ شيءٍ أخذت من أحد الأمرين، نجد معه في الوضع شيئاً من الآخر؛ إذ لا ينفرد أحدهما بوضع عن الآخر، فالذي يقابله هو أنْ يكون ذات هذا متميّزاً في الوضع عن ذات ذلك، فتؤخذ أجزاؤه مباينة لأجزاء ذلك.

(٧) فإنْ قيل إنَّ الهيولى يمتنع عليها التداخلّ – بمعنى السلب الذي هو المعنى الأولّ – فليس كلامنا في ذلك، وذلك مسلّم أنَّ الهيولى في نفسها بهذه الصفة. ولكنْ كلامنا في القسم الثاني؛ وذلك القسم الثاني لا يتصور في الهيولى إلاّ أنْ تُجعل ذات وضع، ولا تصير كذلك إلاّ بالعرض، بسبب البُعْد الذي يعرض لها، فحينئذ تتعرض للتجزؤ والإنقسام. فيكون استعداد الهيولى لأنْ تُحمل عليها هذه المقابلة، وهي التداخل وغير

the other opposite ([that is,] noninterpenetration), follows as a concomitant of the interval. The interval causes this opposite to follow as a concomitant of [the material] and be conceptualized as such. It is because of the interval that the material does not interpenetrate other material (even if the interval is the sort of thing to which that might belong), while it is not in the nature of the material alone to be an obstacle that opposes interpenetration and so to preclude the material's being something that interpenetrates. How could this material that possesses the interval in such as way as not to preclude itself from having the corporeal interval possibly preclude that very [interval] from receiving another corporeal interval? The material has nothing to do with whether something is insusceptible to the nature of the interval when it encounters it, nor has it anything to do with whether it is insusceptible to a given interval and increase, whereas it will be revealed that [the material] is susceptible to rarefaction once we investigate and confirm that.[8] So, if the interval in itself does not preclude the interpenetration of some other interval, whereas the material is what is prepared such that it does receive the interval, and it is not in its nature *qua* material that it be unique to one space such that it would oppose interpenetration,[9] then the interpenetration of two bodies must be possible. So, if there is something [that is] composed of two things and [that] is itself nothing but the composite of the two (without there occurring a certain alteration and affection [so as to produce] a third form or factor different from the two), then, when each one of them is judged to be possible, the whole is [also judged] to be possible, and when one after another is not prevented, the whole is not prevented. The whole body, however, is precluded from interpenetrating another body! So what prevents that is owing to its parts. It is not, however, that every part of it prevents that, since the material is not a cause preventing that, whether by some proper action or affection. So it remains that it is the nature of the interval that does not suffer interpenetration; and if, additionally, the material informed by the interval cannot interpenetrate the interval, then the body cannot enter into an interval at all.

---

8. The reference is to *Kitāb fī al-kawn wa-l-fasād,* although there is also a brief discussion at 2.9.17 and 20–21.

9. Reading *al-mudākhilah* with **Z**, **T**, and the Latin (*infusio*), which **Y** (inadvertently) omits.

التداخـل المقابـل له أمراً يلحقها من البُعْد ؛ والبُعْد هو السـبب في أنْ تلحقها هذه المقابلة وتصور فيها ، وهو السبب في أنْ صارت الهيولى لا تداخل الهيولى الأخرى لأجل البُعْد ، وأنْ كان البُعْـد جائزاً له ذلك . وليـس في طبيعة الهيولى وحدها مَنْعٌ يقابل المداخلة ، فلا يمتنع على الهيولى المداخلة . وكيف يمكن أنْ تمانع هذه الهيولى ذات البُعْد لنفسها لا لامتناع البُعْد الجسـماني ، أنْ تلقى ذاته البُعْد الجسماني الآخر؟ وليست الهيولى ممّا لا يقبل طبيعة البُعْد ويلاقيه ، ولا أيضاً ممّا لا يقبل بُعْداً أو زيادةً ؛ ويكشـف قبولها التخلخل ، وذلك حين نحقّقه ونصحّحه . فإنْ كان البُعْد لا يمتنع عن مداخلة بُعْدٍ آخر في نفسه ، والهيولى مستعدة لأن يلقاها البُعْد ، وليس في طباعها - بما هو هيولىّ - أنْ تنفرد بحيّز فتتقابل المداخلة ، فواجب أن يكون التداخل في الجسمين جائزا . فإن كلّ مؤلف من شيئين ، ليس المؤلف إلاّ نفس مؤلفهما ، من غير أنْ حدث هناك اسـتحالةٌ وانفعالٌ ، هي صورةٌ ثالثة ومعنى ثالث غيرهمـا . فإنَّ الحكم إذا كان جائزاً على كل واحدٍ منهمـا كان جائزاً على الجملة ، وإذا لم يمنعه واحد واحد منهما لم تمنعه الجملة . لكن جملة الجسـم تمانع مداخلة جسمٍ آخر ، بسـبب أنَّ في أجزائه ما يمنع ذلك ، وأنّه ليس كل جزءٍ منه غير مانع لذلك ، إذ ليسـت الهيولى سبباً يمنع ذلك ، ولا بسبب فعلٍ خاص وانفعالٍ خاص . فبقيَ أنْ تكون طبيعة البُعْد لا تحتمل التداخل ، فإنْ كان مع ذلك يجب للهيولى المتصورة بالبُعْد أنْ لا تُداخل البُعْد ، لم يجزْ أنْ يدخل الجسم في بُعْدٍ البتة .

(8) Moreover, the matter and material of the thing placed in the container that has been filled must either encounter that naturally disposed interval or not. Now, on the one hand, if [that interval] remains independent and separate from [the material], then the body possessing the material will have neither filled the container nor entered into it. [That is] because that naturally disposed interval is something that subsists independently, which does not encounter the matter of the body entering into it, whereas the body entering into it is itself never devoid of its matter. On the other hand, if that interval permeates the matter itself along with the interval in the matter, then two equal intervals agreeing in nature will have permeated the matter. It is well known, however, that things agreeing in [their] natures that are not divided into species by differences in their substance do not make up a multiplicity of individuals except through the multiplicity of the matters that underlie them, whereas, when there is only one matter for [the nature], there simply is no multiplicity, and so there will not be *two* intervals. Also, were we to assume that in the matter the interval has become many, then there would be two intervals; but, then, which intervallic property would the matter have owing to one of the intervals that permeates it, and which other property would it have owing to the other interval that permeates it? [That] is because we find in the matter one [intervallic property] that corresponds with the continuous and another that corresponds with being divisible; and, accordingly, were there but one interval in [the matter], then the form would be that form.

(9) So this is what we have to say in refuting the existence of this naturally disposed interval. During the refutation of that, something was said that will provide a basis for [showing that] it is impossible for intervals to exist within intervals *ad infinitum*. At this point, however, we have not reached a full understanding of that according to a true sense that commands confidence; but later, we, or someone else, will reach it.[10]

10. The reference may be to his doctrine expounded in 3.2.8 and elsewhere that within a continuous magnitude, there is not an infinite number of actual, potential, or even latent half-intervals; rather, the potentially infinite divisibility of a continuous magnitude refers to the fact that nothing in the material precludes the estimative faculty from imagining as many divisions and as small an interval as one wants.

(٨) ثم لا يخلو؛ إذا كان المتمكن في الإناء قد ملأه من أنْ تلقى مادته وهيولاه ذلك البُعْد المفطور أو لا يلاقيها، فإنْ انفرد عنها وفارقها فلا يكون الجسم ذو الهيولى قد ملأ الإناء ولا دخل فيه، إذ يكون ذلك البُعْد المفطور قائماً على حياله، ليس ملاقياً لمادة الجسم الداخل فيه. والجسم الداخل فيه لا تكون ذاته خالية عن مادته؛ وإنْ سرى ذلك البُعْد في ذات المادة مع البُعْد الذي في المادة، فتكون المادة قد سرى فيها بُعْدان متساويان متفقا الطبيعة. وقد عُلم أنَّ الأمور المتفقة في الطباع، التي لا تتنوع بفصولٍ في جوهرها، لا تتكّثر في هوياتها إنّما تتكّثر بتكّثر المواد التي تحملها، وإذا كانت المادة لها واحدة لم تتكّثر البتة، فلا يكون بُعْدان. ولو أنّا فرضنا البُعْد قد تكّثر في المادّة لكان فيها بُعْدان. فأيّة خاصية بُعْدية تكون للمادة بسبب سريان أحد البُعْدين فيها؟ وأيّة خاصية أخرى تكون لها بسريان البُعْد الآخر فيها؟ فإنّا لا نجد في المادة إلاّ نحواً من الاتصال واحداً ونحواً من الانقسام واحداً، وعلى ما لو كان فيها بُعْد واحد فقط، لكانت الصورة تلك الصورة.

(٩) فهذا ما نقوله في إبطال وجود هذا البُعْد المفطور. وقد قيل في إبطال ذلك شيء مبني على استحالة وجود أبعادٍ في أبعادٍ بلا نهاية، ونحن لم نحصل إلى هذه الغاية فهم ذلك على حقيقةٍ يُوجب الركون إليها، وسندركها بعد؛ أو يدركها غيرنا.

Chapter Eight

The inconsistency of those who defend the void

(1) The first thing we must do is to explain to the proponents of the void that it is not absolutely nothing, as many people suppose and imagine. Indeed, if the void is simply nothing, there would be no dispute between them and us. So let the void be nothing that has any determinate reality, and we shall happily grant them this.[1] The descriptions applied to the void, however, demand that it be some existing thing—namely, that it is a certain quantity, and that it is a certain substance, and that it has a certain active power. [That is] because *nothing* cannot exist between two things to a greater and lesser extent, whereas the void might exist between two bodies to a greater and lesser extent, for the void measured between Heaven and Earth is greater than that occurring between two cities on the Earth. In fact, it will have a certain ratio, and, indeed, each one of the two will exist as some measured distance having a magnitude, such that one void would be a thousand cubits and the other ten. Also, one void would terminate at a certain occupied place, while the other would go on infinitely. These states simply cannot be predicated of pure nothing.

(2) Now, since [the void] has these properties, and these properties essentially belong to quantity (and, by means of quantity, to whatever else [has them]), the void must have them either primarily and essentially or accidentally. If it has them essentially, then it is quantity [which will

1. For a discussion of Avicenna's argument, which runs through paragraphs 1–4, see Jon McGinnis, "Logic and Science: The Role of Genus and Difference in Avicenna's Logic, Science, and Natural Philosophy," *Documenti e Studi* 18 (2007): 165–86, esp. §4.

<الفصل الثامن>

في مناقضة القائلين بالخلاء

(١) وأمّا القائلون بالخلاء؛ فأول ما يجب علينا هو أنْ نعرّفهم أنَّ الخلاء ليس لا شيء مطلقاً، كما يظنّ ويتوهم قومٌ كثير، وأنّه إنْ كان الخلاء لا شيء البتة؛ فليس ها هنا منازعة بيننا وبينهم، فليكن الخلاء لا شيئاً حاصلاً؛ ولنسلّم هذا لهم. لكن الصفات التي يصفون بها الخلاء تُوجب أنْ يكون الخلاء شيئاً موجوداً، وأنْ يكون كمّاً، وأنْ يكون جوهراً، وأنْ تكون له قوة فعّالة، فإنَّ اللاشيء لا يجوز أنْ يكون بين شيئين أقل وأكثر، والخلاء قد يكون بين جسمين أقل وأكثر. فإنَّ الخلاء المقدّر بين السماء والأرض أكثر من المتحصل بين بلدين في الأرض، بل له إليه نسبة ما، بل وكل منهما يوجد ممسوحاً مقدّراً لمقدارٍ؛ فيكون خلاء ألف ذراع وخلاء آخر عشرة أذرع، وخلاء يتناهى إلى ملاء، وخلاء يذهب إلى غير نهاية. وهذه الأحوال لا تحمل البتة على اللاشيء الصرف.

(٢) ولأنّه يقبل هذه الخواص، وهذه الخواص بذاتها للكمّ وبتوسط الكم ما يكون لغيرهّ - فلا يخلو إمّا أنْ يقبلها الخلاء قبولاً أولياً بالذات، أو قبولاً بالعرض. فإنْ كان قَبلَها

be considered shortly]. If it has them accidentally, then it is something possessing a quantity, doing so as either a substance or an accident. Now, the accident possesses a quantity only because it exists in a substance that possesses a quantity; and so the void would need to be essentially joined to a substance and a quantity, where that quantity would be nothing but the continuous quantity that is divisible in three dimensions. Now, if the substance and the quantity [together] internally constitute it, and every substance having this description [namely, having three dimensions] is a body,[2] then the void will be a body. If the two are joined to it from without and do not constitute it, then the states of [the void] reduce to that of an accident in a body; but a body does not enter into the accident in the body, and so a body would not enter into the void. On the other hand, if that [namely, the aforementioned properties] belongs to it essentially, then it must be essentially quantity. Now, it belongs to the nature of what is essentially a three-dimensional extended quantity that it be impressed into matter and that it be either a part or a configuration of the sensible body. If it is not impressed into the matter, then [its not being so] is not because it is a quantity, but because of some accidental factor. Now, that accidental factor must be of the character that it either subsists without being in a subject or not. On the one hand, if it is of the character that it subsists without being in a subject[3] while being joined to the interval, then this interval inherently subsists as joined to something else that subsists without being in a subject. So that to which [the void] is joined and by which it subsists (namely, what subsists in itself) is a subject by which the void's interval subsists. So the subject of the interval will, in fact, be nothing but a thing that, in

2. Cf. 1.2.2.

3. Reading with **Z**, **T**, and the corresponding Latin *aw yakūnu laysa min sha'nihi dhālika. Fa-in kāna min sha'nihi an yuqūma lā fī mawḍū'* (. . . or not. On the one hand, if it is of the character that it subsists without being in a subject), which **Y** (inadvertently) omits.

بالذات فهو كمٌّ، وإنْ كان قبلها بالعرض فهو شيء ذو كمّ؛ إمّا عرضٌ ذو كمّ، وإمّا جوهرٌ ذو كمّ. والعرض لا يكون ذا كمّ إلاّ لوجوده في جوهرٍ ذي كمّ، فيلزم أنْ يكون الخلاء ذاتاً مقارنة لجوهرٍ وكمّ، وليس ذلك الكمّ إلاّ الكمّ المتصل القابل للقسمة في الأقطار الثلاثة. فإنْ كان كل واحدٍ من الجوهر والكمّ داخلاً في تقويمه - وكل جوهرٍ بهذه الصفة - فهو جسم؛ فالخلاء جسم. وإن كانا مقارنين له من خارج غير مقومين له، فأقلّ أحواله أنْ يكون عرضاً في جسمٍ، والعرض في الجسم لا يدخله جسم، فالخلاء لا يدخله جسم. وإنْ كان قَبِلَ ذلك بالذات فهو لا محالة كمٌّ بالذات، ومن طباع الكمّ بالذات، الذي له ذهاب في الأبعاد الثلاثة، أنْ تنطبع به المادة وأنْ يكون جزءاً أو هيئة للجسم المحسوس. فإنْ لم تنطبع به المادة فلا يكون لأنّه كمٌّ؛ بل لأمرٍ عارض، وذلك العارض لا يخلو إمّا أنْ يكون من شأنه أنْ يقوم لا في موضوع أو يكون ليس من شأنه ذلك. فإن كان من شأنه أن يقوم لا في موضوع وقد قارن البُعْد؛ فهذا البُعْد لا يخرج عن أنْ يقوّم مقارناً لقائم لا في موضوع غيره. فما يقارنه البُعْد ويقوم بهّ - وهو قائم في نفسهّ - فهو موضوع يقوم به بعد الخلاء، فإنّ الموضوع للبُعْد ليس إلاّ شيئاً هو في نفسه لا في موضوع ويقارنه بُعْدٌ ويكمّمه. وإنْ

itself, is not in a subject and with which a certain interval is conjoined and that gives to [the interval] a certain quantity. On the other hand, if that [accidental] factor is of the character that it subsists with a subject, then the only existence it and whatever accompanies it will have is in a subject. So, then, how does the interval come to be something that subsists without a subject, when it needs a subject? If it is said that its subject is the void and that when it is in its subject, it makes its subject be without a subject, the sense of the claim is that what has no subsistence in itself is accidental to what has no subsistence in itself except in a subject, and then it makes it subsist in itself without a subject. Now one of the things would be in its nature an accident, while accidentally being a substance, and so the substantiality would be something accidental to one of the natures—which is impossible, as will become clear, particularly in First Philosophy.[4] In short, the indicated interval admits of both situations,[5] [and yet] it is numerically one nature and so is itself ordered only to one genus, in which case that nature [must] fall under either what exists in a subject or what does not.

(3) Moreover, if [that nature] is sometimes in itself a substance and at other times in itself a nonsubstance, then, when it becomes a nonsubstance, it itself will have been corrupted absolutely such that its highest genus—namely, substantiality—will have passed away. In that case, [that nature itself] will no longer remain, for if its species below its highest genus were corrupted, its substance would no longer remain. So, when its highest genus is corrupted, how can you think that its specific nature, by which it is a substance, will remain? If this factor underlying the interval is something inseparable that does not cease, then either it is inseparable

4. Cf. *Ilāhīyāt* 2.1.
5. That is, it can either exist in a subject or not exist in a subject.

كان ليس من شـأن ذلك المعنى أنْ يقوم لا في موضوع؛ فيكون لا وجود له مع ما هو معه إلّا فـي موضوع. فكيف يصير البُعْـد قائماً لا في موضوع وهو يحتاج إلى موضوع؟ فإنْ قيل إنَّ موضوعه هو البُعْد، وأنّه إذا حصل في موضوعه جعل موضوعه لا في موضوع؛ كان معنى هذا الكلام أنَّ ما لا قوام له بنفسه يعرض لما لا قوام له في نفسه إلّا في موضوع، فيجعله قائماً في نفسـه لا في موضوع. ويكون بعض الأشـياء هـو في طبيعته عرض، ويعرض له أنْ يكون جوهراً، فتكون الجوهرية ممّآ يعرض لبعض الطباع وهذا الاسـتحالة، وخصوصاً في الفلسفة الأولى. وبالجملة فإنَّ البُعْد المشار إليه؛ القابل للأمرين، هو طبيعة واحدة بالعدد، فلا تترتب هي بعينها إلّا في جنسٍ واحدٍ، فتكون تلك الطبيعة إمّا تحت ما وجوده في موضوع، أو تحت ما وجوده لا في موضوع.

(٣) وأيضـاً إنْ كانت تـارة هي بعينها جوهراً، وتارة هـي بعينها لا جوهراً، فإذا صارت لا جوهراً فقد فسـدت منها ذاتها فسـاداً مطلقاً، حتى زال أعلى أجناسها وهي الجوهرية، فلا تكون باقية لا محالة. فإنّها لو كان يفسد نوعها دون جنسها الأعلى، لكان جوهرها لا يبقى؛ فكيف إذا فسـد جنسـها الأعلى، فترى تبقـى نوعيتها التي هي بها جوهـر؟ وأمّا إنْ كان هذا المعنى الموضوع للبُعْـد ملازماً غير زائلٍ، فلا يخلو إمّا أنْ يلزم

from the void because it is a certain dimensionally extended interval (in which case every interval will be separate from matter) or it is inseparable from it because of a certain thing upon which it follows as a consequence that an interval is dimensionally extended (in which case the account about that thing will be the same as the former account, going on *ad infinitum*). Also, this consequence [namely, that an interval is dimensionally extended] will be unlike the species difference belonging to a genus, since (given that the nature of the interval is such as to be divided into three dimensions) the specific nature [of being dimensionally extended] will [equally] belong to magnitude, as well as the natures of a line and a surface. [That is] because the distinction between the specific nature upon which accidents follow and the generic nature upon which differences follow is that the generic nature is divided into different [species] by differences that follow upon the [specific] nature as such. Even when they do not follow as consequences, the intellect needs them to follow at least to the extent that they are perfectly conceptualized in the intellect and so might be given some determinate existence in [the intellect]. In short, on the grounds that [something] is, it should have some difference. So, when it is said that there is an absolute void—that is, some indeterminate thing susceptible to continuous division—the difference upon which this follows as a consequence is that (whether in one, two, or all directions) there should be some difference that qualifies the intelligible concept of the interval and gives it some determinate existence in reality and in the intellect, and that the intellect needs in order to determine whether it is some existing thing or [simply] a vain intelligible.

الخلاء لأجل أنه بُعْدٌ ذاهبٌ في الأقطار ، فيكون كل بُعْدٍ مفارقاً للمادة أو يلزمه لمعنى يلحقه بعد كونه بُعْداً ذاهباً في الأقطار ، ويكون الكلام في ذلك المعنى هو ذلك الكلام بعينه ؛ ويذهب إلى غير نهاية . وليس هذا اللحوق ، كلحوق المعنى الفصلي للمعنى الجنسي ، إذ طبيعة البُعْدّ - إذا كان بحيث ينقسم في الأبعاد الثلاثة - فهي طبيعة نوعية للمقدار . وكذلك طبيعة الخط ، وكذلك طبيعة السطح ؛ لأنَّ التمييز بين الطبيعة النوعية على ما يلحقها من العوارض ، والجنسية على ما يلحقها من الفصول ‹هو› أنَّ الطبيعة الجنسية تنفصل بفصولٍ تلحق الطبيعة بما هي . وإذا لم تلحق يكون العقل مقتضياً للحوقها حتى يُستكمل في العقل تصوّرها ، ويجوز عنده تحصيل وجودها . وبالجملة قد يكون فصلاً له لأنّه هو ؛ فإنّه إذا قيل بُعْدٌ مطلقاً ، أي أمر يقبل الانقسام المتصل بلا تحصيل ؛ كان الفصل الذي يلحق هذا أنّه في جهةٍ أو جهتين أو جميع الجهات فصلاً يكيّف المعنى المعقول من البُعْد ويحصلّه مقرّراً في الوجود وفي العقل ، ويفتقر إليه العقل في تحصيله موجوداً أو معقولاً مفروغاً عنه .

(4) As for part of the interval's coming together with white or black and part of it needing matter while another part subsists without matter, [such considerations] neither qualify its intervallic nature nor does it need them in order to be an interval and subsist. In fact, they follow upon it inasmuch as [the interval] is in matter, and, inasmuch as it exists, its existence is qualified by some external factor. Species differences are those things by which something's essence is qualified, whether it is assumed to exist in concrete particulars or that [assumption] is not taken into account. (The complete scientific account of this is given in another discipline.)[6] In fact, the nature of the interval is completed in[7] its essence as an interval in that it determinately has some manner of division and extension. Everything else simply are concomitants that follow upon it, which we do not need in order to establish that there is a given interval that is correctly assumed to exist. [Other than the interval's having some manner of division and extension,] the intellect [simply] does not require anything else upon which [the interval] follows as a consequence so as to believe that the interval determinately exists—[unlike] the way that [the intellect], when it believes that color or animal exists, requires that there needs to be some species in a certain state and having some description in order that [either color or animal] exist. Therefore, the intellect cannot allow that the true difference is absent from the species while its share of its genus remains, which was explained elsewhere.[8] Consequently, this differentiation between an interval that is in matter and one that is not in matter is not such as to

6. The reference may be to *Kitāb al-madkhal* 1.2, if Avicenna means that the existence of essences is assumed to be either in concrete particulars or in conceptualization. Alternatively, if he is referring to the role of species differences, the reference may be to *Kitāb al-madkhal* 1.13.

7. Reading *fī* with **Z**, **T**, and the Latin (*in*), which is (inadvertently?) omitted in **Y**.

8. Cf. *Kitāb al-burhān,* 2.10, where Avicenna argues that the existence of the genus is simply the existence of its species.

(٤) وأمّا كون البُعْد بعضه ملاقياً للبياض أو السواد، وكون بعضه ملازماً للمادة وبعضه قائماً بلا مادة، فليس يكيّف بُعْديته ولا يُحتاج إليها في تحصيل أنّه بُعْدٌ وتقويمه، بل هي أمورٌ تلحقه من حيث هو في مادة، ومن حيث وجوده، ويكيّف وجوده أمرٌ من خارج. والفصول هي التي تتكيّف بها ماهية الشيء سواء فرض موجوداً في الأعيان أو لم يلتفت إلى ذلك؛ وهذا العلم يستتم من صناعةٍ أخرى. بل طبيعة البُعْد تستتم بُعْداً في ماهيته بأنْ يكون له نحوٌ من أنحاء الانقسام والامتداد محصلاً، ويكون ما سواه لواحق تلحقه لا نحتاج إليها في تقرير كونه بُعْداً ما يصح أنْ يفرض موجوداً. ولا يقتضي العقل لحوق شيء آخر به يجعله محصل البُعْد؛ كما يقتضي إذا جُعل اللون موجوداً أو الحيوان موجوداً أنْ يكون صار بحالٍ ووصفٍ نوعاً متى وجد، لذلك لا يجوّز العقل أنْ يكون الفصل الحقيقي يبطل عن النوع وتبقى حصة جنسه له، وهذا موضحٌ في مواضع أُخر. وإذا كان كذلك، فلا يكون هذا الانفصال بين بُعْدٍ في مادة وبُعْدٍ لا في مادة انفصالاً بفصلٍ

be through a species-making difference, but is a differentiation by certain accidents necessarily external to the subsistence of the interval's nature as a species. Now, it is possible for the estimative faculty to imagine that each one of the things sharing the same nature has the accident that belongs to the other, even though that is almost always impossible owing to a certain obstacle, a [difference of] time, or some other external cause.

(5) It would seem that we just now got carried away with something other than our intended topic of discussion when there is a way more in keeping with the arguments of natural philosophy. So we say: If there is a separate interval [namely, a void], then it must be either finite or infinite. In the opinion of all of those who require the existence of the void, however, its nature is such that it is not finite save [where it terminates] at the interval of some plenum, and that, if the plenum is finite, then it similarly terminates at the void. According to their view, it would necessarily follow that an infinite interval is either a void alone, or a plenum alone (which limits the void), or a composition of void and plenum. Now, it is absurd that there be an infinite interval having this description, as we shall explain soon,[9] and so it is absurd that there be a void according to what they say.

(6) Moreover, if there were a void, then necessarily either the plenum would enter into it, or it would not. On the one hand, if the plenum enters it, then either the void interval continues to exist simultaneously with the interpenetration, or it ceases to exist. Now, if it ceases to exist, it cannot be called a *place;* rather, the place is the void surrounding the body that is joined with it. That is because it is *in* that only, since the void interval between that has ceased to exist. Moreover, [the place] is not

9. See 3.8.

منوّع؛ بل انفصالاً بأعراض لازمة خارجةً عن تقويم طبيعة البُعْد نوعاً . والأشـياء المتفقة بالطبيعة لا يستحيل أنْ يتوهم لكل واحدٍ منها العارض الذي للآخر ، لكنه ربما استحال ذلك لعائقٍ ولزمانٍ ولسببٍ من خارج.

(٥) وكأنّا أمعنا الآن في غير النظر الذي من غرضنا أنْ نتكلم فيه وهو النمط الأشبه بالكلام الطبيعي فنقول: إن كان بُعْدٌ مفارقٌ فلا يخلو إمّا أنْ يكون متناهياً ، وإمّا أنْ يكون غير متناهٍ. لكن طبيعة الخلاء ، عند جميع مَنْ يُوجب وجوده ، هي بحيث لا تنتهي إلاّ إلى بعد ملاء ، وأنّه إنْ كان الملاء متناهياً انتهى أيضاً إلى الخلاء . فيلزم أنْ يكون عندهم بُعْدٌ غير متناه، إمّا خلاء وحده أو ملاء وحده يتحدّد به الخلاء ، أو تأليف خلاء وملاءٌ - ومحالٌ أنْ يكون بُعْدٌ غير متناه، إمّا خلاء وحده أو ملاء وحده يتحدّد به الخلاء ، أو تأليف خلاء وملاءٌ - ومحالٌ أنْ يكون بُعْدٌ غير متناهٍ على هذه الصفة كما نوضّحه بعد ، فمحالٌ أنْ يكون خلاء على ما يقولون.

(٦) وأيضاً إنْ كان خلاء ؛ فلا يخلو إمّا أنْ يدخله الملاء أو لا يدخله ، وإنْ دخله الملاء فلا يخلو إمّا أنْ يبقى بعد الخلاء مع المداخلة موجوداً أو معدوماً . فإنْ كان معدوماً فلا يجوز أنْ يسمّوه مكاناً ، بل يكون المكان هو ما يحيط بالجسم من الخلاء المقارن له وذلك لأنّه في ذلك لا غير ، إذ قد عدم ما بين ذلك من بُعد الخلاء ولا يكون أيضاً جميع

the whole of that [void], but only its limit that is adjacent to the placed thing, because, if the estimative faculty were to imagine the whole of that [void] as not existing except this limit, then the placed thing would be *in* something that, if it is moved, would depart so as to make way for what succeeds it. Also, many bodies might be at rest in what is beyond that [limit], but the place of the body does not contain other bodies together with it. If, despite that, this interval sometimes does not exist and sometimes does, then sometimes it will be in potentiality and sometimes in actuality. Now, as such, its being in potency means that, before its existence, there is something existing in a certain nature that is receptive to the existence of [the interval]. (Let the natural philosophers concede this as a posited principle.)[10] In that case, the void is a composite of an interval and a matter that is informed by that interval so as to have a certain position, and [a form that] can be pointed at. This, however, is a body, and so the void is a body. On the other hand, if [the void interval] continues to exist with the [plenum's] interpenetration, then one interval will enter another; but we have already undermined the possibility of this.[11]

(7) We also say that there can be neither motion nor rest in the void, while there [can] be motion and rest in every place, and so the void is not a place. There cannot be motion in it because every motion is either natural or forced. Now, we argue that there cannot be natural motion in the void because it will either be circular or rectilinear.

---

10. For Avicenna's full account concerning the nature of potency, see *Ilāhīyāt* 4.2.
11. See 2.7.6–7.

ذلك، بل نهايته التي تلي المتمكن، لأنَّ جميع ذلك لو تُوهم معدوماً إلاّ هذا الطرف لكان المتمكن في شيءٍ إنْ تحرّك فارقه مهيئاً لعاقبٍ يخلفه، وأيضاً ما وراء ذلك قد تسكنه أجسامٌ كثيرة. ومكان الجسم لا يسعه معه جسمٌ آخر، ومع ذلك فإنْ كان هذا البُعْد تارة يُعدم وتارة يوجد؛ فيكون تارة بالقوة وتارة بالفعل، وكل ما كان كذلك فإنَّ كونه بالقوة معنى موجود قبل وجوده في طبيعةٍ قابلةٍ لوجوده. ليسلّم الطبيعيون هذا على سبيل الأصل الموضوع، فيكون الخلاء مؤلفاً من بُعْدٍ ومن مادةٍ تتصور بذلك البُعْد، فيصير ذا وضع وتكون إليه إشارة؛ وهذا هو الجسم، فيكون الخلاء جسماً، وإنْ كان يبقى مع المداخلة فيكون بُعْدٌ يدخل في بُعْدٍ، وهذا قد أبطلنا إمكانه.

(٧) ونقول: إنّه لا يجوز أن يكون في الخلاء حركة ولا سكونٌ - وكل مكان ففيه حركة وسكونٌ - فالخلاء ليس بمكان. وأمّا أنّه لا حركة فيه؛ فلأنَّ كل حركةٍ إمّا قسرية وإمّا طبيعية. ونقول إنَّ الخلاء لا تكون فيه حركة طبيعية، وذلك لأنّها إمّا أنْ تكون مستديرة، وإمّا أنْ تكون مستقيمة.

(8) There cannot be circular motion in the void because the void is of the character that it neither comes to an end nor is exhausted unless there is a certain finite body beyond it and that body prevents it from extending infinitely. So let us posit a certain body that is rotated so as [to describe] a circle ABCD, and let us also stipulate that the circle itself is being moved.[12] Now let [the circle's] center be E, and let us posit outside of [the circle] an infinite rectilinear extension, GF, parallel to AD (whether [that extension] be in a void, a plenum, or both together).

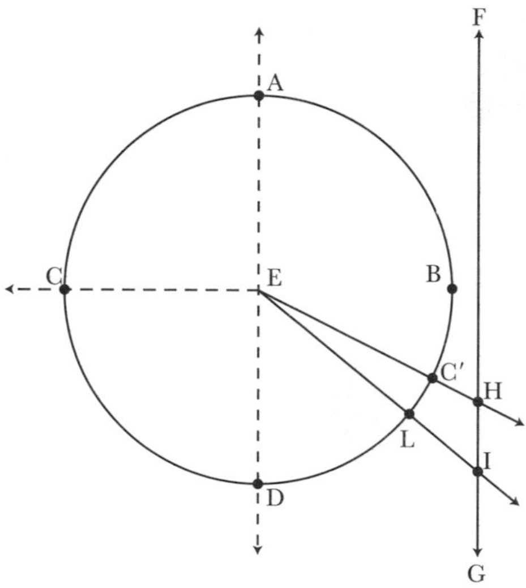

Now let a line, EC, which connects the center and the point C, rotate. Because the line EC is perpendicular (or nearly so) to the line AD in a direction other than FG, when [EC] is extended infinitely in the direction of C, it will not cross FG, since there will undoubtedly be some

12. Avicenna is probably imagining a sphere that is fixed in place but rotating, much like the way that he and other ancient and medieval philosophers imagined the various spheres of the planets and fixed stars. If one then imagines some point on this fixed, rotating sphere, that point will describe the circle Avicenna has in mind.

(٨) ولا يجوز أنْ تكون في الخلاء حركة مستديرة؛ وذلك لأنَّ الخلاء من شأنه أنْ لا يقف ولا يفنى إلاّ أنْ يكون وراءه جسمٌ متناهٍ؛ فذلك الجسم يمنعه أنْ يمتد إلى غير نهاية. فلنفرض جسماً يتحرّك على الاستدارة؛ على دائرة (أ ب جـ د) ولنجعل الدائرة نفسها تتحرّك، وليكن مركزها (طّ) - ولنفرض خارجاً عنها إمتداد (ر هـ) المستقيم بلا نهاية موازياً لـ(أ د) إما في خلاء وإمّا في ملاء أو فيهما جميعاً. وليكن خط (ط جـ) يصل بين المركز وبين نقطة (جـ) المنتقلة كيف كانت الاستدارة، فلأنَّ خط (ط جـ) عمود أو كالعمود على (أ د) في غير جهة (هـ ر)، فإذا أخرج من جهة (جـ) إلى غير النهاية؛ لم يلاق (هـ ر) إذ لا شك أنَّ لنقطة (ط) جهة لا تلي بعد (هـ ر) وما ينفذ فيها لا يصل

direction from the point E that does not lie next to the interval FG, and whatever passes in [that direction] does not contact [FG]. [That] is because, otherwise, the interval FG would be finite, circumscribing the circle ABCD in every direction; but it was not posited as such. So let EC be a certain interval or line that does not cross FG as long as it is in that direction until it coincides with the line AED and then crosses through it. At that time it, [EC′] will intersect FG, for when it is in the direction of FG and is perpendicular (or otherwise) to AD, then when it is extended infinitely, it inevitably intersects FG and crosses some point on it. Now, [the point where EC′ crosses FG] will not always be one and same point. [That] is because, along the line FG, you can posit many points and connect them to the center by many lines, [all] of which will come to be along the projected path of the intersecting point from which that line is produced whenever EC′ corresponds with it. Now, since there will be the projection toward [FG] after not projecting toward [it], there must be a first instant of time of the projection that divides the two times, which [corresponds with] a certain point along some projected path. Let the point be H. Now, let us take a point I that is before the point H and connect E and I along a line ELI, [where L is a point on the circumference of the circle ABCD]. In that case, when the line EC′ reaches the place where C′ would correspond with the point L during the rotation, there would have been a projection to point I on FG that was before [the projection to] point H. It was said, however, that H[13] is the first point on the line FG to which there is a

13. **Y** has *jim* (=C), but the sense of the argument clearly demands *hā* (=H), which, in fact, is what occurs in **Z**, **T**, and the Latin.

إليــه ، وإلاّ فبعــد (هـ ر) متناهٍ يطيف بدائرة (أ ب جـ د) من كل جهةٍ ، ولم يفرض كذلك . فليكــن (ط جـ) بُعْداً أو خطّاً لا يلاقي (هــ ر) ما دام في تلك الجهة إلى أنْ ينطبق على خــط (أ ط د) ثم يجــاوزه ، فهناك لا محالة يقاطع (هـ رّ) - فإنّه إذا صار في جهة (هـ ر) وكان عمــوداً على (أ د) أو غيــر عمود ، فإذا أُخرج إلى غير النهاية قاطع (هـ ر) لا محالة ولاقى نقطة منه ، وليست نقطة واحدة بعينها . فإنّك يمكنك أنْ تفرض في خط (هـ ر) نقطــاً كثيرة وتصلها بمركز (ط) بخطــوط كثيرة كلما انطبق خط (ط جـ) على خطٍ منها صار في ســمت مقاطعة النقطة التي جاء منها ذلك الخط . ولمّا كانت المسامتة بعد لا مسامتة ، فيجب أنْ يكون أول آن زمان المسامتة هي فصل بين الزمانين في سمت نقطةٍ ولتكن نقطة (ح) ، ولنأخذ نقطة (ك) قبل نقطة (ح) ، ولنا أنْ نصل بين (ط) و(ك) على خط (ط ل ك) فيكون خط (جـ ط) إذا بلغ في الدور حتى يلقى (جـ) نقطة (ل) كان مســامتاً لنقطــة (ك) في خط (هـ ر) قبل نقطة (حــ) - وقيل إنَّ (ح) أول نقطة

projection. This is a contradiction.[14] In fact, [the line] will always be projecting toward [FG] and always away from it, which is absurd.[15] So, then, there can be no circular motion in the void as they suppose.

(9) We also say that there would be no natural rectilinear motion. That is because natural motion departs from one direction and is directed toward another, and that from which it naturally departs must be different from that toward which it naturally tends. So, if that which it departs is in every respect like that toward which it tends, there will be no sense that the natural thing naturally departs it so as to acquire its like. [That] is because to depart naturally is to avoid naturally, but it is absurd that what something naturally avoids be what it naturally tends toward.

(10) Let us, instead, take it from the top: Natural motion must either be naturally directed in some direction or not. Now, it is absurd that motion should not be directed in some particular direction; so, if it is directed in some direction, the direction must be either some existing

---

14. Avicenna offers a condensed version of this argument in the *Kitāb al-najāt,* 241–43, where he argues thus: "We might be able to explain that [i.e., that the void is not infinite] quickly. So, we say, let there be circular motion in an infinite void (if an infinite void is possible) and let the moved body, for instance a sphere ABCD, be moved around a center, E. Now let us imagine a line GF in the infinite void and let there be a line EC [extending] from the center toward C away from the line GF. When the sphere rotates, this line [EC] will be such as to cut, pass through, and then depart from [GF] such that there will inevitably be two points where the projected line [EC] encounters and then departs [from GF]; let them be I and H. Yet there will also be a point, J, whose projection [on to GF] is before point I, and [yet] the point I was the first point at which the line [EC] projected [on to GF]. This is a contradiction; however, there is circular motion, and so the void is not infinite." Avicenna generates the contradiction by having us assume any point as the *first* point of contact with the line GF. A structurally similar argument, albeit employed for a radically different end (namely, to show that an infinite could be crossed in a finite period of time) was given by Abū Sahl Bijān al-Qūhī (or al-Kūhī); see Roshdi Rashed, "Al-Qūhī vs. Aristotle: On Motion," *Arabic Sciences and Philosophy* 9 (1999): 7–24.

15. In this sentence, Avicenna extends the general reasoning of the first argument to create another one, which seems to be the following: On the basis of the first argument, *mutatis mutandis,* there cannot be any last moment of projecting toward FG, since there will always be some point after that purported last point to which his initial argument applies. Consequently, there will always be another point toward which the rotating line EC must project before it ceases projecting toward FG. Equally, however, one could construct a line KL paralleling FG that is on the opposite side of the circle ABCD and, since the above argument will hold for the line EC when it projects toward KL, EC will always project away from FG. Clearly, this outcome is absurd, just as Avicenna concludes.

تسـامت من خط (هـ ر) ؛ وهذا خُلفٌ ، بل يلزم أنْ يكون دائماً مسـامتاً ودائماً مبايناً ، وهذا محالٌ ، فإذن لا حركة مستديرة في الخلاء الذي فرضوه .

(٩) ونقول ولا حركة طبيعية مستقيمة ، وذلك لأنَّ الحركة الطبيعية تترك جهة وتنحو جهة ، ويجب أنْ يكون ما تتركه بالطبع مخالفاً لما تقصده بالطبع . فإنّه إنْ كان ما تتركه في جميـع أحواله في حال ما تقصده ؛ فلا معنى لأنْ تكون الطبيعة تتركه طبعاً لتأخذ مثله ؛ فـإنَّ الترك الطبيعي نفارٌ طبيعي ، ومن المحال أنْ يكون منفورٌ عنه بالطبع مقصوداً بالطبع .

(١٠) بل نقول من رأس : إنّه لا يخلو إمّا أنْ تكون الحركة الطبيعية تنحو بالطبع جهةً أو لا تنحـو جهةً . ومحال أن تكون الحركـة لا تنحو جهة خاصّة . فإن كانت تنحو جهةً فلا يخلو إمّا أنْ تكون الجهة شـيئاً موجوداً أو غير موجود ، فإنْ كان شـيئاً غير موجود

thing or some nonexisting thing. If it is some nonexistent thing, then it is absurd that it be such that something either departs from or is directed toward it. If it is some existing thing, then either it is an intellectual object, not itself having a position and so not the sort of thing that one can point to, or it has a position [such as up/down, left/right, front/back] and so can be pointed to. Now, it is absurd that [the void] be an intellectual object that has no position, because no motion is directed toward that. So we are left with its having a position, in which case the interval either can or cannot be divided into parts that are reached during the traversal. On the one hand, if it is divided, then there will be one part that is next to the mobile and, once the mobile is there, either it will have become fully realized in [its natural] direction, and so the part is the intended direction and the rest are superfluous, or it will not be fully realized in the direction and, instead, it will need to cross through it, and so [the part] will be an intermediary to the direction and not part of the intended direction, but its status is like that of everything else adjacent to it. On the other hand, if it is not divided into parts that are reached, the absence of such a division is either because, while it itself allows the division by supposition, it is not in its nature to be broken up, or it is in no way divisible into parts (as they say about the celestial sphere). Now, if it is [the sort of thing that] is not divisible into parts by being broken up, while being divisible by supposition, then it is a body other than the void; and so, as long as no body exists in the void, [the void] will have no direction. In that case, there will be no direction in the absolute void taken alone. Moreover, that body must either be proper by nature to the space of the void that it is in or not. If [that body] is proper to it, then one part of the void will

فمحالٌ أنْ يكون متروكاً أو نحواً متوجهاً إليه. وإنْ كان شيئاً موجوداً؛ فإمّا أنْ يكون موجوداً عقلياً لا وضع لذاته فلا يشار إليه، أو يكون له وضعٌ فيشار إليه. ومحالٌ أنْ يكون عقلياً لا وضع له؛ لأنَّ ذلك لا حركة إليه. فبقي أنْ يكون له وضعٌ، وحينئذ لا يخلو إمّا أنْ يكون شيئاً لا يتجزأ من حيث يُصار إليه بالقطع للبعد، أو يكون يتجزأ. فإنْ كان يتجزأ؛ فالبعض منه يكون أقرب من المتحرّك إليه، وإذا وصل إليه المتحرك، فإمّا أنْ يكون قد حصل في الجهة - فالبعض هو الجهة المقصودة والباقي خارج عنه - وإمّا أنْ لا يكون حصل في الجهة، بل يحتاج أنْ يتعداه، فإن كان يحتاج أن يتعداه فهو سبيل إلى الجهة لا بعض الجهة المقصودة، وحكمه حكم سائر ما يليه. وإذا كان غير متجزٍ من حيث يصار إليه؛ فلا يخلو إما أنْ يكون فقدانه التجزي لا لأنّه في نفسه لا يحتمل فرض القسمة؛ بل لأنّه ليس في طباعه الانكسار كما يقولونه في الفلك أو يكون لا يتجزأ أصلاً. فإنْ كان لا يتجزأ بالتفكّك ويتجزأ بالفرض؛ فهو جسم غير الخلاء. فما لم يكن في الخلاء جسمٌ موجودٌ لا تكون له جهة، فيكون حينئذ لا جهة في الخلاء المطلق وحده. وذلك الجسم أيضاً لا يخلو إمّا أنْ يكون مختصاً بالطبع بالحيّز من الخلاء الذي هو فيه أو لا

differ from another in the nature [of the void itself] in order that some bodies be naturally proper to [a given space] to the exclusion of others. If [that body] is not proper [to that space], then it will pass through it when it departs from it. Now, when the body departs that space in the void (assuming that [the body] is moved), it either must be moved naturally [(1)] toward [some space]—whether it is the earlier space of the void that that body was in or the later space that it will reach—or [(2)] it [must be moved naturally] toward the body that was in [the space]. [As for (1)] it cannot be moved to the earlier space [or some later space], otherwise its motion to *that* space is the natural motion and that which is [proper] to it itself.[16] [As for (2)] the motion [to that space] is, then, accidental, and so it could not have been moved naturally to that other space. [That] is because, if that moved body is not in some way aware of the transition of that body from one place to another, then how can it just happen to leave one direction toward which [its] motion tended because that [body] was in it, and its nature spontaneously tend toward another direction? Unless, that is, that body emits to [the nature] a certain influence or power, and[17] that influence and power are a certain principle that triggers the moved body to move naturally toward [the body], as in the case of the magnet and iron, in which case the motion is forced, not natural. If [the moved body] is aware, then a certain perception will have in fact occurred, and [the motion] will have been volitional, not natural—all of which is vacuous. The discussion of that body's transition, whether naturally or not, however, returns us to our enumeration [of the various options]—namely, if what is directed and has a position is not divisible in any way whatsoever into parts that are reached, it is either a point, a line, or a surface. Additionally, either all the

16. In other words, that other space (in fact, either the space it was in or the space it will come to be in) will be the body's proper place, in which case the immediately preceding argument applies—namely, that there will be natural parts in the void—which, on the present horn of the argument is assumed to be false.

17. Reading *wa* with **Z** and **T**, which is omitted in **Y**.

يكـون مختصـاً به ؛ فإنْ كان مختصاً به فبعض الخلاء مخالـفٌ لبعضه في الطبيعة حتى تختص به بعض الأجسـام طبعاً دون بعـض ، وإنْ كان غير مختص جاز فيه مفارقته له. وإذا فـارق ذلـك الحيّز من الخلاء لم يخلُ إمّا أنْ يتحرك الجسـم ، المفروض متحرّكاً إليه ، بحركته الطبيعية إلى الحيّز الأول الذي كان فيه ذلك الجسـم من الخلاء أو يتحرك إليه نحو الحيّـز الآخر الذي صار إليه. ولا يجـوز أنْ يتحرك إلى الحيّز الأول وإلاّ فحركته إلى ذلك الحيّز هي الحركة الطبيعية ، والتي بالذات ، وإمّا إلى ذلك الجسم الذي كان فيه ، فقد كانت ‹الحركة› بالعرض. ولا يجوز أنْ يتحرك بالطبع إلى الحيّز الآخر ، لأنَّ الجسـم المتحرك إنْ لم يشـعر بوجهٍ من الوجوه بانتقال ذلك الجسـم عن حيّز إلى حيّز ؛ كيف يتأتى أنْ يترك جهـة كانت مقصودة بحركته لأنَّ ذلك فيهـا ويقصد جهةً أخرى من تلقاء طباعه؟ إلاّ أنْ يكون ذلك الجسـم يبعث إليها أثراً أو قوة وذلك الأثـر وتلك القوة تكون مبدأ ما لانبعاث حركة الجسم المتحرّك بالطبع إليه ؛ كحال ما بين المغناطيس والحديد ، فحينئذ تكون الحركة قسـرية لا طبيعية. وإنْ شعر ، فقد حصل هناك إدراكٌ وحصلت إرادية لا طبيعية ، وهذا كله باطل! على أنَّ الكلام في انتقال ذلك الجسـم بالطبع أو بغير الطبع ، يرجع إلى ما نحن نسـرده ونقوله. وإنْ كان المتوجه إليه لا يتجـزأ من حيث يُصار إليه بوجهٍ من الوجوه وله وضعٌ ؛ فهو إمّا نقطة وإمّا سـطح. فلا يخلو بعد ذلك إمّا أنْ تكون الجهات كلها متشـابهة في أنها نقط وخطوط وسـطوح ، أو تكون جهة نقطة ، وجهة خطاً وجهة سـطحاً. فإنْ

directions will be the same in that they are points, lines, or surfaces, or one will be a point, another a line, and another a surface. On the one hand, [consider] if all of them are either points, or lines, or surfaces. Now, points, lines, and surfaces are distinguished only by certain accidents that they happen to have—namely, from those things of which they are predicated, whether through what is proper to them as such or foreign to them. All of that, however, belongs to them because of things that vary in shapes and natures that are their extremities, whereas the void is not like that. So, then, on the basis of this description, there would be no specifically different directions in it. If, on the other hand, it is not like that but, instead, one direction is a point and another a surface or line, or according to some other way in which it can be divided, then how is it possible that, in some location within the void, there is only an actual point, and in another, only an actual line or actual surface, or whatever? The void is one and continuous with no discontinuity in it because it has no matter on account of which it would be susceptible to these states; and, since it is self-evident, we have stipulated that that is not because of a body. So there will not be different directions in the void; but when there is no difference of directions and places, then it is impossible that there be some place that is naturally left behind and another that is naturally tended toward. So, then, there will be no natural rest in the void, since within the void there will be no location that is better suited than another to there being natural rest in it.

(11) Also, we witness bodies being moved naturally toward various directions and, moreover, varying in speed. Now, their variation in speed is either because of some factor in what is moved or something in the medium.[18] The factor in that which is moved is sometimes due to a difference of inclinatory power, since, because of its [inclinatory] power, there is the increase in the heavy thing's descent or the light thing's ascent;

18. The Arabic *masāfah,* which I translate "medium" here, literally means "distance" or "spatial magnitude"; however, since, in paragraph 12, Avicenna will speak of the "distance's" being rarer and denser, he is clearly not thinking of distance in the sense of a kilometer or a mile, but the nature of the traversed medium. Thus, I have preferred "medium" in the present context.

كانت الجهات كلها نقطاً أو خطوطاً أو سـطوحاً ؛ والنقط والخطوط والسطوح لا تختلف إلاّ بعـوارض تعرض لها أي من حواملها ، بما تختص بها من حيث هي كذلك ؛ وإمّا غريبة عنها ، وجميع ذلك يلزمها من جهة الأشـياء المختلفة الأشـكال والطبائع التي هي نهايات لهـا ، والخلاء ليس كذلك ، فإذن لا يجـوز أنْ يكون فيه اختلاف جهات على هذه الصفة بالنوع. وإنْ كان ليس كذلك بل من جهة نقطة وجهة أخرى سطح أو خط ؛ أو على وجهٍ آخر مما تحتمله القسـمة ، فكيف يمكن أنْ يكون في الخلاء في موضع نقطة بالفعل فقط ، وفي موضع خط بالفعل فقط ، أو سـطح بالفعل ، أو وجهٌ آخر؟ والخلاء وأحدٌ متصلٌ لا انقطاع فيه لأنّه لا مادة له ، فيقبل لأجلها هذه الأحوال ، ووضعنا أنَّ ذلك ليس بسبب جسم لمـا بانَ من البيان. فالخلاء ليس فيه اختلاف جهات ، وإذا لم يكن هناك اختلاف جهات وأماكن ، استحال أنْ يكون مكان متروكاً بالطبع ، ومكان مقصوداً بالطبع ، فليس إذن في الخلاء سكونٌ طبيعي ، إذ ليس في الخلاء موضعٌ هو أولى بالسكون فيه بالطبع من موضع .

(١١) وأيضاً فإنّا نشـاهد الأجسام تتحرك بالطبع إلى جهاتٍ ما ، وتختلف مع ذلك في السرعة والبطء ، فلا يخلو اختلافها في السرعة والبطء أنْ يكون إما لأمرٍ في المتحرّك منها ، أو لأمرٍ في المسافة. أمّا الأمر الذي في المتحرّك فقد يكون لاختلاف قوة مَيْلية ؛ فإنَّ الأزيد في الثقل النازل أو الخفّة الصاعدة لقوته أو لزيادة عِظَمه يسـرع ، والأنقص يبطيء .

or it is faster owing to the increase of its bulk and slower owing to the decrease. At other times, [the bodily factor] is due to its shape. So, when, for instance, the shape is a square and it crosses the medium with its surface [projecting forward], it will not be like a cone crossing the medium with its vertex [projecting forward], or even when the square crosses the medium with one of its angles [projecting forward], since, in the former case, it needs to set into motion a larger thing (namely, what it meets first), while in the latter cases it does not need to. So, in every case, the cause of fastness is the greater strength to repel what obstructs and to oppose whatever stands in its way and to penetrate more forcefully. [That] is because the greater the repulsion and penetration, the faster [something] moves, and the weaker they are, the slower it moves; whereas this is indeterminable in the void. Let us, however, set this to one side, since it is not of much use for what we are trying to do now.

(12) [The factor effecting speed] that is due to the medium is that, whenever [the medium] is rarer, it is crossed more quickly, whereas whenever it is denser, it is traversed more slowly (assuming that what is undergoing the natural motion is one and the same). In general, the cause of it is the strength and weakness to resist pushing and being penetrated, for the rarer is affected more readily by what pushes and penetrates, whereas the thickly dense opposes it more forcefully. Thus, something does not pass through earth and stone as it passes through air, while its passing through water will be between the two. Now, rarity and density increase and decrease in varying degrees, which, we have confirmed, causes that opposition. So, whenever the opposition is less, [the mobile] moves faster, whereas whenever the opposition increases, it moves slower;

وقد يكون لاختلاف شكله؛ فالشكل مثلاً إذا كان مربعاً وقطع المسافة بسطحه، لم يكن كمخروطٍ يقطع المسـافة برأسه، وكذلك المربع إذا قطع المسافة بزاويته، إذ ذلك يحتاج أن يحرّك شيئاً أكبر؛ وهو الذي يلاقيه أولاً، وهذا لا يحتاج إلى ذلك. فيكون سبب السرعة في كل حال الاقتدار على شدة دفع ما يمانع الشيء ويقاومه مقاومة ما وعلى شدّة الخرق. فإنَّ الأدفع والأخرق أسرع، والأعجز عنهما أبطأ، وهذا لا يتقرّر في الخلاء، بل لنترك هذا الوجه فإنّه لا كثير نفع لنا فيما نحاوله منه.

(١٢) وأمّا الذي يكون من قبل المسافة؛ فهو أنّها كلّما كانت أرق كان قطعها أسرع، وكلما كانت أغلظ كان قطعها أبطأ، وذلك بحسـب المتحرّك بالطبع الواحد، وبالجملة، السـبب فيه الاقتدار على مقاومة الدافع الخارق والعجز عنه. فإنَّ الرقيق شـديد الإنفعال عن الدافع الخارق، والغليظ الكثيف شـديد المقاومة لـه، ولذلك ليس نفوذ المتحرّك في الهـواء كنفوذه في الأرض والحجـارة، ونفوذه في الماء بين الأمرين. والرّقة والغلظ تختلف في الزيادة والنقصان، ونحن نتحقّق أنَّ السـبب في ذلـك المقاومة؛ فكلّما قلّتْ المقاومة زادت السـرعة، وكلّما زادت المقاومة زاد البُطء، فيكون المتحرك تختلف سرعته وبطؤه

and so the variation of the mobile's fastness and slowness are commensurate with the opposition. Whenever we posit less opposition, the motion must be faster, while whenever we posit more opposition, the motion must be slower. So, when the mobile undergoes motion in the void, it will cross the void[19] medium either in a certain amount of time or in no time. Now, it is absurd that that would take no time, because it crosses some of the medium *before* crossing all of it, and so it must take some time. Now, inevitably, that time will have some proportion to the time of the motion in a plenum that offers opposition, and it will be just as much as an amount of time of a certain opposition (were there such) whose proportion to the plenum's opposition is the proportion of the two times, but slower than an amount of time of opposition that is smaller in proportion to the posited opposition than the proportion of the time. Now, it is absurd that the proportion of the motion's time, where there is absolutely no opposition, should be like a proportion of some amount of time of a motion during a certain opposition (were it, in fact, to exist), let alone slower than some amount of time of some other opposition (were it imagined to be even less than the initial smaller opposition).[20] In fact, whatever requires any opposition imagined as existing for some time cannot have any proportion to a time having absolutely no opposition. So, then, the motion will neither take some time nor not take some time. This, however, is absurd.

---

19. Reading *khālīyah* with **Z**, **T**, and the Latin (*vacuum*) for **Y**'s *ḥālīyah* (present).

20. In more concrete terms, the argument may be expressed thus: Assume that it takes one minute to cover a given distance in a void that offers no opposition, while it takes two minutes to cross the same distance in a certain medium—say, water—that does offer some opposition. Thus, it takes half the time to cross through the void that it does to cross through the water. Now, imagine that there is another opposing medium that is half as rare as water—say, air. Since Avicenna has argued that, ceteris paribus, fastness and slowness of some mobile is proportional to the opposition offered by the medium through which it travels, something traveling through the air would travel twice as fast as it does through the water; however, the same object likewise traveled twice as fast through the void as it did the water, and so the mobile traveling through the air, which offers a certain amount of opposition, would cover the same distance in the same time as it would traveling through a void that offers no opposition. Moreover, were there some medium even rarer than air, the mobile would move through it faster than through a void, while being opposed in the former medium but not being opposed in the void.

بحسب اختلاف المقاومة، وكلّما فرضنا قلّة مقاومةٍ وجب أنْ تكون الحركة أسرع، وكلّما فرضنا كثرة مقاومة وجب أن تكون الحركة أبطأ. فإذا تحرّك <المتحرّك> في الخلاء لم يخلُ إمّا أنْ يقطع المسافة الخالية بالحركة في زمان أو لا في زمانّ - ومحالٌ أنْ يكون ذلك لا في زمانّ - لأنّه يقطع البعض من المسافة قبل قطعه الكل. فيجب أنْ يكون في زمانٍ، ويكون لذلك الزمان نسبة إلى زمان الحركة لا محالة، في ملاء مقاوم، ويكون مثل زمان مقاومة لو كانت نسبتها إلى مقاومة الملاء نسبة الزمانين، وأبطأ من زمان مقاومةٍ هي أصغر في النسبة إلى المقاومة المفروضة من نسـبة الزمان. ومحالٌ أنْ تكون نسبة زمان الحركةّ - حيث لا مقاومة البتّة - كنسبة زمان حركةٍ في مقاومةٍ مّا لو صحّ لها وجود، فضلاً عن أنْ تكون أبطـأ من زمان مقاومةٍ أخرى لو توهمت أقلّ من المقاومـة القليلة الأولى. بل يجب أنْ لا يكون لما توجبه أي مقاومة تُوهمت موجودة من الزمان نسـبة إلى زمان لا مقاومة أصلاً، فيجب إذن أن تكون الحركة لا في زمان ولا ليست في زمان؛ وهذا محال!

(13) We do not have to stipulate in our proof of this whether this opposition (which is according to the noted proportion) must exist or not, because our claim is that the amount of time for this motion in a void would be equal to the amount of time during a certain opposition *were it to exist,* and this premise is true in the way we explained. Also, every motion in a void is a motion in what does not oppose, and this premise is likewise true. Also, any motion in what does not oppose will not at all be equal to some motion during a certain opposition having a given proportion, were it to exist. So, from these premises, it necessarily follows that, in a void, there is no motion whose time is equal to some amount of time of a motion during a certain opposition, were there such. From these [premises] and the first, it necessarily follows that none of the motions in the void is a motion in the void, which is a contradiction.[21]

(14) Now, one thing that could be said against this is that every motive power is in a body, and so, as a result of the body's magnitude with respect to its bulk and [that power's] magnitude with respect to its strengthening and weakening, [the motive power] requires a certain amount of time, even if there were no opposition at all. Besides that, the times might increase in accordance with an increase of certain oppositions, while it does not follow that every given opposition will produce some influence on that body. [That] is because it does not necessarily follow, when a given opposition has some influence, that half of it, or half of half of it, would have an influence. Indeed, when a certain number of

---

21. The reasoning is clear once the proportion is given. Assume two finite periods of time, t_1 and t_2, and two magnitudes of opposition, M_1 and M_2. Also, let M_1 be the amount of opposition imposed by the void, which, by assumption, is 0, while the other variable will be some finite magnitude and so be expressed by some finite number. Consequently, the proportion must be this: $t_1:0::t_2:M_2$, which can equally be expressed as $t_1/0 = t_2/M_2$. Since the ratio on the left-hand side involves division by 0, however, it will be infinite, whereas, since the ratio on the right-hand side involves two finite numbers, it must be finite. In that case, however, an infinite will equal a finite, which, as Avicenna observes, is a contradiction. What is important to note is that the contradiction is generated simply by assuming (1) that there is some proportion, expressed as, $t_1/M_1 = t_2/M_2$, between the time it takes to cross through a given distance in a void and a plenum, and (2) that the void offers no (i.e., 0) opposition.

(١٣) ولا نحتـــاج، في بياننا هذا، إلى أنْ نجعل لهذه المقاومةّ - التي على النســبة المذكورةّ - اســتحقاق وجودٍ أو عدم، لأنّا نقول إنَّ زمــان هذه الحركة في الخلاء يكون مســاوياً لزمان حركــةٍ في مقاومةٍ مّا لو كانت موجودة، وهــذه المقدمة صادقة أوضحنا صدقها. وكل حركة في الخلاء فهي حركة في عدم مقاومة، وهذه المقدمة أيضاً صادقة، وكل حركة في عدم مقاومة فليســت مساوية البتة لحركةٍ في مقاومةٍ مّا على نسبةٍ مّا، لو كانت موجودة. فيلزم من هذه المقدمات أنْ لا حركة في الخلاء هي مســاوية الزمان لزمان حركةٍ في مقاومةٍ مّا لو كانت، ويلزم منها ومن الأول أنْ لا شــيء من الحركات في الخلاء حركة في الخلاء، وهذا خُلْف.

(١٤) وممّا يمكن أنْ يقول القائل على هذا؛ أنَّ كل قوة محرّكة تكون في جســم فإنّها تقتضي، بمقدار الجسم في عِظَمه ومقدارها في شدتها وضعفها، زماناً لو لم تكن مقاومة أصلاً. ثم بعد ذلك، فقد تزداد الأزمنة بحســب زيادة مقاوماتٍ مّا، وليس يلزم أنْ تكون كل مقاومةٍ مّا تؤثّر في ذلك الجسم، فإنّه ليس يلزم إذا كانت مقاومة مّا تؤثّر أنْ يكون نصفها يؤثّــر ونصف نصفهــا يؤثّر. فإنّه ليس يلزم إذا كان عدّة يحركــون ثقلاً وينقلونه؛ أنْ يكون

men move and transport some heavy object, it does not necessarily follow that half the number would move a given thing; nor does it follow that, when many drops of water fall on something and erode a hole in it, a single drop would have an influence. So the opposition whose time corresponds with the proportion of the opposition of the void might not have any influence at all, and only some other opposition (were it to exist) would do so. The response to this is that we took *opposition* on the condition that it should be an opposition having an influence, [with respect to] which (were it to exist) the amount of time for it would [equal] a certain amount of time for some unopposed motion. We did not need to say "opposition having an influence" simply because to say that the opposition has no influence is like saying that the opposition does not oppose, for the meaning of *to oppose* is nothing but *to have some influence.* Now, having this influence is taken in two ways, one of which is to break down the inclination's fierceness and power, and the second [of which] is a certain rest that we suppose that the opposition produces such that, as a result of certain interceding oppositions, rests are continually produced that are imperceptible individually while they are perceived collectively as slowness.[22] (Later you will learn that what has an influence is only according to one of the two.)[23] Be that as it may, the mobile is, in its nature, susceptible to lesser [degrees of influence] (should there be some agent to produce such an influence on it), from which it is necessary that some of those oppositions that the nature of the body experiences be equal in their time to what is unopposed. This, however, is absurd.

---

22. This second position is, in fact, that of certain Atomists, who explained differences in speeds by appealing to the number of purported intervallic rests a mobile makes during the course of its motion.

23. See book 3.4.13–14 where Avicenna argues against the notion of atomic motion with its accompanying intervallic rests.

نصف العدّة يحرّك شيئاً ، أو كانت قطرات كثيرة تثقب المقطور عليه ثقباً ، أن تكون قطرة واحدة تؤثّر أثراً . فيجوز أنْ تكون المقاومة ، التي زمانها على نسبة مقاومة الخلاء ، لا تؤثر شيئاً وإنّما تؤثّر مقاومة أخرى لو كانت موجودة. فالجواب عن هذا ؛ إنّا أخذنا المقاومة على أنّهاّ - لو كانت موجودة مقاومة مؤثّرة - لكان زمانها زمان حركة في لا مقاومة، وإنّما لم نحتج أنْ نقول مقاومة مؤثّرة؛ لأنَّ المقاومة إذا قيل إنّها غير مؤثّرة كان كما يقال: مقاومة غير مقاومة، فمعنى المقاومة هو التأثير لا غير . وهذا التأثير على وجهين : أحدهما الكسر من الحمية ومن قوة الميل ، والثاني ما نظنّ من إحداث المقاومة سكوناً ، فلا تزال تحدث سكونات عن مقاومات متشافعة لا يُحسّ بأفرادها وتُحسّ بالجملة كالبُطءّ - وأنت ستعلم بعد ، أنّه ما من تأثير على أحد الوجهين إلاّ وفي طباع المتحرّك أن يقبل أقل منه ، لو كان مؤثّرٌ يؤثره. فيجب من ذلك أنْ يكون بعض تلك المقاومات التي تحتملها طبيعة الجسم مساوياً في زمانه لغير المقاومة ، وهذا محال.

(15) Since it has become apparent that there is no natural motion whatsoever in a void, we say that there is also no forced motion. That is because forced motion is the result of [something] that is either joined with or separate from the mover. So, on the one hand, if it is by being joined with the mover and the mover is moved, then [that mover] likewise will be moved as a result either of a force, or a soul, or a nature. If it is by force, the discussion will continue until it terminates at either a soul or a nature. If it is moved by a soul, the soul will cause motion by producing a certain inclination that also varies in strength as well as weakening to the point that that [inclination of the soul] is seen together with the coming to rest that is brought about by what opposes the motion, just as is seen in what is moved naturally when it is opposed and prevented from moving. In other words, the inclination varies in power and strength, and [all the problems] accompanying natural inclination [in a void] will accompany it.[24] If it is [moved] by nature, then it entails what was argued [earlier]. When there can be neither psychic nor natural [inclination] in the void, neither can there be forced motion in the void, where the agent producing the motion in it is necessarily moved [either by the soul or by nature].

(16) If the mover is separate from the motion when it produces it, then [the motion] frequently entails some difference of direction with respect to which it is moved, and the same thing that we said about natural motion will necessarily follow.[25] Moreover, the forced motion that is produced by the separate mover might exist while the mover's production of the motion has ceased. Now, it is absurd that the motion that is continuously being renewed should exist while its cause does not exist. So there must be some cause that preserves the motion, and that

---

24. In other words, Avicenna believes that the argument of pars. 11–13 can be expressed not only in terms of natural inclination, as it was above, but also, *mutatis mutandis,* in terms of psychic inclination.

25. See par. 10.

(١٥) فقد ظهر أنّه لا تكون في الخلاء حركة طبيعية البتة؛ نقول ولا حركة قسرية، وذلك لأنَّ الحركة القسرية إمّا أنْ تكون بمقارنة المحرّك أو بمفارقته، فإنْ كان بمقارنة المحرّك فالمحرّك متحرّك، فهو أيضاً إمّا متحرك عن قاسر أو عن نفس أو عن طبع. فإنْ كان عن قاسرٍ لزم الكلام إلى أنْ ينتهي إلى نفسٍ أو طبيعة. وإنْ كان عن نفسٍ، فالنفس تحرّك بإحداث ميلٍ مّا يختلف أيضاً في الشدّة والضعف، حتى أنَّ ذلك ليُحسّ مع التسكين المقاوم للحركة، كما يحسّ في المتحرك طبعاً إذا قووم فمُنعت حركته. وذلك الميل يختلف بالقوّة والشدة، ويلزمه ما يلزم الميل الطبيعي، وإنْ كان طبيعياً لزم ما قيل. فإذا كان النفسي والطبيعي لا يصح في الخلاء، لم يصحّ أنْ يكون في الخلاء تحريك قاسر يلزم المحرّك فيه المتحرّك.

(١٦) وإنْ كان المحرّك يفارق عند إيجاد الحركة، فقد يلزمها الاختلاف من جهة ما تتحرك فيه، ويلزم ما قلنا في الحركة الطبيعية بعينها. وأيضاً، فإنَّ الحركة القسريّة - المفارقة للمحرّك - قد تكون موجودة وتحريك المحرّك قد زال، ومحالٌ أنْ يكون ما يتجدّد على الاتصال من الحركة موجوداً وسببه غير موجود. فيجب أنْ يكون هناك سببٌ

cause will exist in the mobile, producing an effect on it. Now, that [cause] will either be some accidental power that is incorporated into the mobile from the mover, like the heat in water as a result of fire, or a certain influence that the mobile encounters resulting from that which it passes through. The latter influence is understood in two ways:[26] On the one hand, when, as a result of something's being moved, the mover contacts and pushes the first part of the [medium] in which there is the motion, then that [first part] will push what is next to it, and so on to the last parts, and the projectile placed into this medium will be moved as a necessary result of [the medium's being moved] (on the assurance that those pushed parts are moved faster than the projectile thrown by the mover, since that [medium] is more easily pushed than this projectile).[27] On the other hand, what pushes that body might push through that medium, and so what is pushed would be something forced to contract and then collectively curve around behind [the projectile], and that contraction will necessarily push the body forward.[28] All of this, however, is inconceivable in the void.

(17) These were the only options, since this motion is either by some power [whether from the soul or from nature] or by some body that causes motion by contact, where the body that causes motion by contact does so either in that it [itself] is carried along (where the status of what is attracted by contact is like that of what is carried along) or in that it is pushed by contact.

(18) If there is forced motion in the projectile as a result of some power in the void, then [the motion] must continue and never abate or discontinue. That is because, when the power is in the body, it either remains or there is a privation of its existence. If it remains, then the

---

26. The two ways mentioned here are discussed at greater length later at 4.4, where both are, in fact, rejected.

27. This account of how the medium can bring about motion is roughly Aristotle's account of projectile motion found at *Physics* 8.10.266b28–267a20.

28. The theory presented here might be the doctrine of *antiperistasis,* or mutual replacement, which Aristotle mentioned (*Physics* 8.10.267a16–19) as one possible, albeit ultimately rejected, account of projectile motion. The general idea is that the medium moves the mobile in a way similar to that of contractions pushing a baby out of a womb during parturition. This interpretation is further confirmed at par. 18, where Avicenna's use of the expression *ʿalá sabīl ḥamlin wa-waḍʿin* (or *wuḍʿin*) evokes the image of childbirth.

يستبقي الحركة ، وأنْ يكون ذلك السبب موجوداً في المتحرّك <و> يؤثّر فيه ؛ فذلك إمّا قوة عرضية ارتبكت في المتحرّك من المحرّك كالحرارة في الماء عن النار ، وإمّا تأثيرٌ ممّا يلاقي المتحرك ممّا ينفذ فيه . وهذا التأثير معقولٌ على وجهين : إمّا أنْ يكون الجزء الأول من الشيء الذي فيه الحركة ، لمّا دفعه المحرّك بالمتحرّك وهو يلاقيه ، دفع ذلك ما يليه ، واستمر إلى آخر الأجزاء ، وكان هذا المرمي المقذوف موضوعاً في ذلك المتوسط فيلزمه أنْ يتحرك ، في ضمن تلك الأجزاء المتدافعة المتحركة ، أسرع من حركة المرمي الذي دفعه المحرّك ، لأنَّ ذلك أسهل إندفاعاً من هذا المرمي . وإمّا أنْ يكون خرق الدافع لذلك الجسم المتوسط ، فالمدفوع يلجيء الشيء إلى أنْ يلتئم فينعطف من ورائه مجتمعاً ، ويلزم ذلك الاجتماع دفع الجسم إلى قدّام ؛ وهذا كلّه لا يتصور في الخلاء .

(١٧) وإنّما كانت الأقسام هذه ، إذ كانت هذه الحركة إمّا تكون عن قوةٍ أو عن جسم يحرّك بالملاقاة . والجسم المحرّك بالملاقاة فإمّا أن يحرّك بأنّه يُحمل أو أنّه يدفع بالملاقاة . وأما الذي يجذب بالملاقاة ؛ فحكمه حكم الحامل .

(١٨) فإنْ كانت الحركة القسرية في المرمي عن قوةٍ في الخلاء فيجب أنْ تبقى فلا تفتر البتة ، ولا تنقطع البتة ، وذلك لأنَّ القوة إذا وجدت في الجسم فلا يخلو إمّا أنْ

motion would always continue. If there is a privation of its existence, or it even weakens, the privation of its existence or weakening is either from a cause or is essential to it. The discussion concerning the privation of its existence will provide you the way to proceed with respect to weakening. We say: It is impossible for the privation of [the power's] existence to be essential, for whatever is essentially a privation of existence necessarily cannot exist at any time. If the privation of its existence is by a cause, then that cause is either in the moved body or in something else. If [the cause of the privation of the motion] is in the moved body and, at the beginning of the motion, it had not actually been causing that [privation] but had in fact been overpowered, and then later became a cause and dominated, then there is another cause for its being such, in which case an infinite regress results. If either the cause or the auxiliary cause, which assists the cause that is in the body, is external, then the agent or auxiliary cause acts either by contact or not. If it acts by contact, then it is a body in contact with the mobile, but this cause would not exist in a pure void, and so the forced motion would neither abate nor stop in the pure void. If it does not act by contact but is something or other that produces an effect at a distance, then why did it not do so initially? The discussion is just like the one about the cause if it were in the body. The fact is that the most appropriate [explanation] is that the continuous succession of opposing things is what causes this power to decrease and corrupt, but this is possible only if the motion is not in the pure void—that is, if the cause of the motion is a power. If the cause is a contacting body that produces motion in the manner of bearing forth and delivering, then the discussion returns to the separated[29] cause, and what was said there will be said here. So, clearly, in a pure void, there is no forced motion, whether conjoined with or separate from the mobile.

---

29. There is some confusion in **Y**'s text, which reads *al-sabab al-mufāraq al-muqāran* (the separated, conjoined cause). Not only does such a reading involve a contradiction, but it is not confirmed by **Z**, **T**, or the Latin text, which all have "the conjoined cause." Still, both **Y** and **Z** note that there are manuscripts with "the separated cause" (sigla H and M in **Y**'s edition, and, while neither is the earliest MS consulted, both are among the earliest, thirteenth and fouteenth century, respectively). Since the present language here in terms of childbirth is similar to that of how a separate cause might impart motion to a medium such that the medium causes motion (par. 16), "the separated cause" seems preferable.

تبقى ، وإمّا أنْ تُعدم. وإنْ بقيت ، فالحركة تبقى دائماً ، وإنْ أعدمت أو إنْ ضعفت فلا يخلو إمّا أنْ تكون تعدم أو تضعف عن سبب ، أو تعدم أو تضعف لذاتها . والكلام في العدم يعرّفك المأخذ في الكلام في الضعف ، فنقول : ويستحيل أنْ تُعدم لذاتها ، فإنَّ ما يستحق العدم لذاته يمتنع وجوده زماناً . وإن عُدمت لسببٍ ، فإمّا أنْ يكون ذلك السبب في الجسم المتحرّك أو يكون في غيره ، فإنْ كان في الجسم المتحرّك ، وقد كان غير سببٍ لذلك بالفعل عند أول الحركة ، بل كان مغلوباً ثم صار سبباً وغالباً ، فلكونه كذلك سببٌ آخر ، والأمر في ذلك يتسلْسل إلى غير نهاية. فإنْ كان السبب خارجاً عن الجسم ، أو كان المعين للسبب الذي في الجسم خارجاً ، فيجب أنْ يكون الفاعل أو المعين ممّا يفعل بملاقاة ، أو يكون يفعل بغير ملاقاة. فإنْ كان يفعل بملاقاةٍ فهو جسم يلاقي المتحرّك فلا يكون في الخلاء المحض هذا السبب ، فالحركة القسرية لا تفتر في الخلاء المحض ولا تقف. وإنْ كان لا يفعل بملاقاة ، بل يكون شيئاً من الأشياء يؤثّر على المباينة ، فما باله لم يؤثّر في أول الأمر؟ ويكون الكلام عليه كالكلام في السبب لو كان في الجسم ، بل الأولى أنْ يكون تواتر المقاومات على الاتصال هو الذي يسقط هذه القوة ويفسدها ؛ وهذا لا يمكن إلاّ أنْ لا تكون الحركة في الخلاء الصرف ، هذا إذا كان سبب الحركة قوة. فإنْ كان السبب جسماً ملاقياً تحرّك على سبيل حَمْلٍ ووضْع ، فيرجع الكلام إلى السبب المفارق ، وقد قيل فيه ما قيل. فبيّنٌ أنْ لا حركة قسرية مفارقةً للمتحرّك أو مقارنة إياه في خلاء صرف.

(19) Since our argument has made it clear that there is no motion in the void, whether natural or forced, we say that neither is there rest in it. That is because, just as that which is at rest is what is not moving but such that it can be moved, so likewise that *in* which there is rest is that *in* which there is no motion, but it is such that there can be motion *in* it; whereas the void is such that there cannot be motion in it.

(20) The defenders of the void, however, were at their most outrageous when they gave it a certain attractive or motive power, even if in some other way, such that they claimed that the cause of water's being retained in the vessels called clepsydrae and its being attracted into the instruments called siphons[30] is nothing but the attraction of the void and that it first attracts what is denser and then what is more subtle. Others have said instead that the void moves bodies upward—namely, when the body becomes rarefied by a greater amount of void entering it [and] then becomes lighter and moves upward more quickly.

(21) We say that, if the void were to have some attractive power, there could be no differences in strength and weakness in it, since the way that each part of the void would attract would be like any other. So, necessarily, something's being attracted into it is no more fitting than another's,[31] nor is one thing's being retained in it any more fitting than another's. Also, if what retains the water in the clepsydra is the void that became filled by it, then why does [the water] descend when it is free of the instrument? The fact is that [the void] by itself should retain and hold onto the water and not let go of it such that it leaves. Additionally, it should not let the container that [the water] was in descend, since, [if the void] retained that water there, it should equally retain the container [there]. What would they say, then, about a container that is assumed to be lighter than the water?

---

30. The term *zarāqāt al-mā*ʾ (or perhaps *zurāqāt al-mā*ʾ) is not found in the *Physics* of the Arabic Aristotle nor in any of the Arabic commentaries that I have consulted, most notably the Arabic paraphrase of John Philoponus. The skeleton *z-r-ā-q-ā-t* could be vocalized either as *zarāqāt,* "a short javelin," or *zurāqāt,* "an instrument made of copper, or brass, for shooting forth naphtha" (see E. W. Lane, *Arabic-English Lexicon,* s.v. Z-R-Q). Neither is particularly helpful, but perhaps the Greek siphōn, "a reed or tube," was confused with a short javelin, and hence the tentative *siphon.*

31. **Y** has (inadvertently) omitted the phrase *minhu awlá min al-injidhāb ilá shay*ʾ, which appears in **Z**, **T**, and the Latin and completes the thought.

(١٩) فقد وضح مّما قلناه أنَّ الخلاء لا حركة فيه لا طبيعية ولا قسرية، فنقول ولا سكون فيه. وذلك لأنّه كما أنَّ الذي يسكن هو عادم الحركة ومن شأنه أن يتحرّك، كذلك الذي يسكن فيه هو الذي تعدم فيه الحركة ومن شأنه أن يتحرك فيه، والخلاء ليس من شأنه أنْ يتحرّك فيه.

(٢٠) وقد بلغ من غُلو القائلين بالخلاء في أمره، أنْ جعلوا له قوة جاذبة أو محرّكة ولو بوجهٍ آخر، حتى قالوا إن سبب احتباس الماء في الأواني التي تسمى سراقات الماء، وانجذابه في الآلات التي تسمى زراقات الماء؛ إنّما هو جذب الخلاء، وأنّه يجذب أول شيء الأكثف ثم الألطف. وقال آخرون؛ بل الخلاء محرّك للأجسام إلى فوق، وانّه إذا تخلْخل الجسم بكثرة خلاءٍ يداخله صار أخف وأسرع حركة إلى فوق.

(٢١) فنقول: لو كان للخلاء قوة جاذبة لما جاز أنْ يُختلف في أجزاء الخلاء بالأشد والأضعف، إذ سبيل كل جزء جذّابٍ من الخلاء سبيل الآخر، فما كان يجب أنْ يكون الإنجذاب إلى شيءٍ منه أولى من الانجذاب إلى شيء آخر، ولا الاحتباس في شيء منه أولى من الاحتباس في شيءٍ منه آخر. وسراقة الماءّ - إنْ كان حابس الماء فيها هو الخلاء الذي امتلأ بهّ - فلِمَ إذا خُلّي عن الآلة نزل؟ بل كان يجب أنْ يحبس الماء في نفسه، ويحفظه ولا يتركه يفارقه، ولا يدع الإناء الذي فيه أنْ ينزل أيضاً، لأنَّ ذلك الماء احتبس هناك، فيحبس الإناء أيضاً. فما يقولون في إناء يُتخذ أخفّ من الماء؟

(22) The same holds for their claim about the void's raising bodies, for either one of two situations must be the case. On the one hand, the interstitial void that belongs to the parts of the rarefied body might be what is required for its upward motion, and, being something required, it would be inseparable from it. In that case, the void will be inseparable from the rarefied body during its motion and so will locally move with it and will also need some place, when it is something locally moved, having a distinct interval with respect to position. On the other hand, it might not be inseparable, and, instead, during [the rarefied thing's] motion, one void after another [might] be continually replaced. If that is the case, then, for any void that we care to take, it will encounter it for an instant; but one thing does not move another in an instant, and after the instant it no longer is something being encountered in it. Perhaps, however, [the void] gives to [the rarefied body] some power that is of such a character as to remain in it and cause it to move—as, for example, it heats it or produces some other effect on it that remains in it. The mover would be that effect; and, as a result of that effect, each new void would produce some effect; and so that effect would continually strengthen and the motion would accelerate. Be that as it may, it would require that there be a certain direction in the void to the exclusion of some other that also belongs to that effect, whereas the void is homogeneous, making [the suggestion] necessarily impossible.

(23) Also, the void's being dispersed throughout the parts of a plenum miraculously necessitates a certain state in the sum of the parts without its being necessitated in each one of the parts. Indeed, it is absurd that each one of the discontinuous parts should not be moved by some motive cause, but that the whole be moved by it. The fact is that the whole that is composed of separate parts that are touching should undergo locomotion precisely because the local motion produced in each one of the parts

(٢٢) وكذلك قولهم في رفع الخلاء للأجسام، فإنّه لا يخلو إمّا أنْ يكون الخلاء المخلّل لأجزاء الجسم المتخلخل هو الذي يوجب حركته إلى فوقّ - وموجب الشيء ملازمٌ لهّ - فيكون ذلك الخلاء يلازم المتخلخل في حركته؛ فيكون متنقلاً معه ويحتاج إلى مكانٍ أيضاً إذا كان متنقلاً ذا بُعْدٍ متميّزٍ في الوضع، أو لا يكون ملازماً له، بل لا يزال يستبدل بحركته خلاء بعد خلاء. فإنْ كان كذلك، فأي خلاء نفرضه تكون ملاقاته له في آنٍ - وفي الآن لا يُحرّك شيءٌ شيئاً - وبعد الآن لا يكون ملاقياً فيه. بل عسى أن يعطيه قوة من شأن تلك القوة أنْ تبقى فيه وتحرّكه؛ مثلاً أنْ تسخّنه أو تؤثّر فيه أثراً آخر يبقى فيه، ويكون المحرّك ذلك الأثر، ويكون كل خلاء جديد يؤثّر فيه من ذلك الأثر. فلا يزال ذلك الأثر يشتد والحركة تسرع، إلاّ أنَّ إيجاب جهةٍ من الخلاء لذلك الأثر أيضاً من دون جهةٍ - والخلاء متشابهٌ - إيجابٌ مستحيل.

(٢٣) ومن العجائب أنْ يصير انبثاث الخلاء بين أجزاء الملاء موجباً حكماً في الجملة من الأجزاء دون أنْ يوجب في واحدٍ واحدٍ من الأجزاء. فإنه محالٌ أنْ تكون أجزاء منفصلة لا يتحرّك واحدٌ واحدٌ منها عن سببٍ محرّك، ولكن الجملة تتحرّك عنه؛ بل من الواجب أنْ تكون الجملة المركّبة عن أجزاء متباينة ومماسة إنّما تنتقل لوجود انتقال يحدث

exists. Now the rarefied thing whose parts are separated by the void is moved only as a result of the void, in which case each one of its parts will arrive above first; but when we take the simple finite parts in it, there is no void in any one of those parts, and so its ascending is not owing to the dispersal of the void [in it], but, rather, because the void surrounds it. In that case, it seems that, when [the rarefied thing's parts] are joined together and many, they are not acted upon by the void, while, when its parts are separated and few, its fewer parts will be acted upon as a result of the void, and it will just so happen that the whole moves upward. Despite that, not all the parts will be acted on in this way, but only certain bodies having specific natures; and it is their natures that require that the rarefaction come to be in this way, by the void. The reality of this would be that there is something that belongs to the bodies whose nature requires that some of its parts be at a certain distance from others, which produces the volume for that instance of rarefaction, while other bodies require a greater distance than that.

(24) It is also bizarre to picture some of these homogeneous parts running away from others until certain well-defined distances are completed between them, whereas that flight is in ill-defined directions haphazardly—one part fleeing upwards and another downwards, one to the left, another to the right—until the rarefaction is created. In this case, you will see either all the parts undergoing a mass retreat, or one standing its ground and being fled from, while others make a hasty retreat. Now, on the one hand, it would be amazing for one part of them to run away while the others do not, when their parts as well as the void that they are in are both homogeneous. On the other hand, however, it would be equally amazing for one part to take off to the right and another to the left, when the two parts are one and the same with respect to [their] nature and there is no variation in that in which there is the motion. From these things, then, it is clear that there is nothing to the void.

فـي واحدٍ واحدٍ من الأجزاء . فيكون المتخلخل المتباين الأجزاء بالخلاء ، إنّما يتحرك عن الخــلاء فيبلغ أولاً إلى فوق جزء جزء منه ، وكل جــزء من تلك الأجزا لا خلاء فيه ، إذا أخذنا البسـيط من الأجزاء المتناهية فيه ، فيكون ليس صعوده لانبثاث الخلاء ، بل لأجل إحاطة الخلاء به . فحينئذ يشبه أنْ يكون إذا اجتمع وكثر لم ينفعل عن الخلاء ، وإذا تفرّق وصغرت أجزاؤه انفعلت أجزاؤه الصغار من الخلاء ، وعرض فيه أنْ يتحرّك الكلّ إلى فوق. ويكون مع ذلك ليس كل الأجسام تنفعل هذا الانفعال ، بل أجسامٌ مّا لها طبائع مخصوصة ، وطبائعها توجب أنْ تتخلخل هذا التخلخل الكائن بالخلاء ، فتكون حقيقة هذا أنَّ شـيئاً من الأجسـام تقتضي طبيعته أنْ تتباعد أجزاؤه بعضها عن بعض بُعْداً مّا يفعل حجم ذلك التخلخل ، وأجسـام أُخر تقتضي ما هو أشـدّ من ذلك بُعْداً . ومن العجائب تصوّر هرب هذه الأجزاء المتجانسة بعضها عن بعض حتى يتمّ بينها أبعادٌ محدودة ، وكون ذلك الهرب إلـى جهـاتٍ غير محدودة كيف كانت ؛ فجزء يهرب بالطبع إلى فوق وجزء إلى أسـفل ، وجزء يمنة وجزء يسرة حتى يحدث التخلخل ، فترى أنَّ كل واحدٍ من هذه الأجزاء يعرض له الهرب ، أو يكون واحدٌ قارّاً مهروباً عنه ، والبواقي هاربة غير قارّة.

(٢٤) ومن العجائب أنْ يكون جزء منها واحدٌ لا يهرب والبواقي تهرب ، وأجزاؤها متشـابهة والخلاء الذي هي فيه متشابة ! ومن العجائب أيضاً أنْ يكون جزء واحدٌ يأخذ يمنة ، وجزء واحدٌ يأخذ يسـرة ، وحكم الجزئين فـي الطبيعة واحد ؛ وما فيه الحركة غير مختلف ! فمن هذه الأشياء يتبين أنَّ الخلاء لا معنى له .

(25) Also, in the clepsydra and siphon, certain things outside the natural course occur because of the *impossibility* of the void and the necessity that the flat surfaces[32] of bodies adhere to one another. In the case where there is a forcible separation, however, there results from the separation a certain replacement together with a change of a contacting surface, without there being any time at which one surface is free of some other contacting surface. So, when the flat surface of the water in the clepsydra naturally adheres to the flat surface of some contacting body, such as the surface of the finger, [the water] must be kept from falling as long as that surface accompanying it is prevented from falling, and so it necessarily stays put [in the clepsydra]. Were a void possible, however, and were the surfaces separated without a replacement, then [the water] would fall [from the clepsydra]. Also because of that, the water's attraction in the siphon turns out to be due to the adhesion of something having two extreme limits, where [that thing] has fallen to the second limit, [coupled with] the impossibility that, when the things being sucked give way to the suction, there should be a discontinuity in between [the two extreme limits] that would result in the existence of a void. That is why it is possible to raise a great weight by a small bowl snugly fitted to it and other amazing devices that are achieved as a result of the impossibility of the void.

32. Omitting *quwā* (powers) after *ṣafāʾiḥ* with **Z**, **T**, and the Latin.

(٢٥) وأنَّ هـذه الآلات السـرّاقة والزرّاقة إنّما تكون فيها أمـورٌ خارجةٌ عن المجرى الطبيعي لأجل امتناع وجود الخلاء، ووجوب تلازم صفائح الأجسام، إلاّ عند افتراقٍ قسري يكون مع بدلٍ ملاقٍ عوضاً عن المفارق بلا زمان يخلو فيه سـطحٌ عن سطح يلاقيه. فإذا كانت صفيحة الماء الذي في السرّاقة تلزم بالطبع صفيحة جسم يلاقيه، كسطح الأصبع، فيلزم أنْ يكون محبوسـاً عن النزول عند احتباس ذلك السـطح معوقاً عن النزول معه. فلزم أنْ يقف ضرورة، ولو جاز أنْ يكون خلاء وافتراق سـطوحٍ لا عن بدلٍ، لنزل. ولذلك ما صحّ انجذاب الماء في الزرّاقة للزوم ما قد نزل من طرفيه للطرف الثاني، وامتناع الانقطاع في البين المؤدي إلى وجود الخلاء وطاعة الممتصات للمصّ، ولذلك ما أمكن رفع ثقلٍ كبير بقِدْحٍ صغير مهندم عليه، وأشياء أُخر من الحيل العجيبة التي تتم بامتناع وجود الخلاء.

Chapter Nine

The essence of place and its confirmation and the refutation of the arguments of those who deny and are in error about it

(1) [Let the following be taken as given:] *Place* is that in which the body alone exists, and no other body can exist together with it in it (since [place] is coextensive with [body]). It can be entered anew and departed, and a number of placed things can successively enter into one and the same [place]. These descriptions (whether all or some) exist only because of a certain material or form or interval or some contacting surface, however it might be. Now, not all of them exist in the material and form, whereas the [absolute] interval has no existence (whether as void or not). Also, the noncontaining surface will not be a place, and only that which is the limit of the enclosing body contains. [Given all this,] *place* is itself nothing but the surface that is the extremity of the containing body. So it is what is proved to contain and be coextensive with the things subject to local motion, and which the locally moved thing fully occupies, and from which and to which the thing subject to local motion departs and arrives during motion, and in which it is impossible that two bodies exist simultaneously. So the existence and essence of place have become apparent.

<الفصل التاسع>

في ماهية المكان وتحقيق القول فيه ونقض حجج مبطليه والمخطئين فيه

(١) فإذا كان المكان هو الذي فيه الجسم وحده، ولا يجوز أنْ يكون فيه معه جسم آخر غيره؛ إذ كان مساوياً له، وكان يستجد ويفارق، والواحد منه تتعاقب عليه عدّة متمكنات، وكانت هذه الصفات كلّها أو بعضها لا توجد إلاّ لهيولى أو صورة أو بُعْد أو سطح ملاق كيف كان، وجميعها لا توجد في الهيولى ولا في الصورة، والبُعْد لا وجود له خالياً ولا غير خالٍ، والسطح غير الحاوي ليس بمكان، ولا حاوٍ منه إلا الذي هو نهاية الجسم الشامل. فالمكان هو هو السطح الذي هو نهاية الجسم الحاوي لا غير. فهو حاوٍ ومساوٍ ثابت للمتنقلات، ويملؤه المنتقل شغلاً، ويفارقه المنتقل بالانتقال عنه، ويواصله بالانتقال إليه. ويستحيل أنْ يوجد فيه جسمان معاً؛ فقد ظهر وجود المكان وماهيته.

(2) Sometimes the place coincides with a single surface, while at others it coincides with a number of surfaces from which a single place is formed (like the water in a river). Also, sometimes some of these surfaces happen to be moved accidentally, while others remain at rest; and at still other times all of them happen to rotate around the moving thing, while the moving thing remains at rest. What surrounds and what is surrounded might even move away from each other in some extremely complex way, as is the case with much of the Heavens.

(3) Here is something we should consider: When, for example, water is in a jar, and in the middle of the water there is something else that the water surrounds, and we now know that the water's place is the concave surface of the jar, then is it alone its place, or is the water's place [the concave surface] together with the outward convex surface of the body existing in the water? It would be as if the water had a figure that is surrounded by a concave surface and a convex surface and two other surfaces, having this form [i.e. Figure 1]: [In this case,] its place would not be the concave surface of what surrounds alone, but instead the sum of the surfaces that are in contact with all of its sides. So it would seem that its place is the sum of the surfaces that are in contact with the water on all sides: one as concave (belonging to the jar), one as convex (belonging to the body in the water) [see figure 2]. The earlier [figure, 1], however, is a single thing that the latter is not: namely, that the concave surface of the figure that we drew [that is, figure 1] does not alone surround, but, rather, the surfaces as a whole surround like one thing, whereas the latter [figure 2] is not something like that and, instead, the

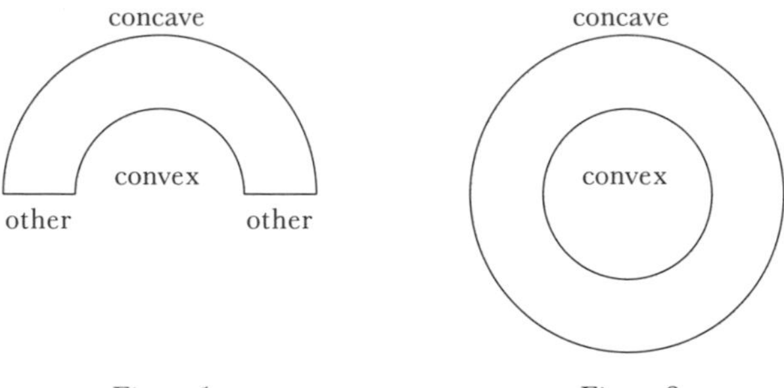

Figure 1 *Figure 2*

(٢) وقد يتفق أنْ يكون المكان سطحاً واحداً، وقد يتفق أنْ يكون عدّة سطوح يلتئم منها مكان واحدّ - كما للماء في النهرّ - وقد يتفق أنْ تكون بعض هذه السطوح متحركة بالعرض وبعضها ساكنة، ويتفق أنْ تكون كلّها متحرّكة بالدور على المتحرّك والمتحرّك عليه ساكن. وربّما كان المحيط والمحاط متخالفي المفارقة؛ كما في كثيرٍ من السماويات.

(٣) ويجب أنْ ننظر هنا: إذا كان ماء مثلاً في جرّة، وفي وسط الماء شيء آخر يحيط به الماء، وقد علمنا أنَّ مكان الماء هو السطح المقعّر من الجرّة؛ فهل هو وحده مكانه أو هو والسطح المحدّب الظاهر من الجسم الموجود في الماء مجموعين مكان الماء؟ كما لو كان الماء على شكلٍ يحيط به سطح مقبّب وسطح مقعّر، وسطحان آخران على هذه الصورة، ولم يكن السطح المقعّر من المحيط به وحده مكانه، بل جملة السطوح التي تلاقي جميع جهاته. فيشبه أنْ تكون جملة السطوح التي تلاقي الماء من جميع جهاته مقعّراً من الجرّة ومحدباً من الجسم الذي في داخل الماء، هو المكان له. ولكن ها هنا شيء واحد ليس هناك، وهو أنَّ المقعّر من الشكل الذي صورناه ليس يحيط به وحده، بل تحيط به السطوح جملة كسطحٍ واحدٍ، وهناك ليس الأمر كذلك، بل بالمقعّر كفاية في

concave surface is enough to surround, whether there is or is not the convex surface. Moreover, in the latter [figure] there are two distinct surfaces from which no single thing, which would be the place, is composed. As for the earlier figure, a single surface, which contacts as a single surface, is composed from the sum of the contacting surfaces. So it would seem that, inasmuch as a single thing comes to be from the sum, then the sum would be a single place, whereas the parts would be the place's parts; but none of them would be a place for the whole, and inasmuch as [nothing] makes up [that place], it will not exist.

(4) As for the arguments denying place, against the first one[1] it is said that place is an accident, and, from it, the name of that in which it is an accident [that is, the substance of the accident] might be derived. It has not been so derived, however, because it has not been common practice. Instances of this are frequent.[2] Even when it is derived, that name need not be the term *placed*. [That] is because the term *placed* is derived from being placed, but to be placed is not that something possessing an accident *is* something's place. A certain accident might be in something, while the name for something else is derived from it, like *begetting* in the *begetter*, and *knowledge* in the *knower*, where the name for the *object of knowledge* is derived from [the *knower*], but knowledge is not in [the object of knowledge]. So it might be that from *place* the name *placed thing* is derived, where the place is not in it, but it is in the place. The fact is, being the body [*x*] that surrounds another body [*y*] so that [*x*'s] internal surface is a place for [*y*] is an intelligible concept from which one might derive the name for that surrounding thing if the infinitive[3] of it were derived from it; but *place* is not an infinitive, and an infinitive has not turned out to be derived from it in this way. It does not necessarily follow from this, however, that place is not an accident.

1. See 2.5.2.

2. For example, it is an accident belonging to me that I am in my office right now, but most frequently we would not refer to me as the *officed one*. Similarly (to take an example from outside the category of *where*), it is accidental to me that I ate steak last night, but we probably would not normally call me the *having-eaten-steak one*.

3. The *maṣdar,* translated here *infinitive,* is the *nomen verbi* or abstract substantive, and it is the simple idea from which most Arab grammarians derive the compound idea of the finite verb. See W. Wright, *A Grammar of the Arabic Language,* 3rd ed., (Cambridge, UK: Cambridge University Press, 1967), vol. 1, §195.

الإحاطة بهّ - كان السطح المحدّب أو لم يكنّ - وهناك أيضاً سطحان متباينان ليس يأتلف منهما شيء واحد يكون مكاناً . وأمّا في هذا الشكل فإنّه يأتلف من جملة السطوح المتلاقية سطحٌ واحدٌ يلاقي سطحاً واحداً . فيشبه أنْ يكونّ - حيث يحصل من الجملة واحدّ - فإنَّ الجملة تكون مكاناً واحداً ، وتكون الأجزاء أجزاء المكان ، ولا يكون شيء منها مكاناً للكل ، وحيث لا يأتلفه لا يكون.

(٤) وأمّا حجج نفاة المكان ؛ فالحجّة الأولى يقال عليها أنَّ المكان عرضٌ ، ويجوز أنْ يشتق منه الاسم لما هو عرض فيه ، لكنه لم يشتق ، لأنّه لم يوقف عليه بالتعارف ، ومثل هذا كثير . وإذا اشتق فلا يجب أنْ يكون ذلك الاسم هو لفظ المتمكن ، فإنَّ لفظ المتمكن مشتقٌ من التمكّن ، وليس التمكّن هو كان الشيء ذا عرض هو مكان لشيءٍ ، ويجوز أن يكون في الشيء عرض ويشتق منه الاسم لغيره كالولادة ؛ فهي في الوالد ، والعلم فهو في العالم ، ويشتق منه للمعلوم الاسم وليس العلم فيه . فيجوز أنْ يشتق من المكان إسم المتمكن ولا يكون المكان فيه ؛ بل هو في المكان . ولكن كون الجسم محيطاً بجسم آخر حتى يكون سطحه الباطن مكاناً له هو معنى معقول ، يجوز أنْ يشتق منه إسم لذلك المحيط لو كان اُشتق له منه مصدر ، والمكان ليس بمصدر ، ولم يتفق أنْ يُشتق منه على هذه الجهة مصدر ؛ فليس يجب من هذا أنْ لا يكون المكان عرضاً .

(5) The response to the second skeptical puzzle[4] is that place is neither a body nor what coincides with body; rather, it is what surrounds in the sense that it primarily applies to its extremity. Our saying [that], "place is coextensive with the placed thing" is just a figure of speech, by which I intended that place is unique to the placed thing. So it is imagined to be really coextensive with it when that is not the case, and it is, in fact, coextensive only with its extremity. It is unique since, in the innermost containing extremity, there cannot be a body other than the body whose outermost extremity is coextensive with that extremity. When what was said about place's coinciding and being coextensive with the placed thing does not have to be accepted as literal, and neither is it a first principle evident in itself nor is it something that we need to prove, then the skeptical doubt does not necessarily follow.

(6) The third skeptical doubt[5] would follow only were we to say that *every* instance of locomotion (however it might be, whether essential or accidental) requires affirming a place. We never said that; rather, we only said that what must be affirmed as having a place is the thing essentially subject to locomotion. That is, it essentially leaves what encompasses and surrounds it as a result of itself, not because it follows upon [something else]. Now, the surface, line, and point follow upon the body that they accompany, and they simply do not leave it, although perhaps the body leaves behind all that accompanies it and all that it encircles, such that the line would leave a line and the surface a surface. So, [only] if the surface, line, and point were such that they could leave by themselves and by their very own motion [and, again, they are not], then the outcome would be what was said. (Their claim that the point is a privation[6] deserves consideration, but this is not the proper place; nor does the resolution of the doubt depend upon it, since it could be resolved without it.)

---

4. See 2.5.3.
5. See 2.5.4.
6. See 2.5.5.

(٥) وأمّا التشكيك الثاني فالجواب عنه أنَّ المكان ليس بجسم ولا مطابقاً للجسم، بل محيط به؛ بمعنى أنّه منطبقٌ على نهايته انطباقاً أولياً. وقولنا إنَّ المكان مساوٍ للمتمكن قولٌ مجازي أُريد به كون المكان مخصوصاً بالمتمكن فيُخيّل أنّه مساوٍ له بالحقيقة وليس كذلك، بل هو مساوٍ لنهايته بالحقيقة، وهو مخصوص به بالحقيقة. إذ لا يجوز أنْ يكون في باطن النهاية الحاوية جسم غير الجسم الذي تساوي نهايته الظاهرة تلك النهاية، وإذا لم يكن ما قيل حقاً من مطابقة المكان ومساواته للمتمكن واجباً تسليمه ولا أولياً بيّناً بنفسه؛ لا يحتاج إلى أنْ ندل عليه، لم يكن التشكيك لازماً.

(٦) وأمّا التشكيك الثالث؛ فإنّما كان يلزم لو قلنا: إنَّ كل انتقالٍ - كيف كان بالذات أو بالعرضّ - يوجب أن يثبت المكان؛ ونحن لا نقول ذلك؛ بل نقول إنَّ انتقال الشيء بالذات؛ وهو أُنْ يفارق كل ما يحصره ويحيط به مفارقة عن ذاته لا بسببٍ ملزوم، هو مفارقٌ بذاته، هو الذي يجب أنْ يكون مثبتاً للمكان. وأمّا السطح والخط والنقطة فإنّها تلزم ما هي معه من الجسم ولا تفارقه البتة، لكن الجسم قد يفارق كل ما معه وكل ما يطيف به. فيلزم أن يكون الخط قد فارق خطّاً والسطح سطحاً. فلو كان السطح والخط والنقطة ممّا يجوز أنْ تفارق بذاتها، وبحركةٍ بنفسها، لكان الحكم ما قيل. وأمّا قولهم إنَّ النقطة عدمٌ، ففيه نظر، وموضعه الخاص به غير هذا الموضع ولا تعلّق له بحل الشك، فقد ينحل دونه.

(7) The fourth skeptical doubt[7] would follow only if it were true that whatever is indispensable is a cause. That is not the case, however, for the effects and necessary concomitants of the effects are indispensable for the cause, while they are not causes. Similarly, the effect cannot dispense with the cause as well as the necessary concomitants of the cause, which are not [themselves] causes, nor do they have anything to do with causing the cause. The fact is that the cause that is indispensable is essential, and nothing else is prior to it. So place is something that is inevitable for motion, but it is not something that is causally prior to motion. Instead, it is perhaps prior by nature, such that whenever there is locomotion there is a place, while it is not the case that whenever there is a place there is locomotion. This priority, however, is not causal priority; and there must instead be something together with the existence of this that imparts existence to the effect in order that there be a cause. (This will be proved for you only in another discipline).[8] So place might be something more general than motion that is necessary for the motion, while not being a cause. Moreover, motion's existing in the mobile does not prevent place from also being a material cause of [motion], for a lot of people think that many things depend upon two subjects. Now, motion is a certain type of departure, and so it is quite likely that it depends upon something that departs and something that is departed from, both of which would be like subjects, in which case the motion would exist with respect to the mobile and the place. If this is false, then it is so by some proof other than the simple truth that motion exists in what is undergoing motion. In summary, place is something necessary on account of the subject of the motion, for the subject of motion is inevitably in place—inasmuch as it is something that may actually undergo motion, not simply inasmuch as motion actually exists in it—and its being in place is not its cause. So place is necessary because of motion's material cause.

7. See 2.5.6.
8. Cf. *Ilāhīyāt* 6.1.

(٧) وأمّا التشــكيك الرابع ؛ فإنّما كان يلزم لو كان صحيحاً أنَّ كل ما لا بُدّ منه فهو عِلّــة ؛ وليس كذلك. فإنّه لا بُدَّ أيضاً للعلة من المعلول ومن لوازم المعلول و<هي> ليســت عِللاً. كما لا بُدّ للمعلول من العِلّة ومن لوازم العِلّة التي ليســت <هي> بعللٍ، وليس شيء منها بعلّةٍ للعلّة، بل العِلّة هي التي لا بُدّ منها، وهو لذاته لا لغيره أقدم. فالمكان من الأمور التي لا بُدّ منها للحركة وليس أقدم من الحركة بالعلّية، بل عساه أنْ يكون أقدم منها بالطبع، حتى أنّه إنْ كانت نقلة كان مكان، وليس إذا كان مكان كانت نقلة. لكن هذا التقدم غير تقدّم العِلّية، بل يجب أنْ يكون الشــيء مع وجود هذا، مفيداً لوجود المعلول حتى تكون علّة. وهذا إنّما يتحقّق لك في صناعةٍ أخرى. فيجوز أن يكون المكان أمراً أعمّ من الحركة لازماً للحركة وليس بعلّة. وأيضاً فإنَّ كون الحركة موجودة في المتحرّك ممّا لا يمنع أنْ يكون المــكان أيضاً علّة عنصرية لها، فكثيرٌ مــن الأمور تتعلّق بموضوعين عند كثير من الناس. والحركة مفارقة ما، فلا يبعد أنْ تتعلّق بالمفارِق والمفارَق؛ على أنهما كلاهما موضوعان. فتكــون الحركة موجودة في المتحرّك وفي المــكان. فإنْ بطل هذا بطل ببيانٍ آخر لا لنفس صحــة وجود الحركة في المتحرّك. وبالجملة المكان أمرٌ لازمٌ لموضوع الحركة، فإنَّ موضوع الحركــة - من حيث هو بالفعل - جائز عليه التحرّك لا من حيث هو بالفعل موجودٌ فيه الحركة فقط، هو في مكان لا محالة، وإنَّ كونه في مكان ليس بعلّةٍ له، فالمكان لازمٌ لعلّة الحركة العنصرية.

(8) The fifth skeptical doubt[9] turns out to be true only if the growing thing that is in place must remain permanently in a single place, whereas when it is always exchanging one place for another, like the exchange of one quantity for another, then what was said is not necessary.

(9) Let us now refute the arguments of those who err about [place's] essence. As for the syllogism of those who said that place is subject to replacement and that the material is subject to replacement,[10] [the syllogism] is well known to be inconclusive, unless it is added that whatever is subject to replacement is place; and this we do not concede. [That] is because [only] some of what is subject to replacement is place; namely, it is that which bodies replace by coming to be in it. The same also holds for the claim that place is the first delimiting container, and so is form.[11] In other words, place is not every first container, but only that which contains something separate. Moreover, the form does not contain anything, because what is contained is separate from what contains, while the material is not separate from the form. Again, if by *delimiting* [in the statement "Place is the first delimiting container"] one means the limit by which something becomes delimited, then it is not commonly accepted that place has this description; and, in fact, it has been shown to be not true. If what is intended by *delimiting* is a container, [*delimiting*] is a synonym for *container,* and the meaning of the one is the meaning of the other. Furthermore, place contains and delimits the placed thing, where the placed thing is a body, whereas the form contains the matter, not a body in it.

(10) As for the argument of those advocating the interval that was based on the changing simple [surface], while the place of the placed thing does not change where nothing remains fixed but the interval,[12] our response is not to concede that the placed thing's place is not changing.

9. See 2.5.6.
10. See 2.6.3.
11. Ibid.
12. See 2.6.4.

(٨) وأمّا التشكيك الخامس، فإنّما يصح لو كان النامي الذي في المكان يجب أنْ يلزم مكاناً واحداً. وأمّا إذا كان دائماً يستبدل مكاناً بعد مكان، كما يستبدل كمّاً بعد كمّ، فليس ما قيل بواجب.

(٩) فلنبطل الآن حجج المخطئين في ماهيته. فأمّا قياس من قال إنَّ المكان يتعاقب عليه، والهيولى تتعاقب عليه، فقد عُلم أنّه غير منتج؛ اللّهم إلاّ أنْ يقال: فكل ما يتعاقب عليه مكانٌ، فلا نسلّم حينئذ، لأنَّ المكان هو بعض ما يتعاقب عليه، وهو الذي تتعاقب فيه الأجسام بالحصول فيه. وكذلك ما قيل إنَّ المكان أوّل حاوٍ محدّد؛ فهو الصورة، وذلك أنّه ليس المكان كل أول حاوٍ، بل الذي يحوي شيئاً مفارقاً. وأيضاً الصورة لا تحوي شيئاً، لأنَّ المحوي منفصل عن الحاوي، والهيولى لا تنفصل عن الصورة. وأيضاً فإنَّ المحدّد إنْ عُني به الطرف الذي به يتحدّد الشيء فليس بمشهورٍ أنَّ المكان بهذه الصفة، وأمّا آنّه غير حق فقد بانَ، وأمّا المحدّد الذي يراد به الحاوي فيو إسم مرادف للحاوي ومعناه معناه. وأيضاً المكان حاوٍ للمتمكن ومحدّده، والمتمكن جسم والصورة تحوي المادة لا جسماً فيها.

(١٠) وأمّا الحجة التي لأصحاب البُعْد؛ المبنية على وجود البسيط متبدلاً، والمتمكن غير مستبدلٍ مكانه، وليس هناك شيء يبقى ثابتاً إلاّ البُعْد؛ فنقول: إنّا لا نسلّم

The fact is that it is changing its place, except that it is neither undergoing motion nor remaining at rest. It is not something at rest because, in our opinion, it is not at a single place for a period of time. If by *something at rest* we did not mean this, but instead [meant] that whose relation to certain fixed things does not change, then, in this sense, it would be at rest; or if[13] its place were that which [as a result of its own action] it vacates, departs, and leaves, it would have preserved that place and would not have been changed as a result of itself and so would preserve a single place. At present, we do not mean by *something at rest* either the first or this one, for if we did so, it would be at rest. It is not undergoing motion because the principle of change does not belong to it, whereas the principle of change does belong to what undergoes motion—namely, that which is the first perfection belonging to what is in potency of itself such that, even if everything else were to remain the same vis-à-vis it, its state would change. I mean that if the things that surround and are joined with it were to remain just as they are with nothing happening to them, its relation with respect to them would still happen to change. This case is not like that one. So[14] the body need not necessarily be at rest or undergoing motion, for there are certain conditions belonging to body with respect to which it would neither be at rest nor undergoing motion in place. One of these is that it has no place.[15] Another is that it has a place, but there is a time when it does not have that same place, and it does not [have] the principle that causes it to leave.[16] Again, another is that it has a place that is the same for a period of time; however, we have not considered it during the period of time, but at an instant, in which case the body is neither resting nor being moved.[17]

---

13. Reading *lau* with **Z**, **T**, and the Latin (*si*), which is omitted in **Y**.

14. Reading *fa* with **Z**, **T**, and the Latin (*quia*) for **Y**'s *wa* (and).

15. For instance, according to Avicenna, the body of the universe as a whole has no place, and so it cannot be said to be moved or be at rest with respect to place, albeit it does change with respect to position (see 2.3, pars. 13–16).

16. The present example of change of the containing surface, while the contained body is not *of itself* changed, would be such an instance.

17. It should be noted that Avicenna is not denying that there might be motion at an instant, but only that there will not be motion (or rest) with respect to place at an instant, understood as a change of place at an instant. This claim is compatible with his earlier claim in 2.1 that there is motion at some posited point for only an instant.

أنَّ المتمكن غير مستبدلٍ مكانه، بل هو يستبدل مكانه إلاّ أنّه ليس بمتحرّكٍ ولا ساكن. أمّا أنّه ليس بساكنٍ فلأنّه ليس عندنا في مكان واحد زمان، اللّهم إلاّ أنْ نعني بالساكن لا هذا؛ بل الذي لا تتبدل نسبته من أمور ثابتة؛ فيكون ساكناً بهذا المعنى، أو الذي لو خُلي وحاله وترك على مكانه، حفظ ذلك المكان ولم يستبدل له من نفسه؛ فكان حافظاً لمكان واحد. ونحن لا نريد الآن بالساكن لا الأول ولا هذا، فإنْ أردنا أحد المعنيين كان ساكناً. وأمّا أنّه ليس بمتحركٍ فلأنّه ليس مبدأ الاستبدال منه، والمتحرّك بالحقيقة هو الذي مبدأ الاستبدال منه؛ وهو الذي الكمال الأول لما بالقوة فيه من نفسه، حتى أنّه لو كان سائر الأشياء عنده بحالها لكان حاله يتغيّر؛ أعنى لو كانت الأمور المحيطة به والمقارنة إياه ثابتة كما هي لا يعرض لها عارض، كان الذي عرض له تبدّل نسبته فيها، وأمّا هذا فليس كذلك. فليس بواجب أنْ يكون الجسم لا محالة ساكناً أو متحركاً، فإنَّ للجسم أحوالاً لا يكون فيها ساكناً ولا متحركاً في المكان؛ من ذلك أنْ لا يكون له مكان، ومن ذلك أنْ يكون له مكان، ولكن ليس له ذلك المكان بعينه في زمان ولا هو المبدأ في مفارقته، ومن ذلك أنْ يكون له مكان وهو له بعينه زماناً، ولكن أخذناه فيه لا في زمان؛ بل من حيث هو في آن، فيكون الجسم حينئذ لا ساكناً ولا متحرّكاً.

(11) As for what they mentioned about analysis,[18] analysis is not as they said, but, rather, involves isolating one thing after another of the parts existing in some thing. So analysis indicates the material in that it demonstrates that there is a certain form and that, by itself, [the form] does not subsist but belongs to a certain matter. So it is demonstrated that there is now a certain form and matter in this thing, whereas their alleged interval is not affirmed in this way at all. That is because the interval is affirmed only in the estimative faculty when the placed thing is removed and eliminated. So, perhaps when the placed thing is removed and eliminated, a certain interval is affirmed in the estimative faculty, whereas it is only the affirmation of the form that makes the matter necessary, not [the fact] that the estimative faculty imagined its removal—that is, unless something else is meant by *removal,* in which case the fallacy of equivocation is being committed. That is because, by *removal,* we mean that the estimative faculty imagines something as nonexistent. Now, this act of the estimative faculty with respect to the form would, in reality, make the matter necessarily cease to be, not affirm it; while, with respect to the placed thing, it would neither necessarily make it cease to be nor affirm it. We can dispense with [showing] that it does not necessarily make the interval cease to be, since the opposing party does not maintain as much. [It does not necessarily] affirm it because simply making the placed thing cease to be does not alone entail that, as long as it is not further added that the bodies encircling it are preserved as they are. If there is only a single body that is imagined not to exist, then, from imagining its elimination, there is no need to maintain an interval. Were it not for the estimative faculty's imagining its elimination, there would be no argument;

18. See 2.6.5.

(١١) وأمّـا ذكروا من حديث التحليل؛ فإنَّ التحليل ليس على الوجه الذي ذكروا، بل التحليل هو إفراد واحدٍ واحدٍ من أجزاء الشـيء الموجود فيه، فإنَّ التحليل يدلّ على الهيولـى بأنّه يبرهن أنَّ هناك صورة وأنّها لا تقـوم بذاتها؛ بل لها مادة، فيبرهن أنَّ في هذا الشـيء الآن صورة ومادة. وأمّـا البُعْد الذي يدّعونه، فهو شـيء ليس ثبوته على هذا القبيل، وذلك لأنَّ البُعْد إنّما يثبت في الوهم عند رفع المتمكن وإعدامه، فعسـى إذا رُفع المتمكـن وأُعدم وَجَبَ أنْ يثبت في الوهـم بُعْد. وأمّا المادة فإنّما يوجبها إثبات الصورة لا توهّم رفعها، اللّهم إلاّ أنْ يعني بالرفع معنىً آخر، فتكون المغالطة واقعة باشـتراك الإسم؛ وذلك لأنَّ الرفع نعني به توهّم الشـيء معدوماً. وهذا التوهم في الصورة يوجب بالحقيقة إبطال المادة لا إثباتها، وفي المتمكن لا يوجب لا إبطال البُعْد ولا إثباته. أمّا أنّه لا يوجب إبطال البُعْد فقد استغنينا عنه إذ الخصم لا يقول به، وأمّا إثباته فلأنَّ نفس إبطال المتمكن وحده لا يوجب ذلك ما لم يضف إليه حفظ الأجسام المطيفة به، موجودة على أحوالها. وأمّا إنْ كان جسـم واحد فقط، وتوهم معدوماً، فليس يجب من توهّم عدمه القول ببُعْدٍ لـولا توهـم عدمه لما قيل به؛ بل التوهّم يتبع التخيّل في إثبات فضاء غير متناهٍ دائماً؛ كان

but, as it is, the act of the estimative faculty always follows that of the imaginative faculty in affirming some infinite empty space, whether you remove a body or not. The existence of a certain interval that determines the measurement follows in the estimative faculty only owing to an elimination of some body with the condition that the bodies encircling it, which had measured the delimited interval, be preserved. Were it not for the measurement, there would be no need to eliminate some body in order that the imaginative faculty picture the interval. Despite all of this, let us grant that this interval is assumed in the estimative faculty, when a certain body or bodies are eliminated. How does one know that this act of the estimative faculty is not false, such that what follows upon it is absurd, and whether this assumption is, in fact, even possible, such that what follows upon it is necessary? Perhaps this advocate sets down as a premise that the estimative faculty judges it to be so, and whatever the estimative faculty demands is necessary. The case is not like that, for many existing situations are different from what the estimative faculty imagines. In summary, we should return to the beginning of the argument and say: Analysis distinguishes things whose existence truly is in the combination, but they are mixed in the intellect, in which case some are different from others in their potency and definition. Alternatively, some [of those things that truly exist but are mixed in the intellect] indicate the existence of something else, and so, when one selectively attends to the state of one of them it is carried from it to the other. In this case, *to remove* means *to set to one side and not to consider,* not *to eliminate.*

(12) The response to the argument after this one[19] is their own claim that the body requires place not on account of its surface, but on account of its corporeality—that is, if by [their claim] they mean that the body is in place not on account of its surface alone, but only on account of its

19. See 2.6.6.

جسم فرفعته أو لم ترفعه. وأمّا وجود بُعْدٍ مّا معيّن التقدير فإنّما يكون في الوهم تبعاً لعدم جسم، بشرط حفظ الأجسام المطيفة به التي كانت تقدّر البُعْد المحدود، ولولا التقدير لما احتيج إلى إعدام جسم في تخيّل البُعْد. ومع هذا كلّه، لنسلّم أنّ هذا البُعْد يُفترض عند الوهم إذا أُعدم جسم أو أجسام، فما يدريه أنّ هذا التوهم ليس فاسداً حتى لا يكون تابعه محالاً، وهل صحيح أنّ هذا الفرض ممكن حتى يكون ما يتبعه غير محال؟.. فعسى أنْ يقضي هذا القائل بأنّ الوهم يحكم عليه، وأنّ كل ما يوجبه الوهم واجب، وليس الأمر كذلك، فكثير من الأحوال الموجودة مخالفٌ للموهوم، وبالجملة يجب أنْ نرجع إلى ابتداء الكلام فنقول: إنّ التحليل تمييز أشياء صحّ وجودها في المجتمع ولكنها مختلطة عند العقل، فيفصل بعضها عن بعض بقوته وبحدّه، أو يكون بعضها يدل على وجود الآخر. فإذا تأمّل حال بعضها انتقل منه إلى الآخر، ويكون الرفع حينئذ بمعنى الترْك له والإعراض عنه إلى آخر لا بمعنى الإعدام.

(١٢) وأمّا الحجة التي بعد هذا فجوابها أنّ قول هذا القائل: إنّ الجسم يقتضي المكان لا بسطحه بل بجسميته، إنْ عنى به الجسم بسطحه وحده لا يكون في مكان، بل إنّما

corporeality, or if they mean that because[20] it is a body it can be in place. In this case the claim is true; and it does not necessarily follow from it that [a body's] place is a body,[21] for, when something requires a certain status or relation to something because of some description it has, the required thing need not have that description as well. So it is not the case that when the body needs certain principles (not inasmuch as it exists, but inasmuch as it is a body), its principles also have to be bodies. When the accident needs a subject inasmuch as it is an accident, its subject [does not have to be] an accident. If they mean by ["body requires place on account of its corporeality"] that every corporeal interval requires an interval in which it exists, then it just begs the original question. In summary, when [the body] requires place on account of its corporeality, it does not necessarily follow that it completely encounters the place in all of its corporeality. It is just as if it required a container on account of its corporeality; it would not necessarily follow that it completely encounters the container in all of its corporeality. In general, it is accepted that the body requires a place on account of its corporeality only to the extent that we accept that it requires a container on account of its corporeality. The sense of both claims is that the whole body is taken as a single thing that is described as being either in a place or in a container, where something's being *in* another in its entirety is not that it completely encounters it in its entirety. We certainly say that *all* of this water and the *whole* of it is in the jar, where we do not mean that the whole of it completely encounters the jar. The response to the argument after this that is based upon place's exactly equaling the placed thing[22] is also now completed.

---

20. Reading *li-anna* with two MSS consulted by **Y**, **Z**, **T**, and the Latin (*ex hoc quod*), which **Y** secludes.

21. The sense of *body* here is probably whatever is three-dimensional (see 1.2.1). So the sense is that the place of a body needs not be three-dimensional simply because the body is three-dimensional.

22. See 2.6.6.

يكون في المكان بجسـميته، أو عنى به أنّه لأنه جسـم يصلح أنْ يكون في مكانٍ فالقول حقٌ. وليس يلزم منه أنْ يكون مكانه جسماً، فإنّه ليس يجب إذا كان أمر يقتضي حكماً مّا أو إضافة إلى شيء مّا بسبب وصف له أنْ يكون المقتضى بذلك الوصف. فليس إذا كان الجسـم يحتاج إلى مبادئ – لكونه جسماً لا لكونه موجوداً – يجب أنْ تكون مبادئه أيضاً أجساماً، إذ كان العرض يحتاج إلى موضوع لكونه عرضاً أنْ يكون موضوعه عرضاً. وأمّا إنْ عنى به أنَّ كل بُعْدٍ من جسميته يقتضي بُعْداً يكون فيه، فهو مصادرة على المطلوب الأول. وبالجملة أنّه ليس إذا كان بجسـميته يقتضي المكان يلزم أنْ يلاقي بجميع جسميته المكان، كما أنّه لو كان بجسميته يقتضي الحاوي، فليس يلزم أنْ يكون بجميع جسميته يلاقي الحاوي. وبالجملة فإنّه غير مسـلّم أنَّ الجسم يقتضي بجسميته مكاناً، إلاّ بمقدار ما نسلّم أنّه بجسميته يقتضي حاوياً؛ ومعنى القولين جميعاً أنّ جملة الجسم المأخوذ كشيءٍ واحدٍ يوصف بأنّه في مكانٍ أو في حاوٍ. وليس كون الشيء بكليته في شيء، هو كونه ملاقياً له بكليته. فإنّا نقول: إنَّ جميع هذا الماء وجملته في هذه الجرّة؛ ولا نعني أنَّ جملته ملاقية للجرّة. أمّا الحجة التي بعد هذه، المبنية على مساواة المكان والمتمكن، فقد فُرغ عن جوابها.

(13) The one after that was based upon on the fact that place does not undergo motion.[23] Now, it is conceded that place does not essentially undergo motion, whereas it is not conceded—nor is it even commonly accepted—that it does not undergo accidental motion. Indeed, [just ask] anyone, and they won't deny that a thing's place might move (since they believe that the jar is a place), while inevitably allowing [what is placed in the jar] to move [with it].

(14) As for the next argument,[24] in the first place, it is based upon the biases of the masses, and that is no argument in things intellectual. Second, just as the man in the street does not disallow you from saying that the naturally disposed interval in the jar is empty and full, so likewise he does not disallow us from saying that the concave simple [i.e., the interior containing surface] in the jar is empty and full, provided that the man in the street understands both meanings (for he has no considered opinion about some expression when custom has not decidedly issued for him how to understand its meaning). Now, it seems that he would more readily apply that to the concave simple [surface] than he would to the other. That is because, on his understanding, what is full is that which surrounds something solid on its inside such that it meets it on every side. Don't you see that, in common parlance, he says that the *jar* is full and the *cask* is full? He does not give a thought to the alleged interval in the jar, but, rather, describes the container in this way, and the container is more like the simple [surface] than it is like the interval. In fact, the interval does not surround anything; but, rather, perhaps, should it exist, it is surrounded by the filled thing. Thus, we find that the common man is not averse to saying that the jar is full, while he might give pause to saying that the interior interval is full. Now, *jar* is

23. See 2.6.7.
24. See 2.6.8.

(١٣) وأمّا التي بعد تلك فهي مبنية على أنَّ المكان لا يتحرّك. والمسلّم أنَّ المكان لا يتحرّك بذاته، وأمّا أنّه لا يتحرّك، ولا بالعرض، فذلك غير مسلّم، ولا هو مشهور، فإنَّ الجمهور لا يأبون أنْ يتحرّك مكان الشيء، فإنّهم يرون الجرّة مكاناً ويجوزون لا محالة حركتها.

(١٤) وأمّا الحجة التي بعد هذه، فهي أول شيءٍ مبنية على عادات الجمهور، وذلك ليس بحجةٍ في الأمور العقلية. وثانياً أنّه كما لا يمنع العامة أنْ تقول إنَّ البُعْد المفطور في الجرّة، فارغٌ ومملوءٌ، كذلك لا يمنع أنْ نقول أنَّ البسيط المقعّر الذي في الجرّة؛ فارغٌ ومملوءٌ، على أنْ تفهم العامة المعنيين جميعاً. فإنّهم لا فتوى لهم في لفظٍ لم تجرِ لهم العادة بفهم معناه محصلاً، ويشبه أنْ يكونوا إلى أنْ يطلقوا ذلك في البسيط المقعّر أسرع منهم إلى غير ذلك. وذلك أنَّ المملوء في عرفهم هو الذي يحيط بشيء مصمت في ضمنه حتى يلاقيه من كل جهة. ألا ترى أنّهم يقولون، فيما بينهم، إنَّ الجرّة مملوءة والزّق مملوء، ولا يعرفون حال البُعْد الذي يدعونه في داخل الجرّة، بل يصفون الحاوي بهذه الصفة، والحاوي أشبه بالبسيط منه بالبُعْد. فإنَّ البُعْد لا يحيط بشيء، بل ربّما أحاط به ما يملؤه إنْ كان موجوداً. فلذلك نجد العامة لا يتحاشون أنْ يقولوا إنَّ الجرّة مملوءة، وربّما توقفوا عن أنْ

just a name for the earthenware substance made according to the shape of the interior, surrounding simple [surface]; and, if the simple [surface] were to subsist on its own, it would stand in for this jar, and what he says about the jar he would say about the simple [surface]. So it has become clear that, when he says that the jar is empty and full and deems that to be like saying [that] a certain place is empty and full, he has been led to what surrounds. The fact of the matter is that he disallows saying that the absolute simple [surface] is empty and full only because the absolute simple [surface] is not place; rather, place is a simple [surface] on the condition that it contain, and when a simple [surface] having this description is permitted to replace the absolute simple [surface], then he is not averse to that.

(15) The basis of the argument after this one is that place becomes an interval that provides every body with a place. That is, it is something properly necessary.[25] One of the things eagerly desired, however, is to show that this is *properly* [*necessary*], for if [the idea that] every body is in a place is not necessary in itself, then our attempt to make it necessary would be a fool's errand. Perhaps there is a greater necessity for one of the bodies not to be in a place. Also, if it is necessary, then there would be no need on our part to lay anything out. Now, if this premise were true—namely, that every body is in a place—*and* it were impossible for a container or anything the estimative faculty images to be a place other than the naturally disposed interval, *and* the naturally disposed interval were to exist, then [these] conditions would require us to hold that the interval be place. None of that, however, is necessary. (Oh, how great the twists and turns we undertake in order to contrive some clever way so that we can make all bodies be in place!) Let us even concede

25. See 2.6.9. The argument at 2.6 was that the interval ensures that every body has a place, while the doctrine of the containing surface would preclude certain bodies from having a place. The body in question is almost certainly the universe itself, because there is nothing outside to contain it. The implicit premise in the argument is that every body *necessarily* has a place, and so any account of place must ensure that every body has a place. Avicenna's move here is (1) to undermine the necessity of that premise and (2) to show that, even given the premise, the inferred conclusion that the interval is place does not necessarily follow.

يقولوا إنَّ البُعْد الباطن مملوء ، والجرّة إسم لجوهر الخزف المعمول على شكل البسيط الباطن المحيط . ولو كان البسـيط يقوم بنفسـه لكان مقام هذه الجرّة ، ولكانوا يقولون في البسيط مـا يقولونـه في الجرّة . فقد بانّ أنهم إذا قالوا إنَّ الجرّة فارغة ومملوءة ، وجعلوا ذلك كقولهم مكانٌ مّا فارغ أو مملوء ذهبوا إلى المحيط . نعم ، إنّما يمتنعون أنْ يقولوا في البسـيط المطلق أنّه فارغ ومملوء ، لأنَّ البسـيط المطلق ليس هو المكان ، بل المكان بسيط بشرط الإحاطة ، وإذا جعل بدل البسيط المطلق بسيط بهذه الصفة ، لم يتحاشوا عن ذلك .

(١٥) وأمّا الحجّة التي بعد هذه ، فمبناها على أنْ يصير المكان بُعْداً يجعل لكل جسم مكاناً ، وهو أمر صواب واجب ، وهذا التصويب شـهوة من الشـهوات . فإنّه إنْ لم يكن واجباً أنْ يكون كل جسم في مكانٍ وجوباً في نفسه ، كان سعينا في إيجابه سعياً باطلاً ، وعسى أنْ يكون الأوجب لبعض الأجسام أنْ لا يكون في مكان ، وإنْ كان واجباً لم يحتج إلى تدبير منا . ولو كانت هذه المقدمة صحيحة ؛ وهي أنَّ كل جسـم في مكان ولم يمكن أنْ يوجد لكل جسم حاوٍ أو شيء من الأشياء المتوهمة مكاناً غير البُعْد المفطورّ - وكان البُعْد المفطور موجوداً - كانت الحاجة تمسّنا إلى أنْ نقول بأنَّ البُعْدَ مكانٌ ، وليس شيء من ذلك واجباً . فما أشـد تحريفنا في أنْ نتمحل حيلة فيكون لنا أنْ نجعل كل جسـمٍ في

that all bodies are in place. It is not necessary that the interval be that place. [That] is because this thing might not be place, but a concomitant of the placed thing and common to every body [owing to] the commonality of place. If this claim is meant to be similar to the common man's belief that every body is in a place, then that is not an argument. Indeed, ascribing this belief to the common and average man—not inasmuch as he adopts some school of thought, but instead, as he speaks and acts according to the imagining of the estimative faculty and what is commonly accepted—is like ascribing another belief to him; namely, that whatever exists is in a place and can be pointed to. Both of these beliefs are alike in that the average man would give them up once [he sets] aside instinct and the imaginations of the estimative faculty, and consideration and thought prevail upon him. We have already explained the states of these premises in our discussion on logic[26] and made clear that they are products of the estimative faculty that fall short of those produced by the intellect, and it is not necessary to consider them. Even then, [the common man's] judgment that every body is in place comes with less assurance than his judgment that whatever exists can be pointed to and occupies space; and his understanding of the placed thing is no different from our understanding of position. Once again, even if this [premise that every body has a place] were true, what they say would not necessarily be true, from what we have explained. Place might be something different from the interval, and both of them belong to every body. So the interval's encountering all of a body is no indication that it is its place, since two things might belong to every body, and one of them, to the exclusion of the other, is place.

---

26. For Avicenna's exhaustive classification of the various types of premises, see *Kitāb al-burhān* 1.4.

مكان؟ ولنسلّم أيضاً أنَّ كل جسم في مكان، فليس يجب أنْ يكون ذلك المكان هو البُعْد؛ فإنّه يجوز أنْ يكون هذا المعنى ليس بمكان، لكنه لازم للمكان وعام لكل جسم عموم المكان. فإنَّ عنى بهذا القول أنْ يكون أشبه برأي الجمهور أنَّ كل جسم في مكان، فليس ذلك حجّه. فإنَّ نسبة هذا الرأي إلى الجمهورّ - والذين هم العامّة - من حيث لا يعتقدون مذهباً يذهبون إليه، بل يعملون ويقولون على ما في المشهور أو الوهم، كنسبة رأي آخر إليهم، وهو أنَّ كل موجود في مكان، وأنّه يشار إليه. وهذان الرأيان يتساويان في أنَّ العامة تنصرف عنهما بتبصير وتعريفٍ يرد عليهم بعد الفطرة العقلية والوهمية. وقد عرفنا أحوال هذه المقدمات حيث تكلمنا في المنطق، وبيّنا أنّها وهميات دون عقلية، ولا يجب أن يلتفت إليها. على أنَّ حكمهم في أنَّ كل جسمٍ في مكان ليس في تأكّد حكمهم في أنَّ كل موجود إليه إشارة وله حيّز، ولا هم يفهمون من التمكّن غير ما نفهم من الوضع. ثم لو كان هذا أيضاً حقاً لما وجب، على ما بينا أن يكون ما قالوه حقاً، وكان يجوز أنْ يكون المكان أمراً غير البُعْد، وكل واحدٍ منهما ممّا يوجد لكل جسم. فلا يكون وجود البُعْد ملاقياً لكل جسم دليلاً على أنّه مكانٌ له، إذ كان يجوز شيئان موجودان لكل جسم، وأحدهما دون الآخر مكانٌ.

(16) As for the argument that is after this one,[27] let it be known that there are two ways that seeking the extremity [can be understood]: one that is possible, the other that is not. The impossible way is for something that possesses volume to seek entry into a surface or extremity of a body with its volume. The possible way is that it seeks to encounter completely [the surface or extremity] so that it is surrounded by what surrounds, and this sense can be realized together with the supposition that the extremity is a place. Moreover, it is not the case that, when it seeks the extremity, it necessarily seeks some order among ordered intervals. On the contrary, it might seek only a certain order in the position without every position needing to be in an interval; and, instead, every position is just a certain relation between one body and another that is next to it in some direction, where there are no intervals but those of successive bodies.

(17) As for the arguments of those advocating the void, the response to the one based upon rarefaction and condensation[28] is that condensation might be [understood] in two ways. On the one hand, condensation might be by the coming together of the parts that are spread throughout the intervening air by forcing out the intervening air such that the parts come to replace it without there being some predisposed void. Just the opposite would hold in the case of rarefaction. On the other hand, condensation might not be in that the separate parts come together, but in that the matter itself receives a smaller volume at one time and a larger one at another, since both [volumes] are accidental to it, neither one of which is more fitting than the other. So, when it receives a smaller volume, it is said that there is condensation. Just the opposite would hold in the case of rarefaction. This is something that will be explained in another discipline;[29] and even if it is not explained here, there is no real harm, since at the most this option is false, while the earlier option that I gave as a response remains.

---

27. See 2.6.10.

28. See 2.6.12.

29. The discipline in question is that associated with the tradition surrounding Aristotle's *On Generation and Corruption* and which Avicenna treats in his *Kitāb fī al-kawn wa-l-fasād*. Discussions of condensation and rarefaction are dispersed throughout it, but see especially chapter 9, although he also summarizes many of the important details below in pars. 20–21.

(١٦) وأمّا الحجّة التي بعد هذا، فليعلم أنَّ طلب النهاية على وجهين: طلبٌ ممكن، وطلبٌ محال. فأمّا الطلب المحال فهو أنْ يكون ذو الحجم يطلب أنْ يدخل بحجمه سطحاً ونهايــة جســم، والطلب الممكن يطلب أنْ يلاقيه ملاقاة محــاطٍ به بمحيط، وهذا المعنى يتحقّق مع وضع النهاية مكاناً. ثم ليس إذا لم يطلب النهاية وجب أنْ يطلب ترتيباً في أبعاد مرتّبــة، بل ربّما طلــب ترتيباً في الوضع فقط من غير حاجةٍ أنْ يكون كل وضع في بُعْدٍ، بل على أنْ يكون كل وضعٍ هو نسبة ما بين جسمٍ وجسمٍ آخر يليه في جهة، ولا أبعاد إلاّ أبعاد الأجسام المتتالية.

(١٧) فأمّــا حجــج أصحاب الخلاء، فالجــواب على المبني منها علــى التخلْخل والتكاثــف إنَّ التكاثف على وجهين: تكاثف باجتماع الأجزاء المنبثة في هواء يتخللها، بــأنْ يخرج الهواء عــن الخلل؛ فتقوم الأجزاء مقامه من غيــر أن يكون هناك خلاء معدّ، ويقابله تخلْخل وتكاثف يكون لا بأنَّ الأجزاء المتفرقة اجتمعت، بل بأنَّ المادة نفسها تقبل حجماً أصغر تارة، وحجماً أكبر أخرى، إذ كان كلاهما أمرين عارضين له ليس أحدهما أولــى بــه من الآخر. فإذا قبل حجماً أصغر قيل إنّه تكاثــف ولمقابله تخلخل. وهذا أمرٌ يتبيّن في صناعةٍ أخرى، وإنْ لم يبيَّن في هذا الموضع لم يضر، إذ تكون غاية ذلك أن هذا القسم يبطل، ويبقى ذلك القسم الذي أجيب عنه.

(18) The report about the container of ash[30] is pure fiction. Even if it were true, it would be the whole of the container that is void, not some ash in it. Concerning the wineskin and wine,[31] it might be that the difference in the holding capacity of the wineskin in relation to the cask is not obvious to the senses. Also, the wine might be squeezed such that some vapor or air is expelled from it, and so it becomes smaller. It also might become smaller by a certain natural or forced condensation, as you [just] learned.

(19) As for the report about what grows,[32] the nutrition has the potential to extend between two contiguous parts of the organs and cause them to move apart and so settle between them, and so the volume expands. Now, were the nutrition to extend only into a void, then the volume at the time [the nutrition] is incorporated would be the same as it was before [the nutrition was incorporated], there being no increase.

(20) The response to the account concerning the phial[33] is based upon what was just mentioned about rarefaction and condensation—namely, that the body might be provided with a smaller or larger volume, sometimes occurring naturally and sometimes by force. So, just as there might be both natural and forced heating and cooling, the same situation holds for becoming large and small. Now, if this is possible, then not every decrease in some part of a body requires that the rest [of the parts] retain their original volume such that, when some part of the air filling the phial is taken away, the volume of [the air] must remain the same as it was, such that a void would be left behind. Now, if this is not necessary, then neither is that argument; whereas, if its contrary is possible, then it is possible that an increase in volume is proper to the air by its nature.

30. See 2.6.13.
31. Ibid.
32. See 2.6.14.
33. Ibid.

(١٨) وأمّا حديث إناء الرماد فهو كذب صرف، ولو كان ذلك صحيحاً كان الإناء كله خالياً لا رماد فيه أصلاً. وأما حديث الزّق والشراب، فيجوز أنْ يكون المقدار الذي للزّق لا يظهر تفاوته في الجُبّ حسّاً، ويجوز أنْ يكون الشراب ينعصر فيُخرج منه بخاراً أو هواء فيصير أصغر، ويجوز أنْ يصغر بتكاثفٍ طبيعي أو قسري على ما تعلمه.

(١٩) وأمّا حديث النامي فإنَّ الغذاء ينفذ بقوّةٍ بين متماسين من أجزاء الأعضاء ويحركهما بالتبعيد فيسكن بينهما، فينفسح الحجم. ولو كان الغذاء إنّما ينفذ في الخلاء لكان الحجم في حال دخوله وقبله حجماً واحداً لا زائداً.

(٢٠) وأمّا حديث القارورة. فإنَّ الجواب عن ذلك مبني على المذكور في التخلْخل والتكاثف، وهو أنّه من الجائز أنْ يكون الجسم يستفيد حجماً أصغر وحجماً أكبر، وأنْ يكون من ذلك ما هو طبيعي ومنه ما هو قسري، فكما أنّه يجوز أنْ يسخن ويبرد، ويكون منه ما هو طبيعي ومنه ما هو قسري، فكذلك الحال في العِظم والصغر. وإذا كان هذا جائزاً؛ لم يكن كل انتقاص جزء من جسم يوجب أنْ يبقى الباقي على حجمه الأول حتى يكون إذا أُخذ جزء من هواءٍ ماليءٍ لقارورةٍ، يجب أنْ يبقى على حجمه فيكون ما وراءه خلاء. وإذا لم يجب هذا؛ لم تجب تلك الحجّة، وإذا كان خلافه جائزاً، فجائز أنْ يكون

Moreover, in a certain case it will have to become larger in that some part is forcibly removed from it without providing any way for a body comparable in volume to what was removed to replace it. Now, if the removal of that part is impossible unless a certain expansion[34] is possible (such that what remains becomes the original volume and fills it in order to prevent the occurrence of a void), and the agent acting by force has the power needed to bring this possibility into actuality by its attracting [what remains] to one side, while [what remains] clings to the surface adjoining it on the other side (in other words, by forcibly expanding and enlarging it), then it yields to that agent and so expands as to become larger. [Given that], then part of what expanded will come to be outside the phial—namely, what was sucked out—and the rest will remain filling the phial, after having expanded owing to the necessary attraction through the length of the phial caused by the suction. When that suction ceases and it can return to its original state (in that either the water or air is attracted so as to occupy the place from which it moved when there was the decrease), it reverts to its normal state.

(21) Also, when we ourselves blew into the phial and then inverted it into water, a considerable amount of vapor came out of it and bubbled up in the water before the water came back and entered the [phial]. In this case, we know that it was we ourselves who necessarily forced something into [the phial], and, when the force ceases, it comes out. In other words, what we forced in enters either by extending into a void or by the condensation of what was already in it, so that what is being forced in will have a place, where that condensation will occur in the way that we ourselves maintain. We also see that what is acted on by force reverts to the natural state once the force ceases. So, if it is by extending into

34. **Y** has the perfect *inbasaṭa,* while **Z**, **T**, and the Latin (*nisi si ut dispergatur*) have the subjunctive *yanbasiṭa,* which is adopted here. Neither editor mentions the other's alternative reading in the critical apparatus; but the structure of the argument seems to require an *aw* subjective, and so the verb should be in the subjunctive.

الهـواء بطبعه يقتضي تحجّماً . ثم أنه يضطر ، في حـــالٍ ، إلى أنْ يصير أعظم ؛ بأنْ يُقتطع منه جزءٌ بالقسر من غير أن يجعل له إلى استخلاف جسم بدل ما يقتطع منه وفي حجمه سـبيل . فإذا كان اقتطاع ذلك الجزء منه لا يمكن ، أو ينبسـط انبســاطاً يصيّر الباقي في حجم الأول لامتناع وقوع الخلاء ووجود الملاء ، وكان هذا الانبساط ممكناً ، وكان للقاسر قــوة تحوج إلى خـروج هذا الممكن إلى الفعل بجذبه إياه في جهة ، ولزوم ســطحه لما يليه في جهة ، وذلك ببســطٍ منه وتعظيم إياه بالقسر ، أطاع القاسر فانبسط انبساطاً عظيماً ، وصار بعض ما انبســط واقعاً خارج القارورة وهو الممصوص ، وبقي الباقي ملء القارورة قد ملأها منبسطاً ، لضرورة الجذْب الماصّ بقدر القارورة . فإذا زال ذلك المصّ ، وجاز أنْ يرجع إلى قوامه الأول ؛ بأنْ يجذب ماء وهواء إلى شـغل المكان الذي يتحرّك عنه متقلصاً عاد إلى قوامه .

(٢١) ونحـن إذا نفخنا في القارورة ثم اكبناها علـى الماء خرجت منها ريحٌ كثيرة يبقْبق منه الماء ، ثم عاد الماء فدخل فيها ، فنعلم أنّا قد أدخلنا فيها بالقسر شيئاً لا محالة ، ولمّا زال القسر خرج . وذلك لا يخلو إمّا أنْ يكون دخول ما أدخلناه بالقسر هو بنفوذه في الخلاء ، أو يكون على ســبيل التكاثف من الموجود الــذي كان فيه حتى حصل للمُدخَل بالقســر مكانٌ ، ويكون ذلك التكاثف على ســبيل التكاثف الذي نقوله نحن . ونرى أنَّ للقسري منه أنْ يعود إلى الطبيعي عند زوال القاسر . فإنْ كان على سبيل نفوذٍ في الخلاء ،

a void until it reaches that place belonging to it, and that place neither belongs to it by force nor does [that place] abhor some airy body filling it such that it would reject it and push it out, and it does not belong to the nature of air to descend downward away from some void that it is in so as to be pushed into the water, then there should be no need for the air to leave and escape from it. If [the air] abhors the void, then why doesn't the other air follow suit? If the water abhors it, then why is it that, when the suction is very strong and stops [only] at the point where all the air that can be drawn out has been and then [the phial] is quickly inverted over the water, the water enters into it? If the void abhors the air's occupying it and pushes it out, it would more aptly abhor attracting water. Perhaps the void by its nature abhors air while attracting water; but, then, why does it let water puffed up in the air [in the form of clouds], which occupies the spaces between the existing air, fall? If its heaviness overcomes the attraction of that void, then why doesn't the heaviness of the inverted water of the phial overcome the void? Quite to the contrary, it is attracted! Is it more difficult to hold onto something heavy that is already possessed than to lift something heavy that is not possessed? So, once it is clear that this option [namely, that what we forcibly blew into the phial enters by extending into a void] is impossible, it remains that the cause of it is that the air seeks refuge in a smaller volume owing to compression and then, when [the compression] ceases, it expands to its [original] volume. Now, because there is another cause that requires an increase in volume and attenuation—namely, heating—if,[35] owing to the forced inflation, [the heat] is prevented from what it demands because the condensing pressure is stronger than the attenuation, then, when the obstacle is removed, the accidental heat will make the air have

35. Following **Z** and **T** which have the conditional *in,* which is omitted in **Y**. The Latin has *ideo* (therefore), which suggests that the Arabic exemplar for the text there may have been *fa.*

حتى حصل في ذلك المكان منه، وليس ذلك المكان له بقسري ولا مُبغضاً لجسم هوائي يملؤه فينفيه عنه أو يدفعه، ولا من طبيعة الهواء أنْ ينزل متسفلاً عن خلاء يحصل فيه نزولاً مندفعاً في الماء، فينبغي أنْ لا يحتاج الهواء إلى أنْ يفارقه ويتخلص عنه. فإنْ كان الخلاء هو الذي يأباه؛ فلِمَ لا يأتي الهواء الآخر؟ وإنْ كان الماء يأباه، فلِمَ إذا أحكم المصّ ثم تُرك حتى يخرج من الهواء ما من شأنه أنْ يخرج، وكُبّ سريعاً على الماء دخله الماء؟ فإنْ كان الخلاء يأبى أنْ يشغله الهواء ويدفعه، فلأنْ يأبى جذب الماء أولى. فلعل الخلاء يبغض الهواء بطبيعته ويجذب الماء؛ فلِمَ يترك الماء المنفوش في الهواء الشاغل لخَلل الهواء الخالي ينزل؟ وإنْ كان ثقله يغلب جذب ذلك الخلاء، فلِمَ ثقْل. الماء المكبّ على القارورة لا يغلب الخلاء، بل ينجذب وإمساك الثقيل المشتمل عليه أصعب من إشالة الثقيل المباين؟ فإذا استبانت استحالة هذا القسم، بقي أنّ السبب فيه إلتجاء الهواء إلى حجم أصغر للانضغاط فإذا زال انبسط إلى حجمه. ولأجل أنَّ هناك سبباً آخر يقتضي حجماً أكبر وهو التسخّن والتلطف، بقسر تحريك النفخ، إنْ كان ممنوعاً عن مقتضاه بالضغط الذي يكثّفه، أشدّ من تلطيف هذا، وقد زال العائق فاقتضى السخونة العارضة أنْ يصير الهواء

a larger volume than it had before the inflation. Because that heating is accidental to this, when it ceases, the air contracts to the volume that its nature requires should there have been no heat, and so the water comes back and enters, owing to the impossibility of a void's occurring. It is because of this that we experience the air emerging from what was vigorously inflated as first bubbling up and then beginning to attract the water into itself, just as if a finger unplugged the mouth of the phial. Also, [when] it is heated with a hot fire that does not shatter it and then is inverted over the water, it first happens to bubble, and then the water is sucked into it.

(22) The response to the argument that is after this one[36] is related to this response. That is because what is moved pushes the air that is immediately in front of it, and that continues to wherever it is that the preceding air no longer yields to the push, and the surge [of air] becomes compressed between what is being pushed and what is not and is forced to receive a smaller volume. Just the opposite happens to what is behind [the compressed air], for some of it is attracted along with [the surging air], and some of it resists and so is not attracted, in which case what is between the two rarefies, and there is a larger volume. From that there comes to be an ongoing and balanced normal state.

(23) So let us be content to this extent with the discussion about place, and let us now talk about time.

---

36. See 2.6.15.

أعظم حجماً من الحجم الذي كان قبل النفخ. ومن أجل أنْ تلك السخونة عرضية بهذا وتزول، ينقبض الهواء إلى الحجم الذي اقتضته طبيعته، لو لم تكن تلك السخونة، فيعود الماء فيدخل لا ستحالة وقوع الخلاء. فلهذا ما نشاهد من أنَّ المنفوخ بالقوة أولاً يتبقبق منه هواء يخرج، ثم يأخذ في جذْب الماء إلى نفسه؛ كما لو سُدّ فم القارورة بأصبع، وسخنت بنار حادةٍ لا تكسرها، ثم أُكبّت على الماء، عَرَض أولاً تبْقبقٌ ثم امتصاص منها للماء.

(٢٢) وأمَّا الجواب عن الحجّة التي بعد هذه، فمناسب لهذا الجواب، وذلك لأنَّ المتحرّك يدفع ما يليه من قدام من الهواء، ويمتد ذلك حيث لا يطيع فيه الهواء المتقدم للدفع، فيتلبّد الموج بين المندفع وغير المندفع، ويضطر إلى قبول حجم أصغر، وما خلفه يكون بالعكس؛ فيكون بعضه ينجذب معه، وبعضه يعصي فلا ينجذب، فيتخلْخل ما بينهما إلى حجمٍ أكبر، يحدث من ذلك وقوفٌ معتدل عند قوام معتدل.

(٢٣) فليكفنا هذا القدر من الكلام في المكان، فلنتكلم الآن في الزمان.

Chapter Ten

Beginning the discussion about time, the disagreement of people concerning it, and the refutation of those erring about it

(1) The inquiry about time is akin to the one about place in that it is one of the things that is inseparable from every motion, and the disagreement among people about its existence and essence is just like that about place. Some people have denied that time has any existence, while others believed that it has an existence, but not at all as [existence] occurs in external concrete particulars, but as a product of the estimative faculty. Still others believed that, although it does exist, it is not a single thing in itself; rather, it is in some way a relation that certain things (whatever they might be) have to other things (whatever they might be). So it was said that time is the collection of moments, where the *moment* is some event that happens, which is taken by supposition, [and that event] is simultaneous with some other event[1] and so it is a moment for the other, whatever accidental occurrence it might be. Others have given time a certain existence and subsistent reality, while others yet even made it a substance subsisting in its own right.

1. **T** additionally has the example of Zayd's arriving with the sunrise, but this is almost certainly a later interpolation.

<الفصل العاشر>

فصل في ابتداء القول في الزمان واختلاف الناس فيه ومناقضة المخطئين فيه

(١) إنَّ النظر في أمر الزمان مناسب للنظر في أمر المكان، لأنَّه من الأمور التي تلزم كل حركة، والحال في اختلاف الناس في وجوده وماهيته كالحال في المكان. فمن الناس مَنْ نفى أنْ يكون للزمان وجود، ومنهم مَنْ جعل له وجوداً، لا على أنَّه في الأعيان الخارجة البتّة بوجهٍ من الوجوه، بل على أنَّه أمرٌ متوهم. ومنهم مَنْ جعل له وجوداً لا على أنَّه أمرٌ واحدٌ في نفسه؛ بل على أنَّه نسبة مّا على جهةٍ ما لأمورٍ أيّها كانت إلى أمورٍ أيّها كانت، فقال: إنَّ الزمان هو مجموع أوقاتٍ، والوقت عرض حادث يفرض وجود عرضٍ آخر مع وجوده، فهو وقت الآخر، أي عرض حادث كان ومنهم مَنْ جعل للزمان وجوداً وحقيقة قائمة، ⟨و⟩ منهم منْ جعله جوهراً قائماً بذاته.

(2) Those who have denied the existence of time have relied on certain skeptical puzzles.[2] One of them is that if time exists, it is either divisible or indivisible. On the one hand, if it is indivisible, then it would be impossible for years, months, days, and hours, as well as past and future, to belong to it. If, on the other hand, it it is divisible, then it exists with either all of its parts or some of them. If it exists with all of its parts, the past and future parts of it must exist together simultaneously. If some of its parts exist, while others do not, then whichever part we consider must occur as present, future, and past, or as days, hours, and the like. Those who affirm time in general agree that both the past and the future are nonexistent, whereas, if the present is divisible, the same question necessarily arises about it, while, if it is indivisible, it is what they call the *present instant*[3] and is not a time. Moreover, [the present instant] cannot exist in actuality. Were it to do so, it would either endure or cease to be. If it were to endure, then part of it would be earlier and another part later, and the whole of it would not be the present instant, and also the past and future would simultaneously be in a single present instant; which is absurd. If it ceases to be, then it does so either in some immediately adjacent instant where between the two there is no time, or in some instant where between the two there is some time. If it ceases to be in some instant where between it and [the present instant that is ceasing] there is some time, then it must endure for some time, which we have already refuted. If it ceases to be in some immediately adjacent instant, then the present instant would be immediately adjacent to the instant on the continuum without any period of time being interposed between the two; but this is something that those who affirm time deny.

2. Most of the following puzzles, or at least variations on them, are mentioned by Aristotle in *Physics* 4.10.

3. The Arabic *ān,* like the Greek *to nun,* can mean both *an instant* and *now;* hence the slight overtranslation *present instant* here. This argument and the subsequent one, which are taken from the list of temporal puzzles found in Aristotle's *Physics* 4.10, draw on this double sense of *al-āna.*

(٢) فأمّـا مَـنْ نفى وجـود الزمان، فقد تعلّق بشـكوكٍ منها أنَّ الزمـان، إنْ كان موجوداً، فإمّا أنْ يكون شـيئاً منقسـماً، أو يكون شـيئاً غير منقسـم. فإنْ كان غير منقسـم فمسـتحيل أن يكون منه سنون وشهور وأيام وسـاعات وماض ومستقبل. وإنْ كان منقسماً؛ فإمّا أنْ يكون موجوداً بجميع أقسامه أو ببعضها. فإنْ كان موجوداً بجميع أقسـامه؛ وَجَبَ أنْ يكون الماضي والمستقبل منه موجودين معاً، وإنْ كان بعض أقسامه موجوداً وبعضها معدوماً، فلا يخلو إما أنْ تكون القسمة التي إياها نعتبر واقعة على سبيل الحاضر والمستقبل والماضي، أو واقعة على سبيل الأيام والساعات، وما أشبه ذلك. فأمّا الماضي والمسـتقبل فكل واحدٍ منهما - باتفاقٍ من مثبتي الزمانّ - معدومٌ، وأمّا الحاضر فإنْ كان منقسـماً وجبت المسـألة بعينها، وإنْ كان غير منقسم، كان الأمر الذي يسمونه آنـاً وليـس بزمان. ومع ذلك فإنّه لا يجوز أنْ يوجد بالفعـل، ولو وجد بالفعل لم يخل إمّا أنْ يبقى وإمّا أنْ يُعدم، فإنْ بقي كان منه شيء متقدماً وشيء متأخراً، ولم يكن كله آناً، وكان الماضي والمسـتقبل معاً في آنٍ واحدٍ وهذا محال. وإنْ عُدم لم يخلُ إمّا أنْ يُعدم في آنٍ يليـه لا زمان بينهمـا، وإمّا أنْ يُعدم في آنٍ بينه وبينه زمان. فإنْ عُدم في آنٍ بينه وبينه زمان، لزم أنْ يبقى زماناً، وقد أبطلنا ذلك. وإنْ عُدم في آنٍ يليه، كان الآن يلي الآن على الاتصال من غير تخلل زمان بينهما، وهذا ممّا يمنعه مثبتو الزمان.

(3) Furthermore, how could time in general exist? Any time we care to take might be delimited by two given instants assigned to it by us: one past instant and another instant that, relative to the past, is future. No matter what the situation, the two will not be able to exist together, and, instead, one will not exist. Now, if one does not exist, then how can that which needs a certain limit—that does not exist—in fact exist (for how can something have a nonexistent limit)? In short, how could there be a certain continuous thing [namely, time] between something that does not exist and something that does? This is a powerful sophism upon which those who deny time rely.

(4) They also give an argument [that assumes the following]: A certain amount of time inevitably belongs to motion in that it is a motion. In that it is motion, this motion does not need some other body different from its body also to be moved. (The fact is that it might need that [other body] in some cases, not in that it is a motion, but because, in order to produce the motion, the one bringing about [the motion's] existence needs [another body] to undergo motion; but this is neither a condition of motion *qua* motion nor one of its concomitants.) So, given the above [assumptions], any motion that you posit as existing necessarily entails that a certain time belong to it inasmuch as there is a motion; but [the motion], inasmuch as it is a motion, does not necessarily entail that there is another motion. If that is the case, then consequent upon each motion is a certain private time that applies to no other motion, just as a private place is consequent upon it. Also, there would be one time for [the different motions] only in the way that there is one place for them. That is, *one* [is predicated] by way of being a universal, but our discussion is not about that. So, when motions are together, their times must also be together, where they will be together either with respect to place, subject, rank, nature, or anything else except being

(٣) ثـم بالجملة كيف يكون للزمـان وجود ، وكل زمان نفرضـه فقد يتحدّد عند فارضه بآنين ؛ آن ماض وآنٍ هو بالقياس إلى الماضي مسـتقبل . وعلى كل حال ، لا يصح أنْ يوجـدا معـاً ، بل يكون أحدهما معدومـاً ؟ وإذا كان معدوماً فكيف يصح وجود ما يحتـاج إلى طرف هو معدوم ؛ فكيف يكون للشـيء طرف معـدوم وبالجملة كيف يكون شيء واصلاً بين معدوم وموجود ؟ فهذه هي الشبهة القوية التي يتعلق بها مَنْ ينفي الزمان .

(٤) ويقولـون أيضاً إنّه إنْ كان لا بُـدَّ للحركة في أنْ تكون حركة ، من أنْ يكون لها زمان ، وليس تحتاج هذه الحركة في أنْ تكون حركة إلى أن يكون جسم آخر يتحرك أيضا غير جسمها ، بل ربما احتيج إلى ذلك في بعض الأمور ، لا أن تكون حركة بل لأنَّ موجدها يحتاج في أنْ يحرّك إلى أنْ يتحرّك ؛ وهذا ليس شـيئاً من شـرط الحركة بما هي حركة ولا مـن لوازمها . فإذا كان كذلك ، فأيّة حركةٍ فرضتها موجودة يلزمها ، من حيث هي حركة ، أنْ يكـون لها زمان ، ولا يلزمها من حيث هي حركـة أنْ تكون هناك حركة أخرى . وإذا كان كذلك ، كان كل حركة مسـتتبعة زماناً على حِدة ، غير موقوف على حركةٍ أخرى ، كما يسـتتبع مكاناً على حِدة ، ولا يكون لها زمان واحد إلاّ على نحو ما يكون لها مكان واحـد ، أي الواحـد بالعموم ، وليس كلامنا في ذلك . فإذا كانـت الحركات معاً ؛ كانت أزمنتهـا لا محالـة معاً ، ولا يخلو إما أن تكون معيتها في المـكان أو في الموضوع أو في الشرف أو في الطبع ، أو في شيء آخر غير المعيّة في الزمان . لكن جميع وجوه معاً لا يمنع

together in time. None of the ways of being together, however, precludes some being before and some being after—that is, some existing while others do not. So it remains that their being together is [a case] of simultaneity, where *simultaneity* [means] that many things occur at a single time, or a single instant, or a single limit of time. From that it necessarily follows that many times would have a single time; but the discussion concerning the sum of that [single] time together with [the many times] is, in this sense, just like the discussion about those that were joined together in [that single time], in which case there would necessarily be an infinite number of simultaneous times. Also, in the opinion of you [Aristotelians], times follow upon motions, and so there would necessarily be an infinite number of simultaneous motions. This is something impossible, whose existence you yourselves reject and deny.

(5) Due to these skeptical puzzles and the fact that time must have some existence, many people felt compelled to give time some other manner of existence—namely, the existence that is in the activity of the estimative faculty. Now, the things that characteristically exist in the act of the estimative faculty are those things that are concomitant with the connotational attributes that, when they are grasped by the intellect and correlated with one another, produce there certain forms of relations whose existence is only in the estimative faculty. So this group made time something that is impressed on the mind as a result of a certain relation of what is undergoing motion to the two limits in the spatial magnitude it [is traversing], where [the mobile] is in actual proximity to one of them while not being in actual proximity to the other, since, in concrete particulars, it cannot occur *here* simultaneously with its occurring *there*, but it can in the soul. [That is] because, in the soul, the conceptualization of the two and the conceptualization of what connects them exist simultaneously, but nothing in concrete particulars exists that

أنْ يكـون بعضها قبل، وبعضها بعد؛ أي بعضها يكون موجوداً وبعض معدوماً. فبقي أنْ تكون معيتها المعيّة التي بالزمان، والمعيّة التي بالزمان هي أنْ تكون أشـياء كثيرة في زمانٍ واحد، أو في آنٍ واحد هو ظرف زمانٍ واحد. فيجب من ذلك أنْ يكون للأزمنة الكثيرة زمـانٌ واحد، ويكون الكلام فـي جميع ذلك الزمان معها في هذا المعنى كالكلام في التي هي مجموعة فيه. فيلزم أنْ تكون أزمنة بلا نهاية معاً، وعندكم أنَّ الأزمنة تتبع الحركات، فيلزم أنْ تكون حركات لا نهاية لها معاً، وهذا من المستحيل الذي تدفعونه وتمنعون وجوده.

(٥) فمن جهة هذه الشـكوك، ووجوب أنْ يكـون للزمان وجود، اضطر كثير من النـاس إلى أنْ جعل للزمان نحـواً من الوجود آخر، وهو الوجود الذي يكون في التوهّم. والأمور التي من شـأنها أنْ توجـد في التوهّم هي الأمور التي تلحـق المعاني إذا عُقلت ونوسـب بينها، فتحدث هناك صور نسبٍ، إنّما وجودها في الوهم فقط. فجعلوا الزمان شـيئاً ينطبع في الذهن، من نسـبةٍ للمتحرّك إلى طرفي مسـافته الذي هو بقرب أحدهما بالفعل وليس بقرب الآخر بالفعل، إذ حصوله هناك لا يصح مع حصوله هاهنا في الأعيان لكـن يصح في النفس. فإنّه يوجد في النفس تصورهما وتصوّر الواسـطة بينهما معاً فلا يكون في الأعيات أمر موجود يصل بينهما، ويكون في التوهم أمر ينطبع في الذهن، وأنَّ

connects the two.[4] In the act of the estimative faculty, however, something is impressed on the mind—namely, that, between its existence *here* and its existence *there*, there is a certain thing during the equivalent of which the distance is traversed at this speed belonging to either those motions or the number of combined motions and rests. In this case, this is a certain measurement of that motion that has no [external] existence but is [something that,] in itself, the mind brings about as a result of motion's limits actually occurring in [the mind] simultaneously. Other examples [of products of the estimative faculty] include the predicate, logical subject, premise, and analogous things that the mind requires for intelligible matters and the relations among [such matters], none of which are in [concrete] existing things.

(6) The group that we mentioned at the beginning said that time is nothing but a collection of moments; for when you order successive moments and collect them together, you do not doubt that their collection is time.[5] Consequently, once we define the moments, we define time. Now, the moment is nothing more than what the one fixing the moment needs—namely, he designates a certain starting point of some given event that will happen. So, for instance, we say that such-and-such will occur after two days, meaning that it will occur with the sunrise following two sunrises; and so the sunrise is the moment. If it were replaced with the coming of Zayd, that would be just as fine as the sunrise. So, then, the sunrise becomes a moment only by the speaker's designating it so. Had he wanted, he could have made something else a moment, except that the sunrise is more prevalent, better known, and more commonly accepted. Hence, that and similar things have been chosen to set the moment. So time is the sum of things that either sets moments or can be stipulated as certain fixed moments. They also claim that time has no existence other than in this way, which is recognized from the previously mentioned skeptical puzzles.

---

4. **Y** has (inadvertently) omitted the text *maʿan fa-lā yakūna fī al-aʿyān amr mawjūd yaṣilu baynhumā* (simultaneously, but nothing in concrete particulars exists that connects the two), which is found in **Z**, **T**, and the Latin ([*simul*]; *non habet autem esse in sensibilibus aliquid existens inter illos*).

5. This view was the common one taken by most *mutakallimūn*, or Islamic speculative theologians. See Jon McGinnis, "The Topology of Time: An Analysis of Medieval Islamic Accounts of Discrete and Continuous Time," *The Modern Schoolman* 81 (2003): 5–25.

بين وجوده ها هنا وبين وجوده هناك شيئاً في مثله تُقطع هذه المسافة بهذه السرعة والبطء اللذين لهذه الحركات أو لهذا العدد من الحركات والسكونات المتركبة. فيكون هذا تقديراً لتلك الحركة لا وجود له، لكن الذهن يوقعه في نفسه بحصول أطراف الحركة فيه بالفعل معاً؛ مثل ما انَّ الحمل والوضع والمقدمة وما يجري هذا المجرى، أشياء يقضي بها الذهن على الأمور المعقولة ومناسبات بينها، ولا يكون في الأمور الموجودة شيءٌ منها.

(٦) وقالت الطائفة التي ذكرناها بَدْءاً إنَّ الزمان ليس إلاّ مجموع أوقاتٍ. فإنك إذا رتبت أوقاتاً متتالية وجمعتها لم تشك أنَّ مجموعها الزمان، وإذا كان كذلك، فإذا عرفنا الأوقات عرفنا الزمان. وليس الوقت إلاّ ما يوجبه الموقت، وهو أنْ يعيّن مبدأ عارضٍ يعرض؛ فنقول مثلاً يكون كذا بعد يومين، معناه أنّه يكون مع طلوع الشمس بعد طلوعين، فيكون الوقت طلوع الشمس، ولو جُعل بدله قدوم زيدٍ لصلح في ذلك صلوح طلوع الشمس. فاذن إنّما صار طلوع الشمس وقتاً بتعيين القائل إياه، ولو شاء لجعل غيره وقتاً، إلاّ أنَّ طلوع الشمس قد كان أعمّ وأعرف وأشهر، ولذلك اختير ذلك وما يجري مجراه للتوقيت. فالزمان هو جملة أمور هي أوقات موقّتة، أو من شأنها أنْ تُجعل أوقاتاً موقته. وقالوا إنَّ الزمان على غير هذا الوجه لا وجود له، يُعرف ذلك من الشكوك المذكورة.

(7) Another group said that time is an eternal substance, and how could it not be a substance [they argued,] when it is something whose existence is necessary?[6] Indeed, the necessity of its existence is such that it does not need to be established by proof. In fact, whenever you try to eliminate time, you necessarily establish it. [That] is because you eliminate it either before or after something; but when you do that, a certain before-ness or after-ness shows up together with its elimination, in which case you have established time together with its elimination, since there is no before-ness and after-ness that have this form unless they either belong to time or are a result of time. So time exists necessarily. Now, whatever necessarily exists cannot have its existence eliminated, and whatever cannot have its existence eliminated is not an accident, whereas whatever exists and is not an accident is a substance. When it is a substance that exists necessarily, then it is an eternal substance. They said: Now, when it is a substance that exists necessarily, it would have been impossible that its existence depend upon motion, and so time sometimes exists even when motion does not. So, in their opinion, sometimes time exists together with motion and so measures motion; but at other times it is separate, in which case it is called *everlasting*. These are the skeptical doubts raised concerning time. It would be best if we first indicate the way time exists and [identify] its essence, and only then come back and attack these sophistries and resolve them.

(8) We say that those who affirm the existence of time as some single thing have also had differences of opinions.[7] Some of them made motion time, while others made time the motion of the celestial sphere, to the exclusion of all other motions. Still others made time the celestial sphere's return (that is, a single rotation), and yet others made the celestial sphere itself time. Those who made motion itself time said

---

6. This position is certainly that of Abū Bakr Muḥammad al-Rāzī, who distinguished between relative and absolute time, making the latter a self-subsistent entity, which is sometimes identified in the sources as *dahr* (everlasting). See al-Rāzī, *Opera philosophica fragmentaque quae supersunt,* ed. Paul Kraus (Cairo: Fuʾad I University, 1939), 195–215.

7. For the source of the following (erroneous) opinions about the nature of time, see Aristotle, *Physics* 4.10.218a31–b10.

(٧) وقالت طائفة إنَّ الزمان جوهر أزلي؛ وكيف لا يكون جوهراً وهو واجب الوجود؟ فإنَّ وجوب وجوده بحيث لا يحتاج فيه إلى إثباتٍ بدليل، بل كلما حاولت أنْ ترفع الزمان وجَبَ أنْ تثبت الزمان، لأنك ترفعه قبل شيء أو بعد شيء، ومهما فعلت ذلك فقد أوجدت مع رفعه قبْليةً أو بَعْديةً، فتكون قد أثبتَ الزمان مع رفعه. إذ القبْلية والبَعْدية، التي تكون على هذه الصورة، لا تكون إلاّ للزمان أو بزمان. فالزمان واجب الوجود، وما كان واجب الوجود فلا يجوز أنْ يُرفع وجوده، وما لا يجوز أنْ يُرفع وجوده فليس بعرض، وما كان موجوداً وليس بعرض فهو جوهر، وإذا كان جوهراً واجب الوجود؛ فهو جوهر أزلي. قالوا وإذا كان جوهراً واجب الوجود فقد استحال أن يتعلق وجوده بالحركة، فجائز أنْ يوجد الزمان وإنْ لم توجد الحركة. فالزمان عندهم تارة يوجد مع الحركة فيقدّر الحركة، وتارة مجرّداً فحينئذ يسمى دهراً. فهذه هي الشكوك المذكورة في أمر الزمان. والأولى بنا أنْ ندل أولاً على نحو وجود الزمان وعلى ماهيته، ثم نكرّ على هذه الشُبه فنحلها،

(٨) ونقول: إنَّ الذين أثبتوا وجود الزمان معنى واحداً، فقد اختلفوا أيضاً، فمنهم مَنْ جعل الحركة زماناً، ومنهم مَنْ جعل حركة الفلك زماناً دون سائر الحركات، ومنهم مَنْ جعل عودة الفلك زماناً، أي دورة واحدة، ومنهم مَنْ جعل نفس الفلك زماناً. فأمّا الذين جعلوا الحركة نفسها زماناً فقالوا إنَّ الحركة، من بين ما نشاهده من الموجودات، هي التي

that from among the existing things that we experience, motion is that which includes past and future things, and it is in its nature always to have two parts[8] with this description, and whatever has this description is time. They also said that we believe that there has been time only when we sense motion, such that the sick and afflicted will find a given period of time long that one engrossed in wanton pleasure would find short, because the motions used to measure [the time] are firmly fixed in the memory of the former two, while they vanish from the memory of the one savoring wanton pleasure and rapture. Whoever is not aware of motion is not aware of time, just like the Companions of the Cave;[9] for, since they were unaware of the motions between the instant that they first settled down for a nap and the instant that they awoke, they did not realize that they had slept more than a day. The First Teacher [i.e., Aristotle] also related that something like that happened to a group of godlike men,[10] and history reveals that they were before the Companions of the Cave. These are the early views about time before the maturity of philosophy, but all of them are incorrect.

(9) Motion is not time because, while motion is sometimes fast and at other times slow, no time is faster and slower than another, but, rather, shorter and longer. Also, two motions might be simultaneous,[11] while two times are not. Also, you know that two different motions might occur simultaneously at a single time, while their time does not differ. Also, motion's specific differences are not time's, and the things related to time—as, for example, *now*, *suddenly*, the *present instant*, and *previously*—have nothing to do with the motion itself in something. Again, while it is proper to take time in the definition of *fast motion* as a part of the specific difference, it is not proper to take motion in the same way, but,

---

8. Reading *juzʾāni* with **Z**, **T**, and the Latin (*duas partes*) for **Y**'s (inadvertent) *juzʾ ān* ("a part of an instant," or perhaps "a present part"?).

9. See Quʾrān 18:1–23.

10. See *Physics* 4.11.218b23–27.

11. Reading *maʿan* with **Z**, **T**, and the Latin (*simul*), which is omitted in **Y**.

تشــتمل على شيء ماضٍ وشيء مســتقبل، وفي طبيعتها أنْ يكون لها دائماً جزءان بهذه الصفة، وما كان بهذه الصفة فهو الزمان. قالوا، ونحن إنّما نظنّ أنّه كان زمان إذا أحسسنا بحركة، حتى أنَّ المريض والمغتم يســتطيلان زماناً يســتقصره المتمادي في البطر، لرسوخ الحركات المقاســة في ذكر هذين، وانمحائها عن ذكر الملتهي عنها بالبطر والغبطة. ومَنْ لا يشعر بالحركة لا يشــعر بالزمان كأصحاب الكهف؛ فإنّهم لمّا لم يشعروا بالحركات التي بين آن ابتداء إلقائهم أنفســهم للاســتراحة بالنوم، وآن انتباههم، لم يعلموا أنّهم زادوا على يوم واحــد. فقــد حكى المعلم الأول أيضاً أنَّ قوماً من المتألهين عرض لهم شــبيهٌ بذلك، ودل التاريخ على أنّهم كانوا قبل أصحاب الكهف. فهذه هي الأقوال الســالفة قبل نضج الحكمة في أمر الزمان، وكلها غير صحيحة.

(٩) أما أنَّ الحركة ليســت زماناً، فلأنّه قد تكون حركة أســرع وحركة أبطأ، ولا يكون زمانٌ أســرع من زمانٍ وأبطأ، بل أقصر وأطول. وقد تكون حركتان معاً ولا يكون زمانــان معاً، وأنت تعلم أنّه قد تحصل حركتان مختلفتــان معاً في زمانٍ واحد وزمانهما لا يختلــف. والحركة فصولها غير فصول الزمان، والأمور المنســوبة إلى الزمان مثل هوذا وبغتة والآن وآنفاً ليس هي من ذات الحركة في شــيء. والزمان يصلح أنْ يؤخذ في حدّ الحركة الســريعة جزءاً من الفصل، والحركة لا تصلح أنْ تؤخذ كذلك، بل تؤخذ على أنّها

instead, it is taken as a premised part. So it is fine to say that *fast* is that which covers a longer distance in a shorter *time,* whereas it is not okay to say in a shorter *motion.* The case of the motion of the first celestial sphere is the same as this one, for it can be said about it that it is the fastest of motions because, simultaneously with another motion, it covers a larger [distance], although this is something that we will discuss later.[12] Now, this *simultaneity* indicates something different from the two motions and, rather, indicates a certain thing to which both of them are related and with respect to which the two are equal, while differing with respect to the distance [covered]. That thing is not itself either one of the two, because the second is not common to the other in itself, while it is common to it in the thing with which they are simultaneous.

(10) From this vantage point, it becomes obvious what is wrong with the claim of those who made moments certain events that set the moment for other events. That is because they do not make that passing event itself a moment *qua* motion, generation, black, white, or whatever, but are forced both to say that it becomes a moment by setting a moment, and that setting a moment involves the concurrence of some other thing simultaneous with its existence. Now, concerning this *concurrence* and *simultaneity,* one must understand something different from either one of the two events, when both concurrences concur with respect to something and both instances of simultaneity are simultaneous with respect to a certain thing. When both exist simultaneously (or one of them exists as that which sets the moment in that it is simultaneous with the existence of the other), then what is understood about the simultaneity cannot be what is understood about either one of [those events]. In fact, this simultaneity would have been the opposite of what was meant had either one of the two been earlier or later. It is this thing with respect to which

12. See 4.5.1–4.

جزء متقدم ، فإنّه يصلح أنْ يقال إنَّ السريع هو الذي يقطع مسافةً أطول في زمانٍ أقصر ، ولا يصح أنْ يقال في حركةٍ أقصر . وحكم الحركة الأولى الفلكية هذا الحكم بعينه ؛ فإنَّه يصحُّ أنْ يقال فيها إنّها أسرع الحركات ، لأنّها تقطعّ - مع قطع الحركة الأخرىّ - أعظم مع ما في هذا ، ممّا نتكلم فيه بعد . وهذه المعيّة تدل على أمر غير الحركتين ، بل تدل على معنى تنسبان كلتاهما إليه ويتساويان فيه ويختلفان في المسافة . وذلك المعنى ليس ذات أحدهما ، لأنّ الثاني لا يشارك الآخر في ذاته ويشاركه في الأمر الذي هما معاً فيه .

(١٠) ويمكن من هذا الموضع أنْ يظهر فساد قول مَنْ جعل الأوقات أعراضاً توقّت لأعراض، وذلك لأنّهم لا يجعلون نفس ذلك العرض الحادث من حيث هو حركة أو كون أو سواد أو بياض أو غير ذلك - وقتاً. ولكن يضطرون إلى أنْ يقولوا إنّه يصير وقتاً بالتوقيت، ويضطرون إلى أنْ يكون التوقيت يُقرن وجود شيء آخر مع وجوده. وهذا الاقتران وهذه المعيّة يفهم منها ضرورة معنى غير معنى كل واحدٍ من العرضين. وكل مقترنين يقترنان في شيء وكلّ معيّن، فهما في أمرٍ مّا معاً. فإذا كان وجودهما معاً، أو وجود واحدٍ منهما موقّتاً بأنّه مع وجود الآخر، فالمفهوم من المعيّة هو أمرٌ لا محالة ليس هو مفهوم أحدهما. وهذه المعيّة مقابلة لمعنى أنْ لو تقدّم أحدهما أو تأخّر، وهذا الشيء

there is the simultaneity that is the moment that accounts for the two states of affairs being together, and so it can be made to indicate either one of the two, just as if [that state of affairs] were something else occurring at that moment. Now, were that state of affairs in itself a moment, then, when it persisted for some duration and was one and the same, the duration of the persistence and the start of [the duration] would be one and the same moment. Now, you know that the moment that sets the moment is a certain limiting point between what is earlier and later, and that what is earlier and what is later do not differ [as such], while, *qua* motion, rest, and the like, they do differ. So its being a certain event (for example, a motion or rest) is not like its being earlier, later, or simultaneous; rather, the true nature of earlier, later, and simultaneous is something else—namely, a state of time.

(11) The argument upon which those who make time a motion rely is based on an unacceptable premise—namely, their claim that time is, in its nature, whatever requires something past and something future. This is certainly unacceptable, for many things that are not time are past and future, such as the Flood and the Resurrection. In fact, there must be another condition together with this [premise]—namely, that [time] is essentially what is such that it belongs to it to be the thing that is the very past or the very future so that the nature of [time][13] is the thing that when compared with something else, there is, in that case, essentially something past and future. When motion is past, its very existence as a motion is not that it is past, but that it is linked with the past; and, because of that, it can be said that some motion is in some past period of time, whereas it cannot be said that some motion is in some past motion (that is, unless it is meant that it is in a group of past motions; but that it is not our intention, but, rather, that the thing corresponds with that thing in which it is).

13. Reading *ṭabīʿatuhu al-amr* with **Z** and **T** for **Y**'s *ṭabīʿat al-amr* and the Latin's *natura rei* (the nature of the thing)

الذي فيه المعيّة هو الوقت الذي يجمع الأمرين، فكل واحدٍ منهما يمكن أنْ يجعل دالاً عليه كما لو كان غير ذلك الأمر، ممّا يقع في ذلك الوقت. ولو كان ذلك الأمر في نفسه وقتاً، لكانّ - إذا بقي مدة وهو واحد بعينهّ - وجب أنْ تكون مدة البقاء وابتداؤها وقتاً واحداً بعينه، ونحن نعلم أنَّ الوقت الموقّت هو حدٌّ بين متقدّم ومتأخر، وأنَّ المتقدّم والمتأخر لا يختلف، وبما هو حركة أو سكون أو غير ذلك يختلف. فليس كونه عرضاً ككونه حركة أو سكوناً وكونه متقدماً أو متأخراً أو معاً، بل حقيقة التقدّم والتأخر والمعيّة أمرٌ آخر هو حال الزمان.

(١١) وأمّا الحجّة التي اعتمدها جاعلو الزمان حركة؛ فهي مبنية على مقدمةٍ غير مسلّمة، وذلك قولهم: إنّ كل ما يقتضي أنْ يكون في طبيعته شيء ماض وشيء مستقبل فهو زمانّ - فإنَّ هذا غير مسلَّم، فإنَّ كثيراً ممّا ليس بزمان هو ماض ومستقبل كالطوفان والقيامة، بل يجب أنْ يكون مع هذا شرط آخر وهو أنْ يكون لذاته ما هو؛ بحيث منه الشيء الذي هو نفس الماضي أو نفس المستقبل، حتى تكون طبيعته الأمر، الذي إذا قيس إلى أمرٍ آخر، كان لذاته حينئذ ماضياً أو مستقبلاً. ولاحركة إذا مضتْ لم يكن نفس وجودها حركة هي أنّها ماضية، بل تكون قد قارنت الماضي. ولذلك يصح أنْ يقال حركة في زمانٍ ماضٍ، ولا يجوز أنْ يقال حركة في حركة ماضية؛ اللّهم إلاّ أنْ نعني في جملة الحركات الماضية وليس قصدنا هذا، بل أنْ يكون الشيء مطابقاً لوجود ذلك الذي هو فيه.

(12) As for those who maintain that time is a single rotation of the celestial sphere, its absurdity is evident in that any part of time is a time, whereas a part of a rotation is not a rotation. Even more far-fetched than all of this is the opinion of those who think that time is the celestial sphere by reasoning from two affirmative propositions in figure two.[14] Even then, one of the two premises in it is false—namely, the claim that every body is in the celestial sphere, for that is not the case; and, instead, the truth is that every body *that is not the celestial sphere* is in the celestial sphere, whereas perhaps every[15] body absolutely is in time, in which case the celestial sphere itself would also be in time in just the way that bodies are in time.

(13) Since we have pointed out the false schools of thought concerning time's essence, it is fitting that we point out the essence of time; and then, from there, its existence will become clear, as well as the solution to the sophisms mentioned about its existence.

---

14. Syllogisms in figure two are those whose middle term is the predicate term of both the major and minor premises. Formally, figure two has this form:

> (major premise) PM,
> (minor premise) SM;
> therefore (conclusion) SP.

An example would be (major premise) "no stone is an animal"; (minor premise) "every human is an animal"; therefore (conclusion): "no human is a stone." The only valid forms of syllogism in figure two have one negative proposition.

15. Reading *kull* with **Z**, **T**, and the Latin (*omne*), which is (inadvertently) omitted in **Y**.

(١٢) وأمّا القائلون بأنَّ الزمان هو دورة واحدة من الفلك، فتتبيّن إحالته بأنَّ كل جزء زمان زمان، وجزء الدورة ليس بدورة. وأبعد من هذا كله ظنّ مَنْ ظنَّ أنَّ الزمان هو الفلك بقياسٍ من موجبتين في الشكل الثاني، على أنَّ إحدى المقدمتين فيه كاذبة وهي قوله: «كل جسم في فلك» فإنّه ليس كذلك، بل الحق أنَّ كل جسم ليس بفلكٍ فهو في فلك. وأمّا الذي في الزمان؛ فلعله هو كلّ جسم مطلقاً، فإنَّ الفلك نفسه أيضاً في زمانٍ، على النحو الذي تكون الأجسام في الزمان عليه.

(١٣) وإذ قد أشرنا إلى المذاهب الباطلة في ماهيّة الزمان، فحقيق بنا أنْ نشير إلى ماهية الزمان، فيتضح لنا من هناك وجوده، ويتضح حلّ الشُبه المذكورة في وجوده.

Chapter Eleven

Identifying and affirming the essence of time

(1) It is plainly clear that two moving things might begin and end moving together, of which one will cover a lesser distance and the other more, either because of differences in speed or, as some people think,[1] because of a dissimilarity in the number of intervening rests. Also, two [mobiles] might begin [together] and cover two equal distances, but one of them reaches the end of the distance, while the other has not yet finished and that again because of the aforementioned differences. Now, in every case, there is, from any motion's starting point to its end point, a certain possibility to cover that same distance by that motion that has the same speed (or the same composition of rest); there is also a possibility to cover more than that distance by one faster than [that motion] (or having fewer rests mixed in); and, again, a certain possibility to cover less than it by one slower than it (or having more rests mixed in). Indeed, that simply cannot be disputed. So it has been established that, between the starting point and ending point, there is a certain definite possibility relative to the motion and the speed. Now, when we posit half of that distance, while positing the same speed, there is another

1. The "some people" is certainly the Atomists, who did, in fact, explain differences in velocity in terms of differences in the number of intervallic rests.

<الفصل الحادي عشر>

في تحقيق ماهية الزمان وإثباتها

(١) فنقول إنَّ من البيّن الواضح أنّه قد يجوز أنْ يبتدىء متحركان بالحركة وينتهيا معاً ، وأحدهما يقطع مسافة أقل والآخر يقطع مسافة أكثر ، إمّا لا ختلاف البطْء والسرعة ، وإمّا لتفاوت عدد السكونات المتخلّلة ، كما يراه قوم . ويجوز أنْ يبتدىء إثنان ويقطعا مسافتين متساويتين ، لكن أحدهما ينتهي إلى آخر المسافة ، والآخر بعد لم ينته ؛ وذلك للاختلاف المذكور . ويكون ، في كل حالٍ من الأحوال ، من مبتدأ كل حركةٍ إلى منتهاها إمكان قطع تلك المسافة بعينها بتلك الحركة المعيّنة السرعة والبطء ، أو المعيّنة التركيب مع السكون ، وإمكان قطع أعظم من تلك المسافة بالأسرع منها أو الأقل مخالطة سكونات ، وإمكان قطع أقل منها بالأبطأ من تلك أو الأكثر مخالطة سكوناتٍ ، وأنْ ذلك لا يجوز أنْ يختلف البتة . فقد ثبت بين المُبتَدأ والمنتهى إمكان محدود بالقياس إلى الحركة وإلى السرعة ، وإذا فرضنا نصف تلك المسافة ، وفرضنا السرعة بعينها والبطء بعينه ؛ كان إمكان آخر بين ابتداء

possibility between the start of that distance and the end of half of it in which, at that speed, only half can be covered; and the same holds for [the distance] between that halfway endpoint that was now posited and the original endpoint. In this case, the possibility up to the halfway point and from the halfway point are equal, and each one of them is half of the initially posited possibility, and so the initially posited possibility is divisible. (Do not worry for now[2] whether you make this mobile something really undergoing motion with respect to [the category] of place or some part you posit as undergoing motion with respect to position, which is similar to being moved with respect to place. [That] is because [the mobile] will either leave one state of contiguity for another through continuous states of contiguity, or leave one state of juxtaposition for another through continuous states of juxtaposition. What is traversed is called *distance,* however it might be, and so no conclusion in the course of what we'll say is going to change because of that.)[3]

(2) So we claim that it has turned out that the possibility is divisible, and whatever is divisible is a certain magnitude or has a magnitude, and so this possibility is never stripped of a magnitude. In that case, its magnitude must be either the magnitude of the distance or some other magnitude. On the one hand, if it were the magnitude of the distance, then two things that are equal in the distance [they cover] would be equal in this possibility; but that is not the case, and so it is some other magnitude. In that case, either it is the magnitude of the mobile or not. However, it is not the magnitude of the mobile. Otherwise, the larger the mobile is, the larger it would be in this magnitude; and, again, that

2. Reading *al-āna* with **Z**, **T**, and the Latin (*nunc*) for **Y**'s *illā anna* (except that).

3. It should again be recalled that, unlike others in the Aristotelian tradition, Avicenna takes local motion and positional motion to be distinct kinds of motion involving different categories: the former involves moving from one place to another, which here is explained in terms of being contiguous and then leaving some place along the distance (*masāfah*), while the latter involves moving from one position to another, which here is explained in terms of the juxtaposition of certain positions on the spatial magnitude (*masāfah*) of the rotating object.

تلك المســافة ومنتهى نصفها ، إنما يمكن فيه قطع النصف بتلك الســرعة والبطء ، وكذلك بــين هذا المنتهى المنصّف المفروض الآن وبين المنتهــى الأول ، فيكون الإمكان إلى النصف ومن النصف متساويين ، وكل واحد منهما نصف الإمكان المفروض أولاً . فيكون الإمكان المفروض أولاً منقســماً . ولا عليك الآن أنْ تجعل هذا المتحرّك شــيئاً متحركاً بالحقيقة في المكان ، أو جزءاً تفرضه لمتحرّكٍ بالوضع يشبه المتحرّك في المكان ، فإنّه يفارق مماسّة إلى مماسّــةٍ بمماسّــاتٍ متصلة ، أو موازاةٍ إلى موازاةٍ بموازياتٍ متصلة ، وإنْ يســمى ما يقطعه مسافة كيف كان ، فليس يختلف لذلك حكمٌ فيما نحن بسبيله .

(٢) فنقول : إنَّ هذا الإمكان قد صحّ أنّه منقسم ، وكل منقسم فمقدار أو ذو مقدار ، فهذا الإمكان لا يعرّى عن مقدارٍ ، فلا يخلو إمّا أنْ يكون مقداره مقدار المســافة أو مقدار آخر ، ولو كان مقدار المســافة لكانت المتساويات في المسافة متساوية في هذا الإمكان ، لكــن ليس كذلك ، فهو إذن مقــدار آخر . فإمّا أنْ يكون مقدار المتحرّك أو لا يكون ، لكنه ليس مقدار المتحــرّك وإلاّ لكان المتحرّك الأعظم أعظم في هذا المقدار وليس كذلك . فهو

is not the case, and so it is a magnitude other than that of the distance and the mobile. Now, it is known that motion itself is not this very magnitude itself, nor is that the speed, since motions *qua* motions are the same in being motion[4] and also [might] be going the same speed, while differing in this magnitude. Also, sometimes the motion varies in speed, while being the same with respect to this magnitude. So it has been established that a certain magnitude exists that is some possibility involving motions between what is earlier and later, occurring in such a way as to require certain definite distances; and [this possibility] is not[5] the magnitude of the mobile, distance, or motion itself. Now, this magnitude cannot be something subsisting in itself. How could it be something subsisting in itself when it comes to an end together with that which it measures? Whatever comes to an end is subject to corruption and so is in a subject or what has a subject, in which case this magnitude is something dependent upon a subject. Now, its first subject cannot be the mobile's matter because of what we explained; for, if it were a magnitude of some matter without intermediary, the matter would become larger and smaller as a result of it. So, then, it is in the subject by means of some other disposition. It cannot be by means of some fixed disposition, like white or black; otherwise, the magnitude of that disposition in the matter would occur in the matter as a firmly fixed magnitude. So it remains that it is a magnitude of an unfixed disposition—namely, motion from place to place or from one position to another between which there is some distance through which the positional motion circulates. This is what we call *time*.

---

4. "Being motion" translates the *nisba* adjective *ḥarakīyah* (literally "motion-ness"). Avicenna is probably using it in the sense of motion's very definition. Thus, every motion agrees in being "a first perfection belonging to what is in potency from the perspective of what is in potency" (2.1. 3); and so, if this magnitude in question were identified with motion-ness itself, this magnitude could not vary from motion to motion, which it does.

5. Reading *laysa* with **Z**, **T**, and the Latin (sing. *non est*) for **Y**'s *laysat,* ("are not," which apparently would be referring back to "definite distances").

إذن مقدار غير مقدار المسافة، وغير مقدار المتحرّك. ومن المعلوم أنَّ الحركة ليست نفسها ذات هذا المقدار نفسه، ولا السرعة والبُطْ ذلك، إذ الحركاتّ - في أنّها حركاتّ - تتفق في الحركية، وتتفق في السرعة والبطء، وتختلف في هذا المقدار، وربّما اختلفت الحركة في السرعة والبطء، واتفقت في هذا المقدار. فقد ثبت وجود مقدار لإمكان وقوع الحركات بين المتقدّم والمتأخر وقوعاً يقتضي مسافات محدودة، ليس مقدار المتحرّك ولا المسافة ولا نفس الحركة. وهذا المقدار ليس يجوز أنْ يكون قائماً بنفسه، وكيف يكون قائماً بنفسه وهو مُنْقَضٍ مع مقدّره؟ وكل مُنْقَض فاسد، فهو في موضوع أو ذو موضوع، فهذا المقدار هو متعلق بموضوع، ولا يجوز أن يكون موضوعه الأول مادة المتحرّك لما بيّناه. فإنّه لو كان مقدار مادة بلا واسطة، لكانت المادة تصير به أعظم أو أصغر. فإذن هو في الموضوع بوساطة هيئة أخرى ولا يجوز أنْ يكون بوساطة هيئة قارّة كالبياض والسواد، وإلّا لكان مقدار تلك الهيئة في المادة يحصل في المادة مقداراً ثابتاً قارّاً. فبقي أنْ يكون مقدار هيئة غير قارّة؛ وهي الحركة من مكانٍ إلى مكان، أو من وضعٍ إلى وضعٍ بينهما مسافة تجري عليها الحركة الوضعية، وهذا هو الذي نسميه الزمان.

(3) Now, you know that being divisible into earlier and later parts is a necessary concomitant of motion, and the earlier and later parts are found in it only as a result of [motion's] relation to the earlier and later parts in the distance. Be that as it may, there also comes with that the fact that the earlier part of the motion will *not* exist together with its later part in the way that the earlier and later parts in distance exist together. Also, what corresponds with the earlier part of the motion in the distance cannot become the later part, nor can that which corresponds with its later part become earlier, in the way that it can in distance. So there is some property that belongs to being earlier and later in motion—which necessarily follows the two [states] because they belong to motion—[but] which is not due to their belonging to distance. Also, the two are numbered by motion. [That] is because motion, through its parts, numbers what is earlier and later; and so the motion has a number inasmuch as being earlier and later belong to it with respect to distance. Moreover, it has a certain magnitude by paralleling the magnitude of the distance. Time is this number or magnitude. So time is the number of motion when it is differentiated into earlier and later parts—not by time, but, instead, with respect to distance; otherwise, the definition would be circular. (This is what one of the logicians believed—namely, that a circle occurred in this explanation—but he believed wrongly, since he did not understand this).[6]

(4) Moreover, this time is that which is essentially a magnitude owing to what it is in itself, possessing [the states] of being earlier and later, the later part of which does not exist together with what is earlier, as might be found in other types of [things that might] be earlier and later. This

6. Unfortunately, it is not clear who the logician is to whom Avicenna refers, although it is most likely one of the Baghdad Peripatetics. Among these, the writings of al-Fārābī, Yaḥyá ibn ʿAdī, Ibn al-Samḥ, and al-Sijistānī give no indication that any of them found fault with the Aristotelian definition of time as the number of motion with respect to before and after. Others who commented on Aristotle's *Physics* 4 and thus might have criticized Aristotle's definition of time were Abū Karnīb and his student Abū Bishr Mattá. Abū Bishr Mattá was the teacher of both al-Fārābī and Yaḥyá ibn ʿAdī, and neither suggests that Abū Bishr questioned the Aristotelian definition of time, although this is far from definitive proof that he did not.

(٣) وأنت تعلم أن الحركة يلحقها أنْ تنقسم إلى متقدم ومتأخر، وإنّما يوجد فيها المتقدم ما يكون منها في المتقدّم من المسافة، والمتأخر ما يكون منها في المتأخر من المسافة، لكنه يتبع ذلك أنَّ المتقدم من الحركة لا يوجد مع المتأخر منها، كما يوجد المتقدم والمتأخر في المسافة معاً. ولا يجوز أنْ يصير ما هو منها مطابق المتقدم من الحركة في المسافة متأخراً، ولا الذي هو مطابق المتأخر منها متقدماً كما يجوز في المسافة، فيكون للتقدم والتأخر في الحركة خاصية تلحقهما من جهة ما هما للحركة، ليس من جهة ما هما للمسافة ويكونان معدودين بالحركة، فإنَّ الحركة بأجزائها تعدّ المتقدم والمتأخر؛ فتكون الحركة لها عددٌ من حيث لها في المسافة تقدّم وتأخّر، ولها مقدار أيضاً بإزاء مقدار المسافة، والزمان هو هذا العدد أو المقدار. فالزمان عدد الحركة؛ إذا انفصلت إلى متقدّم ومتأخّر لا بالزمان، بل في المسافة، وإلاّ لكان البيان تحديداً بالدور، والذي ظنّ بعض المنطقيين أنّه قد وقع في هذا البيان دورٌ إذ لم يفهم هذا، فقد ظنَّ غلطاً.

(٤) وهذا الزمان هو أيضاً - الذي هو لذاته - مقدارٌ لما هو في ذاته، ذو تقدّم وتأخر، لا يوجد المتأخر منه مع المتقدم، كما قد يوجد في سائر أنحاء التقدم والتأخر.

is something part of which is essentially before some part and part of which is essentially after some part, whereas everything else will either be before or after on account of it. [That] is because the things in which there is a *before* and an *after*—in the sense that their *before* has passed away, while the *after* does not exist together with the *before*—are not such essentially, but have their existence together with one of the divisions of this magnitude. So it is said of that part that corresponds with a *before* part that it is before, while it is said of that part that corresponds with an *after* part that it is after. Now, it is known that these things undergo change, because there is no passing away or ensuing in what does not change. Also, this thing [that is, time] cannot be before and after on account of some other thing, because, if that were the case, then its *before* would become a *before* only because it existed in some other thing's *before*. In that case, that thing (or something else at which the regress eventually ends) possesses a *before* and an *after* essentially. In other words, it essentially admits of the relation by which there is before and after. That thing is known to be that in which the possibility of changes (in the manner mentioned above)[7] primarily occurs and on account of which [that possibility] occurs in others. So that thing is the magnitude that, in itself, measures the aforementioned possibility and about which we are solely concerned.

(5) We ourselves have made *time* only a name for the possibility noted above and in which that possibility primarily occurs. Clearly, from this, then, the noted magnitude is the same thing that admits of the relation of before and after, or, more precisely, it itself is divisible into before and after. By this I do not mean that time's being before is not through the relation,[8] but, rather, that this relation is inseparable from time and, because of time, it is inseparable from other things. So,

---

7. See par. 1.

8. The Latin reads *Et non dico quod tempus est prius relative* (I am not saying that time is before relatively), and thus drops the second negation in our phrase "*not* through the relation." It is not clear whether the Latin text reproduces an omission of the second negation in the exemplar underlying the Latin and so is a variant reading, or—as is more likely—is an omission on the part of the Latin translator.

وهذا هو لذاته يكون شيء منه قبل شيء ، وشيء منه بعد شيء ، وتكون سائر الأشياء لأجله ، بعضها قبل ، وبعضها بعد هذا ؛ لأنَّ الأشـياء التي يكون فيها قبل وبَعد ، بمعنى أن القبـل منها فائـت والبَعد غير موجود مع القبل ، إنّما تكون كذلك لا لذواتها ، بل لوجودها مع قسـم من أقسـام هذا المقدار ، فما طابق منها جزءاً هو قبل قيل له إنّه قبل ، وما طابق جزءاً هو بَعد قيل له إنّه بَعد ، ومعلومٌ أنَّ هذه الأشياء هي ذوات التغيّر ؛ لأنَّ ما لا تغيّر فيه فلا فائت فيه ولا لاحق . وهذا الشيء ليس يكون قبل وبَعد لأجل شيءٍ آخر ، لأنّه لو كان كذلك لكان القبل منه إنّما صار قبلاً لوجوده في قبل شيء آخر ، فيكون ذلك الشيءّ – أو شيء آخر ينتهي إليه التدريج آخر الأمرّ – هو لذاته ذو قبل وبَعد ، أي لذاته يقبل الإضافة التي بها يكون قبل وبَعد . ومعلومٌ أنَّ ذلك الشـيء هو الذي يقع فيه إمكان التغييرات على النحو المذكور ؛ وقوعاً أولياً ، ويقع في غيره لأجله ، فيكون ذلك الشيء هو المقدار المقدّر للإمكان المذكور تقديراً بذاته ، ويكون ما نحن فيه لا غيره .

(٥) فنحـن إنّما كنّـا جعلنا الزمان إسـماً للمعنى الذي هو لذاتـه مقدار للإمكان المذكور ، ويقع فيه الإمكان المذكون وقوعاً أولياً ، فبيّن من هذا أنَّ المقدار المذكور هو بعينه الشيء الذي هو لذاته يقبل إضافة قبل وبَعد ، بل هو نفسه منقسم إلى قبل وبَعد ، ولست أعني بهذا أنَّ الزمان يكون قبل لا بالإضافة ، بل أعني أنَّ الزمان تلزمه هذه الإضافة ، وتلزم

when something is said to be before and that thing is not time (but, for example, is motion, humans, and the like), [its being before] means that it exists together with a certain thing that is in some state such that when that state is compared with the state of something later, it is inseparable from [that state] if the thing in [the former state] is essentially before. That is, this inseparability belongs to it essentially. So the [state of] being earlier of some earlier thing [*x*] is that it has a certain existence simultaneous with the nonexistence of some later thing [*y*] that has not existed when [*x*] exists. So [*x*] is earlier than [*y*] when [*y*'s] nonexistence is taken, whereas [*x*] is simultaneous with [*y*] only when [*y*'s] existence is taken and is in a given state that is simultaneous with [*x*]. Now, the determinate thing [that is, *x*] occurs in both states, whereas a given state that it has when it is *earlier* is not a state of *being simultaneous;* and so, inevitably, something belonging to it when it was earlier ceases to belong to it when it is simultaneous. So being earlier and before-ness do not belong essentially to this determinate thing, nor do they persist simultaneously with the persistence of that determinate thing. It is simply and essentially impossible that being earlier and before-ness should persist simultaneously with [the determinate thing's] other state, when it is impossible that they become simultaneous and it is known that this existence does not persistently belong to it, since it is after the existence of the other. That is not impossible for the [determinate] thing that has these [states of being earlier, before-ness, and simultaneity], since it sometimes exists and is before. At other times it exists and is simultaneous with, and at still other times it exists and is after, while being one and the same thing. As for the very thing that is essentially[9] before and after, even if by comparison, it cannot persist the same as it is such that it will be after, after it was before. [That] is because that thing by which it was

9. Reading *li-dhātihi,* which is (inadvertently) omitted in **Y** but appears in **Z**, **T**, and the Latin (*per seipsam*).

سائر الأشياء بسبب الزمان. فإنَّ الشيء إذا قيل له قبل، وكان ذلك الشيء غير الزمان، وكان مثل الحركة والإنسان وغير ذلك، كان معناه أنّه موجود مع شيءٍ هو بحال؛ تلك الحال تلزمهاّ - إذا قيست إلى حال الآخر إنْ كان الشيء بها قبل لذاته، أي يكون هذا اللزوم له لذاتهّ - فالمتقدّم تقدّمه أنّه له وجودٌ مع عدم شيء آخر لم يكن موجوداً - وهو موجود، فهو متقدمٌ عليه إذا اعتُبر عدمه، وهو معه إذا اعتُبر وجوده فقط. وفي حال ما هو معه فليس متقدماً عليه وذاته حاصلة في الحالين، وليس حال ما هو له متقدّم هو حال ما هو مع، فقد يبطل منه لا محالة أمرٌ كان له من التقدّم عندما هو مع. فالتقدّمُ والقَبْلية معنى لهذا الذات ليس لذاته، ولا ثابتاً مع ثبات ذاته، وذلك المعنى مستحيلٌ فيه أنْ يبقى مع حاله الأُخرى البتّة استحالة لذاته، ومستحيلٌ فيه أنْ يصير مع، ومعلومٌ أنَّ هذا الوجود لا يثبت له، فإنّه بعد وجود الآخر. وأمّا الشيء، الذي له هذا المعنى والأمر، فلا يستحيل ذلك فيه؛ فإنّه تارة يوجد وهو قبل، وتارة يوجد وهو مع، وتارة يوجد وهو بَعد، وهو واحدٌ بعينه. وأمّا نفس الشيء الذي هو قبل وبعد لذاتهّ - وإنْ كان بالقياسّ - فلا يجوز أنْ يبقى هو بعينه فيكون بَعد، بَعد ما كان قبل؛ فإنّه ما جاء

something after arrived only when that by which it is before ceased, while it is the thing that has this [*after*] factor that persists together with the cessation of the *before* factor. Now, this factor cannot merely be some relation to nonexistence (or existence), since the existing thing's relation to the nonexisting thing might be one of being later just as easily as being earlier (and the same holds with regard to existence). The fact is that its relation to nonexistence is associated with some other factor [with respect to] which, when associated with it, there is the [state of] being earlier, and, if associated with something else, there is the [state of] being later, whereas the nonexistence in both states is simply nonexistence (and the same holds for existence). Likewise [this factor] is associated *mutatis mutandis* with the related thing, because the related thing is equally—[albeit] conversely—related to it and has that status. This factor either is time or [is] some relation to time. So, if it is time, then that is what we claim. If it is some relation to time, then its[10] before-ness is on account of time, and the factor reduces to the before-ness and afterness whose subject primarily is time.[11] So time turns out essentially to have a *before* and an *after;* and so that which turns out essentially to have a *before* and an *after* we call *time,* since we explained that it is the magnitude of the previously indicated possibility.

(6) Since it has turned out that time is not something subsisting in itself (and how could it be something subsisting in itself when it has no fully determinate being, but is coming to be and passing away?), and the existence of whatever is like that depends upon matter, time is material. Now, although it is material, it exists in matter through the intermediacy of motion; and so, if there is neither a motion nor a change, there is no time. Indeed, how could there be time without *before* and *after,* and how could there be *before* and *after* when one thing does not come to be after

10. Reading *hā* with **Z** and the Latin (sing. *eius*) for **Y**'s *humā* ("their," dual); **T** is sufficiently unclear that it might be either *hā* or *humā.*

11. There seems to be some confusion in **Y**'s text in which there is a repetition of this phrase, which reads "and the factor reduces to the before-ness and afterness whose subject primarily is on account of the time." This reading is not found in **Z**, **T**, or the Latin and almost certainly is a result of dittography in **Y**'s text.

المعنى الذي به الشئ بَعد إلاّ <و> بطل ما هو به قبل، والشيء ذو هذا الأمر هو باقٍ، مع بطلان الأمر القبل. وهذا الأمر لا يجوز أنْ يكون نسبة إلى عدم فقط، أو إلى وجودٍ فقط فإنَّ نسـبة وجود الشيء إلى عدم الشيء قد يكون تأخراً مّا، كما يكون تقدّماً، وكذلك في جانب الوجود، بل هو نسبته إلى عدم مقارنٍ أمر آخر إذا قارنه كان تقدماً، وإنْ قارن غيره كان تأخّراً، والعدم في الحالين عدمٌ، وكذلك الوجود، وكذلك نظيره يقارن المنسوب؛ لأنَّ المنسـوب أيضاً منسـوب إليه بالعكس وله ذلك الحكم، وهذا الأمر هو زمانٌ أو نسبة إلى زمان، فإنْ كان زماناً فذلك ما نقوله، وإنْ كان نسـبة إلى الزمان فتكون قبْليتها لأجل الزمـان، ويرجع الأمر إلى أنَّ هذه القَبْليـة والبَعدية أول موضوعهما لأجل الزمان، فالزمان لذاته يعرض له قبل وبعد، فالذي يعرض له قبل وبعد لذاته؛ نسـميه الزمان، إذ قد بيّنا أنّه لذاته هو مقدار الإمكان المشار إليه.

(٦) ولمّـا صحَّ أنَّ الزمان ليس ممّا يقـوم بذاته، وكيف يكون مما يقوم بذاته وليس له ذات حاصلة وهو حادث وفاسد؟ وكل ما يكون مثل هذا فوجوده متعلق بالمادة، فيكون الزمان مادياً. ومع أنّه مادي موجود في المادة بتوسط الحركة، فإنْ لم تكن حركة ولا تغيّر لـم يكن زمان، فإنّه كيف يكون زمان ولا يكون قبل وبعد، وكيف يكون قبْل وبَعد إذا لم

another? Certainly, *before* and *after* do not exist simultaneously; but, rather, something that was *before* ceases inasmuch as it was *before* because something that is *after* inasmuch as it is *after* comes to be. So, if there is no variation or change inasmuch as something ceases or something comes to be—nothing being *after* (since there was no *before*) or nothing being *before* (since there is no *after*)—time will not exist. [In other words, time exists] only together with the existence of the renewal of some state, where that renewal must also be continuous; otherwise, again, there will be no time. [That] is because, when something is all at once and then there is nothing at all until something else is all at once, there must be, between the two, either the possibility for the renewal of some things or not. On the one hand, if there is a certain possibility for the renewal of some things between them, then there is a *before* and *after* with respect to what is between them, but the *before* and *after* are realized only by a renewal of things, whereas we are assuming that there is no renewal of things, which is a contradiction. On the other hand, if this possibility does not exist between them, the two are adjacent, in which case that adjacency must be uninterrupted or not. If it is uninterrupted, then what we supposed results; however, it is an absurdity whose impossibility will be explained later.[12] If it is interrupted, then the argument returns to the beginning. So, if there is time, there must necessarily be a renewal of certain states, either by way of contiguity or continuity. So, if there is no motion, there is no time. Because time, as we said, is a certain magnitude—namely, something continuous that parallels motions and distances—it inevitably has a division that is the product of the estimative faculty, which is called the *present instant.*

---

12. See 3.4.3.

يحدث أمرٌ فأمر؟ فإنّه لا يكون قبْل وبَعد معاً ، بل يبطل الشيء الذي هو قبلٌ - من حيث هـو قبلٌ - لأنّه يحدث الشـيء الذي هو بَعد من حيث هـو بَعد ، فإنْ لم يكن اختلاف وتغيّر مّا ؛ بأنْ يبطل شيءٌ أو يحدث شيءٌ ، لا يكون أمرٌ هو بَعد إذ لم يكن قبْل ، أو أمرٌ هو قبل إذ ليس بَعد . فإذن الزمان لا يوجد إلاّ مع وجود تجدّد حالٍ ، ويجب أنْ يستمر ذلك التجدّد وإلاّ لم يكن زمان أيضاً ؛ لأنّه إذا كان أمر دفعة ، ثم لم يكن شيء البتّة حتى كان شـيء آخر دفعة ، لم يخـل إمّا أنْ يكون بينهما إمكان تجـدّد أُمورٍ أو لم يكن ؛ فإنْ كان بينهمـا إمكان تجدّد أمورٍ ، فيكون فيما بينهما قبْل وبَعـد ، والقَبْل والبَعد إنّما يتحقّق بتجـدّد أمورٍ ، وفرضنا أنّه ليس هناك تجدّد أُمـورٍ ، هذا خُلْف . وإنْ لم يكن بينهما هذا الإمكان فهما ملتصقان ، فلا يخلو إمّا أنْ يكون ذلك الالتصاق مسـتمراً أو لا يكون ، فإنْ كان مستمراً فقد حصل ما فرضناه - على أنّه محالٌ ستتضح استحالته بَعدُ - وإنْ كان منقطعاً عاد الكلام من رأس . فيجب ضرورة ، إنْ كان زمان ، أنْ يكون تجدّد أحوال ، إمّا علـى التلاحق وإمّا على الاتصال ؛ فإنْ لم تكن حركة لم يكن زمان ، ولأنَّ الزمانَ - كما قلنا - مقدارٌ وهو متصلٌ محاذٍ لاتصال الحركات والمسـافات ، فله لا محالة فصل متوهم وهو الذي يُسمى الآن .

Chapter Twelve

Explaining the instant

(1) We maintain that we know the instant from knowing time. [That] is because, since time is continuous, it inevitably has a certain division, which is a product of the estimative faculty and is called the *instant*.[1] Now, the instant does not at all exist as actual in relation to time itself; otherwise the continuity of time would be severed. Instead, its existence is only as the estimative faculty imagines it—namely, as a certain connection in a linear extension. The connection does not exist as actual in the linear extension inasmuch as it is a connection; otherwise, there would be infinitely many connections (as we shall explain later).[2] It would be actual only if time were, in some way, severed; but it is absurd that the continuity of time should be severed. That is because, if one concedes that time is severed, that severance must be either at the beginning of time or at its end. If it is at the beginning of time, it necessarily follows that that time has no *before*. Now, if it had no *before*, it could not have been nonexistent and then existed. [That] is because, when it is nonexistent and then exists, it exists after not existing, and so its nonexistence is before its existence. In that case, it must have a *before*,

1. The Arabic *ān* might mean *instant, present instant,* and *now*. No single translation covers all the senses in which Avicenna uses *ān* in this chapter; and so, to facilitate philosophical clarity, all three translations are used in this chapter as context requires.

2. See 3.3–4.

<الفصل الثاني عشر>

في بيان أمر الآن

(١) نقول إنَّ الآن يُعلم من جهة العلم بالزمان، فإنَّ الزمان لمّا كان متصلاً فله لا محالة فصل متوهم وهو الذي يُسمى الآن. وهذا الآن ليس موجوداً البتة بالفعل بالقياس إلى نفس الزمان، وإلاّ لقطع اتصال الزمان؛ بل إنّما وجوده على أنْ يتوهمه الوهم واصلاً في مستقيم امتداداً، والواصل ليس موجوداً بالفعل في مستقيم الامتداد من حيث هو واصل، وإلاّ لكانتّ - كما نبيّن بعدّ - واصلات بلا نهاية. بل إنّما يكون بالفعل لو قُطع الزمان ضرباً من القطع، ومحالٌ أنْ يقطع اتصال الزمان، وذلك لأنّه إنْ جُعل للزمان قطعٌ؛ لم يخل إمّا أنْ يكون ذلك القطع في ابتداء الزمان أو في انتهائه، فإنْ كان في ابتداء الزمان، وَجَبَ من ذلك أنْ يكون ذلك الزمان لا قبْل له، وإنْ كان لا قبْل له؛ فيجب أنْ لا يكون معدوماً ثم وجد، فإنّه إذا كان معدوماً ثم وجد، يكون وجوده بعد عدمه، فيكون عدمه قبل

and that *before* is something different from the nonexistence describing it, according to what we stated elsewhere.[3] So the thing of which this species of before-ness is predicated would be some existing thing, while not being *this* time. So before *this* time, there would be a time that is continuous with it—that [time] before, this [time] after—where this division would be what unites the two; but it was posited as what divides [the two]. This is a contradiction. Likewise, if it is posited as what divides in the way of an endpoint, then something's existing after it is either possible or not. On the one hand, if it is not possible that something exists after it—not even what exists necessarily, such that it would be impossible for something to exist with the nonexistence that is reached at the endpoint—then necessary existence and absolute[4] possibility would have been eliminated. Necessary existence and absolute possibility, however, are not eliminated. If, on the other hand, there is [the possibility of something's existing] after that, then it has an *after,* and so there is a *before,* in which case the instant is something connecting, not dividing. So time does not have an instant existing as something actual in relation to [time] itself, but only as potential (I mean the potential proximate to actuality). In other words, time is so disposed that the instant can always be posited in it, whether by someone's simply positing [it as such] or the motion's arriving at some common indivisible limiting point—as, for example, the beginning of sunrise or sunset or the like. That does not really create a division in the very being of time itself, but only in its relation to motions, just as it is created from the relational divisions in other magnitudes. For example, one part of a body is divided from another by being juxtaposed against or contiguous with [something else], or by someone's simply positing [it as such], without an actual division occurring in it in itself, but only a certain division having occurred in it relative to something else.

---

3. See 2.11.5.

4. Reading *al-muṭlaq* with **Z**, **T**, and two MSS consulted by **Y**, which **Y** secludes; like **Y**, the Latin omits *absoluta.*

وجوده، فيكون لـه قبْل ضرورة، ويكون ذلك القبْل معنى غير العدم الموصوف به، على النحو الذي قلنا في غير هذا الموضع. فيكون الشـيء الذي به يقال هذا النوع من القبْلية حاصلاً، ولا هذا الزمان، فيكون هذا الزمان قبله زمان يكون متصلاً به، ذلك قبْل وهذا بَعد، وهذا الفصل يجمعهما، وقد فرض فاصلاً، وهذا خُلْف. وكذلك إنْ فُرض فاصلاً على أنّه نهاية؛ لم يخلُ إمّا أنْ يكون بعده إمكان وجود شيء أو لا يكون، فإنْ كان لا يمكن بعده أنْ يوجد شيء، ولا واجب الوجود، حتى يستحيل أنْ يوجد شيء، مع عدم ما انتهى إليه من النهاية؛ فقد ارتفع أنْ يكون وجود واجب وارتفع الإمكان المطلقّ - والوجود الواجب والإمكان المطلق لا يرتفعانّ - وإن كان بعده ذلك؛ فله بَعدٌ فهو قبل فالآن واصلٌ لا فاصل. فالزمان لا يكون له آنٌ بالفعل موجوداً بالقياس إلى نفسه، بل بالقوة، أعني القوة القريبة من الفعل، وهو أنَّ الزمان يتهيأ أن يُفرض فيه الآن دائماً؛ إمّا آنٌ بفرض الفارض، وإمّا بموافاة الحركة، حدّاً مشتركاً غير منقسم كمبدأ طلوعٍ أو غروب، أو غير ذلك. وذلك بالحقيقة ليس إحداث فصل في ذات الزمان نفسه، بل في إضافته إلى الحركات؛ كما يحدث من الفصول الإضافية في المقادير الأُخر، كما ينفصل جزء جسم من جزء آخر بموازاةٍ أو مماسّةٍ أو بفرض فارضٍ من غير أنْ يكون قد حصل فيه بالفعل فصلٌ في نفسه، بل حصل فيه فصلٌ مقيساً إلى غيره.

(2) When this present instant occurs through this relation, then its nonexistence is only in all of the time that is after it. Now, one can say that [the present instant] must cease to exist either in an immediately adjacent instant or an instant that is not immediately adjacent [only] after it is conceded that [the present instant] can begin ceasing to exist at some instant and, in fact, [conceding] that there is a beginning of its cessation—namely, at the limit of all the time during which it does not exist. The fact is that, by *cessation,* nothing more is understood than that something does not exist after existing. Now, the existence of [the present instant] in this situation is that it is the limit of the time that is nonexistent at it (as if you said that [the present instant] exists at the limit of the time that is nonexistent at it), and its cessation does not have a beginning when it ceases that is a first instant at which it ceased to exist. Instead, between [the time's] existing and not existing, there is a certain division that is nothing but the existence of [the present instant]. (You will learn that the things undergoing motion, rest, generation, and corruption also do not have a first instance in which they undergo motion, rest, generation, and corruption, since the time is potentially divisible infinitely.)[5]

(3) It might erroneously be thought that, against this, one can argue either that the present instant ceases to exist gradually (in which case it takes a period of time to reach its end, at which it ceases to exist) or it ceases to exist all at once (in which case its nonexistence is at an instant).[6] The falsity of this argument needs to be explained. So we say that what either exists or does not exist all at once (in the sense that it happens at a single instant) is not necessarily the opposite of that which exists or does not exist gradually; but, instead, it is more specific than that

5. See par. 4 and 3.6.3–6.
6. Cf. Aristotle, *Physics* 4.10.218a11–21.

(٢) وهذا الآن، إذا حصل بهذه النسبة، فليس يكون عدمه إلاّ في جميع الزمان بعده. وقول القائل إنّه إمّا أنْ يفسد في آنٍ يليه أو آنٍ لا يليه؛ هو بعد أنْ يتسلّم أنَّ له فساداً مبتدأ في آنٍ، بل ابتداء فسادهّ - وهو في طرف الزمانّ - الذي هو في جميعه يعدم. فإنّه لا يفهم من الفساد غير أنْ يكون الشيء معدوماً بعد وجوده، ووجوده في هذا الموضع هو أنّه طرف الزمان الذي هو فيه معدوم؛ كأنك قلت إنّه في طرف الزمان الذي هو معدومٌ فيه موجودٌ، وليس لفساده مبدأ فسادٍ هو أول آنٍ فسد فيه، بل بين وجوده وعدمه فصلٌ هو وجوده لا غير. وأنت ستعلم أنّه ليس للمتحرّك والساكن والمتكوّن والفاسد أول آنٍ هو فيه متحرك أو ساكن أو متكوّن أو فاسد، إذ الزمان منقسم بالقوة إلى غير نهاية.

(٣) والذي يظنّ، من أنّه يمكن أن يقال على هذا، إنّ الآن إمّا أنْ يُعدم قليلا قليلا فيمتد آخره آخذا إلى العدم مدة، أو يعدم دفعة فيكون عدمه في آنٍ، هو قولٌ يحتاج أنْ يبيّن فساده، فنقول: إنَّ المعدوم، أو الموجود. دفعة، بمعنى الذي يحصل في آن واحدٍ، ليس لازماً لمقابل الذي يُعدم قليلاً قليلاً، أو الذي يوجد قليلاً قليلاً، بل هو أخص من

opposite. That [proper] opposite is *that which does* not *proceed gradually toward existence, nonexistence, alteration, and the like.* This is true of (1) what occurs all at once, but it is also true of either (2a) the thing that does not exist during an entire period of time, while existing at its limit (which is not time), or (2b) the thing that does exist during an entire period of time, while not existing at its limit (which is not time). Indeed, neither of these latter two exists or does not exist gradually, and the same equally holds for the first (namely, that whose existence or nonexistence is at an instant). The latter sense is distinct from the former first sense because, on the one hand, in the first sense, the status for the instant of time, which is essentially its extremity, is assumed to be the same as that for all of time. In the latter sense, on the other hand, the status for the instant is assumed to be different from that for time—namely, that one instant is not placed after a different one, unless an intervening state occurs between the two instants and that instant is essentially the limit. Now, our discussion of this second sense is not whether it in fact exists or not, for we are not discussing it with an eye to affirming its existence, but with an eye to a certain negation's being predicated of it. Again, that negation is that *it exists or does not exist gradually;* and, with respect to that, it has a certain subclass, and that subclass is more specific than that negation. Now, the more specific does not necessarily entail the more general. Also, it is not the case that, from our conceptualization of something as a subject or predicate, its existence must either be affirmed or not. (This is something learned in the discipline of logic.)[7] So, when our claim "It is not the case that it exists or does not exist gradually," is more general than our claim "It exists or does not exist all at once" (in other words, that state of it is at some beginning instant), then one's claiming that it is either gradual or all at once in this sense

7. See *Kitāb al-madkhal* 1.3.

ذلك المقابل ، وذلك المقابل هو الذي ليس يذهب إلى الوجود أو إلى العدم أو الاستحالة أو غير ذلك قليلاً قليلاً . وهذا يصدق على ما يقع عليه دفعة ، ويصدق على الأمر الذي يكون في جميع زمانٍ مّا معدوماً ، وفي طرفه الذي ليس بزمانٍ موجوداً ، أو الأمر الذي يكون في جميع زمانٍ مّا موجوداً ، وفي طرفه الذي ليس بزمانٍ معدوماً ، فإنَّ هذين ليسا موجودان أو يعدمان قليلاً قليلاً . والأول أيضاً كذلك ، وهو الذي يكون وجوده أو عدمه في آنٍ . لكن هذا الوجه يباين ذلك الوجه الأول ، لأنَّ الوجه الأول قد فرض فيه الحكم في آن الزمانّ - الذي هو نهايته بالذاتّ - كالحكم في جميع الزمان ، وفي هذا الوجه قد فرض الحكم في الآن مخالفاً للحكم في الزمان من غير أنْ يُوضع آنٌ بعد الآن المخالف ؛ وإلاّ لوقعت مشافعةٌ بين آناتٍ ، ولكان ذلك الآن هو الطرف بالذات . وليس كلامنا في أنَّ هذا الوجه الثاني يصح وجوده أو لا يصحّ ، فإنّا لا نتكلم فيه من حيث يصدق بوجوده ، بل نتكلم فيه من حيث هو محمول عليه سَلْبٌ مّا ؛ وذلك السلّب هو أنّه ليس يوجد أو يعدم قليلاً قليلاً ، وله في ذلك شريك ، فذلك الشريك أخصّ من هذا السلّب ، والأخصّ لا يلزم الأعم ، وليس يجب أنْ يكون الشيء ؛ من حيث نتصوره موضوعاً أو محمولاً ، بحيث يصدق بوجوده أو لا يصدق ، ‹و› قد عُلم هذا في صناعة المنطق . فإذا كان قولنا ليس يوجد أو يعدم قليلاً قليلاً ، أعم من قولنا يوجد دفعة أو يعدم دفعة ، بمعنى أنّه يكون حاله ذلك في آنٍ مبتدأ ، فليس قول القائل إنّه إمّا أنْ يكون قليلاً قليلاً ، أو يكون

is not affirmed in the same way as a disjunctive [proposition] whose scope includes two contradictory disjuncts or one contradictory disjunct and what its contradictory disjunct necessarily entails. Likewise, the opposite of what exists all at once is what does not exist all at once—that is, it does not exist in some beginning instant, and it does not necessarily follow that it exists or does not exist gradually, and, in fact, that which corresponds with the previously noted sense [i.e., 2] might be affirmed together with it. [That is so] unless, by *that which exists all at once,* one means that whose existence is not an instant, but when it is at it, the existence is fully realized and there is no instant in which it is still in procession; and, accordingly, the same holds true for that which does not exist all at once. So, if this is meant, then this is necessarily the opposite, and the premise turns out true; but why should its beginning existence or nonexistence be all at once?

(4) There is something here, and even if this is not its proper place, we should mention it in order that there be a way to confirm what we said. Also, being familiar with it is worthwhile for learning [the answer to the following question, namely]: In the instant common to two periods of time, in one of which something is in one state and in the other of which it is in another state, is the thing altogether lacking both states, or does it have one of the states to the exclusion of the other? If the two situations are potentially contradictory, such as being contiguous and not being contiguous, existing and not existing, or the like, it would be absurd that at the assumed instant the thing would be altogether lacking both. So it must inevitably have one of the two [states], and I wish I knew which one! We argue that some thing undoubtedly opposes the exiting thing so as to render it nonexistent, in which case either one of two situations must be the case. That opposing thing might, in fact, do so at an instant.

دفعـــة ، بهذا الوجه ، صادقاً صدق المنفصل المحيط بطرفي النقيض ، أو المحيط بنقيض وما يلـــزم نقيضه . وأيضاً فإنّ مقابل ما يوجـــد دفعة هو ما لا يوجد دفعة ، أي لا يوجد في آنٍ مبتدأ ، وليس يلزمهّ - لا محالةّ - أنّه يوجد أو يُعدم قليلاً قليلاً ، بل قد يصدق معه الذي بحســـب الوجه المذكور . اللّهم إلاّ أنْ يعني بالموجود دفعة ؛ الذي لا يوجد آنٌ إلاّ وهو فيه حاصل الوجود ، ولا يوجد آنٌ هو فيه بعد في السلوك وكذلك في المعدوم دفعة بحسبه ، فـــإنْ كان عني هذا ؛ كان هذا لازم المقابل وصحت القضيـــة . ولكن ، لِمَ يجب أنْ يكون وجوده المبتدأ دفعة ، أو عدمه؟

(٤) وهـــا هنا شـــيءٌ - وإنْ كان لا يليق بهذا الموضـــعّ - فينبغي أنْ نذكره ليكون سبيلاً إلى تحقّق ما قلناه ؛ وهو : أنّه بالحرى أنْ يتعرّف ليعرف هل الآن المشترك بين زمانين ، في أحدهما الأمر بحالٍ وفي الاخر بحال أُخرى ، قد يخلو الأمر فيه عن الحالين جميعاً ، أو يكون فيه على إحدى الحالين دون الأخرى؟ فإنْ كان الأمران في قوة المتناقضين ، كالمماس وغير المماس ، والموجود والمعدوم ، وغير ذلك ، فمحالٌ أنْ يخلو الشيء في الآن المفروض عنهما جميعاً ، فيجب أنْ يكون على أحدهما لا محالة ، فليتَ شعري على أيّهما يكون؟ فنقـــول : إنَّ الأمر الموجـــود لا محالة يرد عليه أمرٌ فيعدمه ، فلا يخلـــو إمّا أنْ يكون ذلك الشـــيء الوارد ممّا يصح وروده في آنٍ ، وهو الشـــيء الذي يتشابه حاله في أي آنٍ أخذتَ

In other words, it is something whose state remains the same during any instant you take during the time that it exists, and it does not need some [other] instant in such a way as to correspond with a period of time in order to exist. As long as that is the case, the thing in the common division is described by [such a state]—as, for example, being contiguous, being square, and the other fixed dispositions whose existence remains the same during every instant of the time that they exist.[8] Alternatively, the thing might be contrary to this description, and so its existence would occur during a period of time, while not occurring at an instant. In that case, its existence would be in the second period of time alone and would not be predicated of the instant dividing the two such that there would be a certain opposition at it. Examples are *to depart* and *to cease being contiguous* or *moving.* The state of some of the latter [types] can remain the same at certain instants during their time, setting aside the instants that they begin to occur, whereas the state of others can in no way remain the same. Those that can are like *not being contiguous,* which is to be separate, since it occurs only by motion and a variation of some state; however, it can remain not contiguous and, in fact, be separate for a period of time, during which it remains the same; and, even if its states do vary from other perspectives, that will not be from the perspective that the two are separate and not contiguous. Those for which that is impossible are like motion, for its state does not remain the same at some instant or other, but, rather, at every instant there is a renewal of a new proximity and remoteness, both of which result from the motion. So, when the thing that is not undergoing motion is moved and what is contiguous stops being such, there will be the instant that divides the two times; [and], since at it there is no beginning separation and motion, there will be contact and an absence of motion at it. Even if this takes us beyond our immediate goal, it does provide some help here, as well as for other questions.

---

8. Adding the phrase *fī kull ān zamān wujūdihā* (during every instance of the time that they exist) with **Z**, **T**, and the Latin, which is (inadvertently) omitted in **Y**.

في زمان وجوده، ولا يحتاج في أنْ يكون إلى آنٍ يطابق مدة، وما كان هكذا، فالشيء في الفصل المشـــترك موصوف به، كالمماسّة والتربيع وغير ذلك من الهيئات القارّة التي يتشابه وجودهـــا في كلّ آن زمان وجودها. وإمّا أنْ يكون الشـــيء بخـــلاف هذه الصفة، فيقع وجـــوده في زمانٍ ولا يقع فـــي آن، فيكون وجوده في الزمان الثاني وحده، والآن الفاصل بينهمـــا لا يحتمله؛ فتكون فيه مقابلة؛ مثل المفارقة وترك المماسّـــة والحركة، فمن ذلك ما يجوز أنْ تتشـــابه حاله في آناتٍ من زمانـــه دون آنات الوقوع ابتداء، ومنه ما لا يجوز أنْ يتشابه حاله البتة. أما الذي يجوز فمثل الامماسّة؛ التي هي المباينة، فإنّها لا تقع إلاّ بحركةٍ واختـــلاف حال، ولكنها تثبت لا مماسّـــة بل مباينة، زماناً مّا تتشـــابه فيه وإنْ اختلفت أحوالهمـــا من جهاتٍ أخرى، فليس ذلك من جهة إنهما مباينة ولا مماسّـــة. وأمّا الذي لا يجوز ذلك فيه فكالحركة، فإنّها لا تتشـــابه حالهـــا في آنٍ من الآنات، بل يكون في كل آنٍ تجدد قربٍ وبعدٍ جديدٍ هما من أحوال الحركة. فالشيء غير المتحرّك إذا تحرّك والمماسّ إذا لم يماسّ، فالآن ‹هو› الفاصل بين زمانيه، إذ لا ابتداء مفارقةٍ فيه ولا حركة، ففيه مماسّة وعـــدم حركةّ – وهذا وإنْ كان خارجاً عن غرضنا، فإنّه نافعٌ فيه وفي مســـائل أخرى.

(5) That which we have discussed is the instant that is bounded by the past and future, as if a certain period of time came to be and then, after its occurrence, it is delimited by this instant. The estimative faculty, however, imagines another instant with a different description. So, just as you assume that, through the motion and flow of the limit of the moving thing (and let it be some point), there is a certain spatial magnitude[9] or, rather, a certain line (as if it—I mean that limit—is what is moving), and then you assume that there are certain points in that line (not that they make up the line, but that they are only what the estimative faculty imagines to be its connections), so it would likewise seem that, in time and motion (in the sense of *traversing* [a distance]), there is something like that and something like the line's internal points, which do not make it up. In other words, the estimative faculty imagines something that is being borne along and a certain limiting point on the spatial magnitude and time. So what is being borne along some continuous spatial magnitude produces a certain continuous locomotion that corresponds with a period of time. So it is as if what is borne along—or, more exactly, its state that necessarily accompanies it during the motion—is a certain indivisible limit, corresponding with a point on the spatial magnitude and some present instant of time that produces a certain continuum through its flow. So, together with [that limit], there is neither a spatial line (for it followed behind it), nor motion in the sense of traversing [the distance] (for it came to an end), nor time (for it is past). Together with it, there is only an indivisible limit of each one [of these], which it divides. So, from time, the present instant is always with it; from traversing [the distance], there is the thing, which we explained, is in reality motion, as long as something is undergoing motion;[10] and from the spatial magnitude, there is the limiting point,

9. The Arabic *masāfah* also means "distance"; however, "spatial magnitude," which the Arabic can happily support, seems more fitting in the present context, since Avicenna's point here, following John Philoponus (*In Phys.* 727.21–29), will be that that just as the motion of a point produces a line (as, for example, when a pen's nib is drawn across a piece of paper), so the flow of a *now* produces time.

10. Meaning, motion at an instant. See 2.1.6.

(٥) فهذا الذي تكلمنا فيه هو الآن المحفوف بالماضي والمستقبل، كأنه حدث زمانٌ فحُدَّ بعد حصوله بهذا الآن. وقد يُتوهم آنٌ آخر على صفةٍ أخرى، فكما أنَّ طرف المتحرّك، وليكن نقطة ما تفرض بحركته وسيلانه مسافة ما، بل خطاً ما؛ كأنّه - أعني ذلك الطرفّ - هو المنتقل، ثم ذلك الخط تفرض فيه نقطٌ، لا الفاعلة للخط بل المتوهمة، واصلة له، وكذلك يشبه أنْ يكون في الزمان وفي الحركةّ - بمعنى القطعّ - شيءٌ كذلك، وشيء كالنقط الداخلة في الخط الذي لم تفعله، وذلك أنّه يتوهم منتقلٌ وحدٌّ في المسافة وزمان، فالمنتقل يفعل نقلةً متصلةً على مسافة متصلة، يطابقه زمانٌ متصل. فكأن المنتقلّ - بل حالته التي تلزمه في الحركةّ - هو طرفٌ غير منقسم فعّالٌ بسيلانه اتصالاً، ويطابقه من المسافة نقطة، ومن الزمان آن، فإنّه لا يكون معه لا خط المسافةّ - فقد خلّفهّ - ولا الحركة بمعنى القطع فقد انقضت، ولا الزمان فقد سلف. إنّما يكون معه من كل واحدٍ طرفٌ له غير منقسم انقسامه، فيكون معه دائماً من الزمان الآن، ومن القطع الشيء الذي بيّنا أنه بالحقيقة هو الحركة ما دام الشيء يتحرك، ومن المسافة الحدّ، إمّا

whether a point or the like. Each one of these is an extremity, and even what is borne along is an extremity in its own right inasmuch as it was borne along. It is as if there is something extending from the beginning of the spatial magnitude up to where it has reached. [That] is because, inasmuch as it is something borne along, there is something extending from the beginning to the end, while it itself—the continuously existing present instant—is a certain limiting point and extremity inasmuch as it had been borne along to this limiting point.

(6) We should investigate whether, just as what is borne along is one and by its flow produced its limiting point and endpoint as well as the spatial magnitude, there is likewise in time something that is the *now* that flows.[11] In this case, [the now] would itself be something indivisible *qua* itself, and it itself would persist inasmuch as it is like that, while it would not persist inasmuch as it is a given present instant, because it is a present instant only when it is taken as something delimiting time, just as the former is what is borne along only when it delimits what it delimits, whether in itself a point or something else. Just as it happens that what is borne along *qua* being borne along cannot exist twice, but passes away when it is no longer being borne along, likewise the now *qua* a given present instant does not exist twice. Still, the thing that, owing to whatever condition, becomes a *now* perhaps does exist several times, just as what is borne along, *qua* a thing (which just so happened to be borne along), perhaps exists several times. So, if something like this exists, then it is rightly said that *the now produces time through its flow;* but this is not the instant that is posited between two periods of time that connects them, just as the point that the estimative faculty imagines producing a spatial magnitude by its motion is different from the point that it imagines in the spatial magnitude at [a connecting instant]. So, if

11. Avicenna's discussion of the "flowing now" relies heavily upon John Philoponus's own discussion from his *Physics* commentary; see especially *In Phys.* 727.21–29.

نقطة وَإما غير ذلك ، وكل واحدٍ من هذه نهاية ، والمنتقل أيضاً نهاية لنفسه من حيث انتقل ، كأنه شيء ممتد من المبدأ في المسافة إلى حيث وصل ؛ فإنّه من حيث هو منتقل شيء ممتد من المبدأ إلى المنتهى ، وذاته الموجودة المتصلة الآن حدٌّ ونهايةٌ لذاته من حيث قد انتقل إلى هذا الحدّ ،

(٦) فحريٌ بنا أنْ ننظر ، هل كما أنَّ المنتقل ذاته واحدة ، وبسيلانه فعل ما هو حدّه ونهايته ؛ وفعل المسافة أيضاً ؟ كذلك في الزمان شيءٌ هو الآن يسيل فيكون ذاتاً غير منقسمة من حيث هي هو ، وهو بعينه باقٍ ، من حيث هو كذلك ، وليس باقياً من حيث هو آن ، لأنّه إنما يكون آناً إذا أُخذ محدّداً للزمان ، كما أنَّ ذلك إنّما يكون منتقلاً إذا كان محدّداً لما يحدّده ، ويكون في نفسه نقطة أو شيئاً آخر . وكما أنَّ المنتقل يفرض لـهّ - من حيث هو منتقلّ - أنّه لا يمكن أنْ يوجد مرتين ؛ بل هو يفوت بفوات انتقاله ، وكذلك الآن ، من حيث هو آن ، لا يوجد مرتين ، لكن الشيء الذي ، لأمرٍ مّا ، صار آناً عسى أنْ يوجد مراراً ، كما أنَّ المنتقلّ - من حيث هو أمرٌ - عرض له الانتقال ؛ عسى أنْ يوجد مراراً . فإنْ كان شيء مثل هذا موجوداً ، فيكون حقاً ما يقال إنَّ الآن يفعل بسيلانه الزمان ، ولا يكون هذا هو الآن الذي يفرض بين زمانين يصل بينهما ، كما أنَّ النقطة المتوهمة فاعلة بحركتها مسافة هي غير نقطة المسافة المتوهمة فيه . فإنْ كان لهذا الشيء

this thing exists, then its existence is joined to the thing that we previously identified with motion[12] without taking what is earlier and later nor coinciding with [the motion]. Now, just as its possessing a *where* when it continuously flows along the spatial magnitude produces motion, so its possessing that thing that we called the *now* when it continues along the earlier and later parts of motion produces time. So this thing's relation to what is earlier and later is in that it is an instant, while, in itself, it is something that makes time.

(7) [The instant] also numbers time by what is produced when we take some instant from among limiting points in [what is earlier and later]. In this case, numbered instances of being earlier and later are produced, just as points number the line through two relations in that each point is shared between two lines. Now, the one that truly numbers is that which, first, provides the thing with a unit, and, [second,] provides it with number and multiplicity by[13] repeating [the unit]. So the instant that has this description numbers time, for, as long as there is no instant, time is not numbered. Now, what is earlier and later numbers time in the second way. That is, it is part of it, whose being a part occurs through the existence of the instant. Because what is earlier and later are time's parts, and [because] each part of it can be divided, like the parts of a line, the instant is better suited to the unit, and the unit [is] better suited to do the numbering. So the instant numbers in the way that the point does, while not being divisible. Also, the motion numbers time in that it makes the earlier and later parts exist by reason of the distance, and so, through the motion's magnitude, there is the number of what is earlier and later. Thus, motion numbers the time in that it makes time's number to exist—that is, what is before and after—whereas time numbers motions in that it is itself a number belonging to [motion].

---

12. Again, Avicenna is referring to his account of motion at an instant see 2.1.6.

13. **Y** (inadvertently) omits the *bi,* which occurs in **Z** and **T**; the Latin has the adverbial phrase *iterum et iterum* (repeatedly), which confirms the presence of *bi.*

وجودٌ؛ فهو وجود الشيء مقروناً بالمعنى الذي حقّقنا فيما سلف، إنّه حركةٌ من غير أخذ متقدم ولا متأخر ولا تطبيق. وكما أنَّ كونه ذا أين؛ إذا استمر سائلاً في المسافة أحدث الحركة، كذلك كونه ذا ذلك المعنى الذي سميناه الآن إذا استمر في متقدّم الحركة ومتأخرها؛ أحدث الزمان. فنسبة هذا الشيء إلى المتقدّم والمتأخر هي في كونه آناً وهو في نفسه شيء يفعل الزمان.

(٧) ويعدّ الزمان بما يحدث، إذا أخذنا آناً من حدودٍ فيهما، فتحدث تقدمات وتأخرات معدودة؛ كالنقط تعدّ الخط بأنْ تكون كل نقطة مشتركة بين خطين بإضافتين، والعادّ الحقيقي هو الذي هو أول معْطٍ للشيء وحدة، ومعطٍ له الكثرة والعدد بالتكرير، فالآن الذي بهذه الصفة يعدّ الزمان. فإنّه ما لم يكن آن لم يُعدّ الزمان، والمتقدّم والمتأخر يعدّ الزمان على الوجه الثاني؛ أي بأنّه جزؤه تحصل جزئيته بوجود الآن، ولأنَّ المتقدّم والمتأخر أجزاء الزمان، وكل جزءٍ منه من شأنه الانقسام، كأجزاء الخط، فالآن أولى بالوحدة، والوحدة أولى بالتعديد، فالآن يعدّ على الجهة التي تعدّ النقطة ولا ينقسم، والحركة تعدّ الزمان بأنْ توجد المتقدّم والمتأخر بسبب المسافة. فبمقدار الحركة يكون عدد المتقدّم والمتأخرّ - فالحركة تعد الزمان، على أنها توجد عدد الزمان وهو المتقدّم والمتأخر، والزمان يعدّ الحركة بأنه عدد لها نفسها.

(8) An example of this is that people, owing to their existence, are the cause of their number, which is, for instance, ten. Because they exist, their being ten occurs, whereas being ten did not make the people to exist or be things, but only to be numbered—that is, to have a number. When the soul numbers the people, what is numbered is not the nature of the humans, but the ten-ness, which, for example, the spatially dislocated nature of the humans brings about. So, through the humans, the soul counts the ten-ness; and likewise, the motion numbers time in the aforementioned sense. Were it not for the motion, through the limiting points of earlier and later that it produces in the spatial magnitude, a number would not belong to time. As it is, however, time measures motion, and motion measures time. Time measures motion in two ways:[14] [the first] one is that it provides it with a determinate measure, while the second is that it indicates the quantity of its measure (where motion measures time as indicating its measure through the earlier and later parts that exist in it); and there is a difference between the two situations. As for indicating the measure, it is sometimes like the measure of wheat that indicates the holding capacity for the wheat, and at other times it is like the holding capacity for the wheat that indicates the measure of wheat [being held]. Similarly, sometimes the distance indicates the measure of motion, and at other times the motion indicates the measure of the distance, and so sometimes it is said that a trip is two parasangs [approximately seven miles] and sometimes it is said that a distance is a stone's throw; however, that which provides the magnitude for the other is just one of them—namely, the one that is in itself a measure.

---

14. Cf. John Philoponus, *In Phys.* 741.21–742.14.

(٨) مثال هذا ؛ أنَّ الناس لوجودهم هم أسباب وجود عددهم الذي هو مثلاً عشرة، ولوجودهم وجدت عشريتهم، والعشرية جعلت الناس لا موجودين وأشياء، بل معدودين أي ذوي عدد، والنفس إذا عدت الناس كان المعدود ليس هو طبيعة الإنسان، بل العشرية التي جعلها افتراق طبيعة الإنسان مثلاً، فالنفس بالإنسان تعدّ العشرية، فكذلك الحركة تعدّ الزمان على المعنى المذكور. ولولا الحركة - بما تفعل في المسافة من حدود التقدّم والتأخّر - لما وجد للزمان عددٌ، لكن الزمان يقدّر الحركة والحركة تقدّر الزمان، والزمان يقدّر الحركة على وجهين: أحدهما أنّه يجعلها ذات قدر، والثاني أنّه يدل على كمّيّة قدرها، والحركة تقدّر الزمان على أنّها تدل على قدره بما يوجد فيه من المتقدّم والمتأخر، وبين الأمرين فرقٌ. أمّا الدلالة على القدر؛ فتارة يكون مثل ما يدل المكيال على الكيل، وتارة يكون مثل ما يدل الكيل على المكيال، وكذلك تارة تدل المسافة على قدر الحركة، وتارة الحركة على قدر المسافة. فيقال تارة مسيرة فرسخين، وتارة مسافة رميةٍ، ولكن الذي يعطي المقدار للآخر هو أحدهما، وهو الذي هو بذاته قدر.

(9) Because time is something continuous in its substance, it is appropriately said to be *long* and *short;* and, because it is a number relative to what is earlier and later in the way explained, it is appropriately said to be *little* and *much*. The same holds for motion, for it accidentally has a certain continuity and discontinuity, and so the properties of what is continuous and of what is discontinuous are attributed to it, albeit it happens to have that from something other than itself, whereas that which is most proper to it is speed. So we have indicated the way [in which] the *now* exists as actual (if it exists as actual) and the way [in which] it exists potentially.

(٩) ولأنَّ الزمـــان متصلٌ في جوهره صلح أنْ يقال طويل وقصير ، ولأنّه عدد القياس إلى المتقــدّم والمتأخرّ – على ما أوضحاهّ – صلح أنْ يقال قليل وكثير . وكذلك الحركة؛ فإنّها يعرض لها اتصالٌ وانفصال؛ فيقال عليها خواص المتصل وخواص المنفصل، لكن ذلك يعرض لها من غيرها ، والذي هو أخص بها السـريع والبطيء . فقد دلّلنا على نحو وجود الآن بالفعلّ – إنْ كان له وجودٌ بالفعلّ – وعلى نحو وجوده بالقوة.

Chapter Thirteen

The solution to the skeptical puzzles raised about time and the completion of the discussion of things temporal, such as being in time and not in time, everlasting, eternity, [and the expressions] suddenly, right away, just before, just after, *and* ancient

(1) Everything that was said to undermine the existence of time and about its not having any existence is based upon its not existing at an instant.[1] Now, there is a distinction between saying that it has no existence absolutely and that it has no existence that occurs at an instant. We ourselves wholeheartedly concede that time does not exist so as to occur according to the latter sense, save in the soul and estimative faculty, whereas it does, in fact, have the absolute existence that opposes absolute nonexistence. Indeed, if [its existence] were not a fact, its negation would be true; and so it would be true that it is *not* the case that between two spatially separated points there is some magnitude [corresponding with] a possibility of some motion to traverse [that distance] at some definite speed. Now, when this negation is false and, in fact, a certain magnitude does belong to the motion at which it is possible to traverse this distance at that definite speed and possible to traverse a different one

1. See 2.10.2–3.

<الفصل الثالث عشر>

في حل الشكوك المقولة في الزمان وإتمام القول في مباحث زمانية مثل الكون في الزمان والكون لا في الزمان وفي الدهر والسرمد وبغتةً وهو ذا وقُبيل وبُعيد والقديم

(١) وأمّا الزمان؛ فإنَّ جميع ما قيل في أمر إعدامه وأنه لا وجود له، فهو مبني على أنْ لا وجود له في الآن، وفرقٌ بين أنْ يقال لا وجود له مطلقاً، وبين أنْ يقال لا وجود له في آنٍ حاصلاً. ونحن نسلّم ونصحّح أنَّ الوجود المحصل على هذا النحو لا يكون للزمان إلاّ في النفس والتوهم، وأمّا الوجود المطلق المقابل للعدم المطلق فذلك صحيح له. فإنّه إنْ لم يكن صحيحاً صدق سلبه، فصدق أنْ نقول: إنّه ليس بين طرفي المسافة مقدار إمكانٍ لحركةٍ على حدّ من السرعة يقطعها. وإذا كان هذا السلب كاذباً، بل كان للحركة على ذلك الحدّ من السرعة مقدارٌ فيه يمكن قطع هذه المسافة، ويمكن قطع غيرها بأبطأ

by going slower or faster (as we explained before),[2] the affirmation that opposes [this negation] is true—namely, that there is the magnitude of this possibility. The affirmation indicates that something exists, even if it does not indicate that it does so as something occurring at an instant or in some other way. Also, it doesn't have this manner [of absolute existence simply] because of the activity of the estimative faculty, for, even if there were no activity of the estimative faculty, this manner of existence and truth would obtain. Still, it should be known that some existing things have a determinate and realized existence, while others have a more tenuous existence. Time seems to exist more tenuously than motion, akin to the existence of things that are relative to other things, even though time *qua* time is not relative, but relation necessarily accompanies it.[3] Since distance and the limiting points in it exist, whatever has some affiliation with [distance]—whether as mapping onto it or traversing it or a magnitude of its traversal—exists in some way such that it is simply false that [that thing] should not exist at all. If it is intended, however, that we provide time with an existence contrary to this way, such that it exists determinately, then it will occur only in the act of the estimative faculty. Thus, the premise used in affirming that time does not exist in the sense of not existing at a single instant is granted, and we ourselves do not deny its nonexistence at an instant. Its existence, rather, is in the way of generation in that, for any two instants that you care to take, there is something between them that is time, while not at all being at a single instant. In summary, we don't need to worry ourselves [about answering] their question, "If [time] exists, then does it exist at an instant or during a period of time? Otherwise, when does it exist?" The fact is that time does not exist at an instant or during a period of time, nor

2. See 2.11.1–2.

3. Following **Z**'s and **Y**'s suggestion to seclude the phrase *min ḥaythu kawnuhu miqdār al-shay᾿ wa-kawnuhu zamānan ghayr kawnihi miqdāran* (inasmuch as it is the magnitude of something and its being time is different from its being a magnitude), which is also omitted in **T** and the Latin.

وأسرعّ – على ما بيّنا قبلّ – فالإثبات الذي يقابله صادقٌ، وهو أنَّ هناك مقدار هذا الإمكان والإثبات دلالة على وجود الأمر مطلقاً، وإنْ لم تكن دلالة على وجوده محصّلاً في آنٍ، أو على جهةٍ ما. وليس هذا الوجه له بسبب التوهم، فإنّه – إنْ لم يتوهمّ – كان هذا النحو من الوجود وهذا النحو من الصدق حاصلاً. ومع هذا، فيجب أنْ يُعلم أنَّ الموجودات منها ما هي متحقّقة الوجود محصلّة، ومنها ما هي أضعف في الوجود، والزمان يشبه أنْ يكون أضعف وجوداً من الحركة، ومجانساً لوجود أمور بالقياس إلى أمور، وإنْ لم يكن الزمان، من حيث هو زمان مضافاً، بل قد تلزمه الإضافةً. ولمّا كانت المسافة موجودة وحدود المسافة موجودة، صار الأمر الذي من شأنه أنْ يكون عليها ومطابقاً لها أو ‹قاطعاً› لها أو مقدار قطع لها نحوا من الوجود، حتى إنْ قيل إنّه ليس له البتة وجود كذبَ. فإنْ أُريد أنْ نجعل للزمان وجوداً لا على هذا السبيل، بل على سبيل التحصيل؛ لم يكن إلاّ في التوهم. فإذن المقدمة المستعملة في أنَّ الزمان لا وجود له ثابتاً؛ معناه لا وجود في آنٍ واحدٍ مسلّمةٌ، ونحن لا نمنع أن يكون له وجود ليس في آن، بل وجوده على سبيل التكوّن، بأن يكونّ – أي آنين فرضتهماّ – كان بينهما الشيء الذي هو الزمان، وليس في آنٍ واحدٍ البتة. وبالجملة، طلبهم أنَّ الزمان إنْ كان موجوداً فهو موجود في آنٍ أو في زمان، أو طلبهم متى هو موجود، ممّا ليس يجب أنْ نشتغل به. فإنَّ الزمان

does it have a *when*. Instead, it exists absolutely and just *is* time, and so how could it exist in time? So [to begin with] their claim is incorrect that either time doesn't exist, or it exists at an instant or as something persisting during a period of time. Moreover, it does not oppose our claim that it does not exist either at an instant or as persisting during a period of time, and yet time exists, while not being one of these two types of existence. That is because it neither is at an instant nor persists during a period of time. This is just like the one who says that either place does not exist, or it exists in place or at some definite point of place. That is because it does not have to exist either in place or at some definite limiting point of place, or [otherwise] not exist. The fact is that some things simply do not exist in place, and some simply do not exist in time. Place falls within the first class, and time within the second (you will learn this later).[4]

(2) [Next is the argument] that maintains that, if time exists, then time necessarily follows every motion, in which case a [private] time is consequent upon each motion.[5] The response is to distinguish between saying that time is a certain magnitude of each motion and saying that its individual existence is dependent upon each motion. Furthermore, there is a distinction between saying that *the time itself* depends upon motion as one of its accidents and saying that time depends upon *the motion itself* such that time is an accident of [that very motion], because the first sense is that certain things accidentally belong to a given thing, while the second is that certain things are consequent upon a given thing. As for the first, it is not a condition of what measures that it be something accidental to *and* subsist with the thing; rather, it might measure something distinct [from itself] by being brought next to and

4. While Avicenna does discuss various "modes" of existence throughout his *Ilāhīyāt* (see especially 1.5), I have not been able to find a discussion corresponding with the one promised here.

5. See 2.10.4.

موجــودٌ لا فــي آنٍ ولا في زمانٍ ولا له متى ، بل هو موجــودٌ مطلقاً ، وهو نفس الزمان ، فكيف يكون له وجودٌ في زمان؟ فليس إذن قولهم إنَّ الزمان إمّا أنْ لا يكون موجوداً ، أو يكون وجوده في آنٍ ، أو يكون وجوده باقياً في زمانٍ قولاً صحيحاً ، بل ليس مقابل قولنا إنَّــه ليس بموجود ، هو أنَّــه موجودٌ في آنٍ أو موجودٌ باقيا في زمان ، بل الزمان موجودٌ ؛ ولا واحــد من الوجودين ، فإنَّه لا في آنٍ ولا باقٍ فــي زمان . وما هذا إلاّ كمَنْ يقول ، إمّا أنْ يكــون المكان غير موجود ، أو يكون موجوداً في مكانٍ أو في حدّ مكان . وذلك لأنّه ليــس يجب أنْ يكون إمّا موجوداً في مكان أو في حدّ مكان ، وإمّا غير موجود ، بل من الأشياء ما ليس البتة موجودا في مكان ، ومن الأشياء ما ليس البتة موجودا في الزمان ؛ والمكان من جملة القسم الأول ، والزمان من جملة القسم الثاني وستعلم هذا بعد .

(٢) والذي قيل ، إنّه إنْ كان للزمان وجود ، وجب أنْ يتبع كل حركةٍ زمانٌ ، فتكون كل حركة تســتتبع زماناً . فالجــواب عن ذلك أنَّه فرقٌ بين أنْ يقــال إنَّ الزمان مقدارٌ لكل حركــة ، وبين أنْ يقال إنَّ إنّيتــه متعلقة بكل حركة . وأيضاً فرقٌ بين أنْ يقال إنَّ ذات الزمان متعلقة بالحركة على ســبيل العروض لها ، وبين أنْ يقــال إنَّ ذات الحركة متعلقٌ بها الزمان ، على ســبيل أنَّ الزمان يعرض لها ، لأنَّ الأول معناه أنَّ أشــياء تعرض لشــيءٍ ، والثاني أنّ أشــياء تســتتبع شيئاً . أمّا الأول فلأنَّه ليس من شــرط ما يقدّر أنْ يكون الشيء عارضاً لــه وقائماً به ، بــل ربمّا قدّر المباين بالموافاة والموازاة لما هو مباين له . وأمّا الثاني فلأنه ليس

juxtaposed with what is distinct from it. As for the second, it is not the case that when *the thing itself* depends upon the nature of a given thing, the nature of the thing must not be devoid of it. Now, what was demonstrated for us concerning time is only that it depends upon motion and is a certain disposition of it, while, concerning motion, it is only that every motion is measured by time. It does not necessarily follow from these two [propositions] that a given time is dependent upon and peculiar to each motion, nor that whatever measures something is an accident of it, such that time is, essentially in itself, an accident belonging to each motion. The fact is that time is not dependent upon the motions that have a beginning and ending. How could time be dependent upon them? If time were to belong to them, then it would be divided by two instants, and we precluded that. Certainly, when time exists by a motion having a certain description upon which the existence of time is truly dependent (where this motion is continuous and does not have actual extreme limits that delimit it), the rest of the motions will be measured by it.

(3) One might ask: Do you think that if that motion did not exist, time would vanish such that the other motions different from it would be without an *earlier* and a *later,* or is it as you said in the skeptical puzzles[6]—namely, that the body, in order[7] to exist as undergoing motion, does not need a motion of another body such that it might undergo motion, whereas it cannot not have a time? The response to that (and, in fact, it will be explained [later])[8] is that, if a certain circular body did not have a circular motion, then you could not impose directions on rectilinear [motions], and so there would be no natural rectilinear [motion], and so no forced [motion]. So it might be that a motion of some one body alone without any other bodies is impossible, even though the impossibility is

6. See 2.10.4.

7. Reading *fī an* with **T** and the Latin (*ut*) for both **Z**'s and **Y**'s *fī ān* (at an instant). First, orthographically the two phrases are practically identical, and, again, both **T** and the Latin are in agreement with me; second, philosophical sense seems to require reading *fī an,* for, while Avicenna does countenance motion at an instant, it is not clear what it would mean for a body to be at an instant, which seems to be an obvious category mistake.

8. See 3.13 (all).

إذا تعلّقت ذات شيء بطبيعة شيء يجب أنْ لا تخلو طبيعة الشيء عنه. ونحن إنّما تبرهن لنا من أمر الزمان أنّه متعلّق بالحركة وهيئةٌ لها، ومن أمر الحركة أنَّ كل حركة تقدّر بزمانٍ. وليس يلزم من هذين أنْ تكون كل حركةٍ يتعلق بها زمانٌ يخصّها، ولا أنَّ كل ما قدّر شيئاً فهو عارضٌ له، حتى يكون لكل حركةٍ زمانٌ عارض لذاته بعينه، بل الحركات التي لها ابتداء وانتهاء لا يتعلق بها الزمان، وكيف يتعلق بها الزمان؟ ولو كان لها زمانٌ لكان مفصولاً بآنين، وقد منعنا ذلك. نعم، إذا وجد الزمان بحركةٍ على صفةٍ، يصلح أنْ يتعلق بها وجود الزمان، تقدّر به سائر الحركات، وهذه الحركة حركة يصح عليها الاستمرار ولا يتحدّد لها بالفعل أطراف.

(٣) فإنْ قال قائل: أرأيتَ؛ إنْ لم توجد تلك الحركة، أكان الزمان يُفقَد، حتى تكون حركات أُخرى غيرها بلا تقدم ولا تأخّرٍ؟ أو قيل ما ذكرناه في الشكوك، إنَّ الجسم في أن يوجد متحركاً غير محتاج إلى حركة جسم آخر؛ فيجوز أنْ يتحرّك، ولا يجوز أنْ لا يكون له زمان؟ فالجواب عن ذلك أنّه سيتبين لك أنَّه إنْ لم تكن حركة مستديرة لجرم مستدير لم تفترض للمستقيمة جهات، فلم تكن حركات مستقيمة طبيعية، فلم تكن قسرية، فيجوز أنْ تكون حركة جسمٍ من الأجسام وحده - ولا أجسام أخرى - مستحيلة؛ وإنْ لم يكن

not [self-]evident. It happens that not every absurdity wears its impossibility on its sleeve; but, rather, there are many absurdities that are not obvious, and their impossibility becomes clear only though proof and demonstration. If we rely on the activity of the estimative faculty, then, when by its act of imagining we eliminate circular [motion] and affirm in our estimative faculty that a finite rectilinear [motion] is possible, a finite time is affirmed without objection in the estimative faculty. Our concern is not with this, however, but with what really exists. Hence, the existence of time is dependent upon a single motion that it measures and, equally, the rest of the motions whose existence would be impossible without the motion of the body that, through its motion, produces time (except in the act of the estimative faculty). That is like the measure existing in some body that measures [that body] as well as whatever is parallel and juxtaposed to it. Its being a measure—that is, its being one and the same thing for two bodies—does not require that it depend upon the two bodies. It might depend on only one of them, measuring it as well as the other one that it is not dependent upon. Now, the continuity of motion is only because of the continuity of distance; and, because of distance's continuity, there comes to be a cause for the motion's being earlier and later, by which [that is, being earlier and later] the motion is a cause of its having a number, which is time. So motion is continuous in two ways—owing to the distance and owing to the time—whereas it in itself is only a perfection of what is in potency. Moreover, neither some continuity nor [the fact] that it is a measure enters into the essence of this account; for it is not understood, concerning a perfection of what is in potency or a transition from one thing to another or a passage from potency to actuality, that there is a certain interval between the start

بيّن الاستحالة. وليس كل محال يعرض يكون بيّن عروض الاستحالة، بل كثير من المحالات لا يظهر ولا تستبين استحالتها إلاّ ببيانٍ وبرهان. وأمّا إنْ اعتمدنا التوهم، فإذا رفعنا المستديرة بالتوهم وأثبتنا المستقيمة المتناهية في الوهم أمكن وثبت في التوهم زمانٌ محدودٌ لا يستنكره التوهم، وليس نظرنا في هذا، بل فيما يصح في الوجود. فالزمان إذن وجوده متعلق بحركةٍ واحدة يقدّرها ويقدّر أيضاً سائر الحركات التي يستحيل أنْ توجد دون حركة الجسم الفاعل بحركته الزمانّ – إلاّ في التوهمّ – وذلك كالمقدار الموجود في جسم يقدّره ويقدّر ما يحاذيه ويوازيه. وليس يوجب تقديرهّ – وهو واحدٌ بعينه للجسمينّ – أنّ يكون متعلقاً بالجسمين، بل يجوز أنْ يتعلق بأحدهما، ويقدّره ويقدّر أيضاً الآخر الذي لم يتعلق به. والحركة اتصالها ليس إلاّ لأنَّ لامسافة متصلة، ولأنَّ اتصال المسافة يصير علّة لوجود تقدّمٍ وتأخرٍ في الحركة، تكون الحركة بهما علّة لوجود عددٍ لها هو الزمان. فتكون الحركة متصلة من جهتين، من جهة المسافة ومن جهة الزمان، وأمّا هي في ذاتها فليست إلاّ كمالٌ ما بالقوة. وليس يدخل في ماهية هذا المعنى أيضاً اتصال أو ‹تقدير›، فإنّه لا يفهم، من كمالٍ ما بالقوة أو انتقالٍ من شيءٍ إلى شيء ومن خروجٍ من قوةٍ إلى فعل، أنَّ هناك بُعْداً ما

and end that is continuous and is susceptible to the division to which the continuous is susceptible. The fact is [that] this is something known by a kind of reflection by which you learn that this account applies solely to continuous magnitude. So, if our estimative faculty were to imagine three atoms and something undergoing motion while it was at the middle one, then, during its motion from the first to the third, there would be at [the middle atom] a perfection of what is in potency; and yet it would not apply to something continuous. So the very fact that it itself is a perfection of what is in potency does not require that it be divisible. Thus, as long as certain other things are not known, neither will we know the necessity of that—namely, that [motion] applies only to something continuous that is susceptible to such-and-such a division. So continuity is clearly something that accidentally accompanies motion owing to either distance or time, and is not included in its essence. In summary, if we do not consider distance or time, we do not find that motion is continuous. Hence, whenever we need to measure motion, we need to mention distance or time.

(4) The proximate cause of time's continuity is the continuity of the motion through the distance, not the continuity of the distance alone; for, as long as there is no motion, the continuity of the distance alone will not necessitate the continuity of time. Similarly, there may be a certain distance over which the mobile is moved, pauses, and then begins again from there and moves until it is done—in which case the distance's continuity exists, but the time is not continuous. The fact is that the cause of time's continuity must be the distance through the intermediacy of the motion, because the continuity of time is the continuity of the distance on the condition that there be no rest in it. So the cause of time's continuity is one of motion's two continuities *qua* motion's continuity and is nothing but the continuity of the distance relative to the

بين المبتدأ والمنتهى، متصلاً قابلاً للقسمة التي يقبلها المتصل، بل هذا يعلم بنوع من النظر؛ تعلم به أنَّ هذا المعنى يكون به على المقدار المتصل لا غير. فلو أنّا توهمنا ثلاثة أجزاء لا تتجزأ، وكان المتحرّك، حين يتحرك في الأوسط منها، لكان فيه عند حركته من الأول إلى الثالث كمالٌ ما بالقوة ولم يكن على متصل. فنفس كونها حقيقة كمال ما بالقوة لا يوجب أنْ تكون منقسمة، ولذلك ما لم تعرف أشياء أخرى؛ لا نعرف وجوب ذلك، وأنّها لا تكون إلاّ على متصلٍ قابلٍ للقسمة كذا. فبيّنٌ أنَّ الاتصال أمرٌ عارض يلزم الحركة من جهة المسافة أو من جهة الزمان لا يدخل في ماهيتها، وبالجملة فإنا لو لم نلتفت إلى مسافة أو إلى زمان لم نجد للحركة اتصالاً، ولذلك متى احتجنا إلى تقدير الحركة، احتجنا إلى ذكر مسافةٍ أو زمان.

(٤) وأمّا اتصال الزمان؛ فعلّته القريبة اتصال الحركة بالمسافة لا اتصال المسافة وحدها. فإنَّ اتصال المسافة وحدها، ما لم تكن حركة موجودة، لا توجب اتصال الزمان وحدها، كما تكون مسافة يتحرك فيها المتحرّك ويقف، ثم يبتدىء من هناك ويتحرّك حتى يفنيها. فيكون هناك اتصال المسافة موجوداً؛ ولا يكون الزمان متصلاً، بل يجب أنْ تكون علّة اتصال الزمان اتصال المسافة بتوسط الحركة. ولأنَّ اتصال الزمان اتصال المسافة - بشرط أنْ لا يكون فيها سكونٌ، فعِلّة اتصال الزمان أحد اتصالي الحركة من جهة ما هو اتصال الحركة، وليس هذا إلاّ اتصال المسافة مضافة إلى الحركة، وهذا لا يكون وهناك

motion, where there will not be this [continuity relative to the motion] wherever there is rest. This continuity is not the cause of time's becoming continuous, but for making time exist. [That] is because time does not accidentally have the continuity proper to it, but it is itself that continuity. So, if there were something that gave time continuity (not meaning that time itself is one and the same as the continuity), then the continuity would be accidental to time and not its substance. Just as we say that a certain color was a cause of color or a certain heat was a cause of heat—by which we mean that they are a cause of a color or heat's existing, not of the quality's being [color or] heat—so, likewise, we say that a continuity is a cause of a continuity's existing, not that it is a cause of that thing's becoming a continuity. The fact is that it is essentially a continuity, just as the former is essentially heat.

(5) One should not say: We understand motion's having continuity only because of distance or time. Now, you yourselves denied that spatial continuity is a cause of time; and you cannot say that temporal continuity is a cause of time and then go on and say that the continuity of motion is a cause of time, when there is no continuity other than the former two. Our response is to say that we do, in fact, make spatial continuity a cause of time, but just not absolutely, but only inasmuch as there is some motion;[9] and so, by means of [the spatial continuity], the motion is continuous, where considering the continuity of the distance by itself is one thing and considering it joined to motion is another. So know now that distance's continuity inasmuch as it belongs to motion is a cause of time's existence that is itself something continuous or a continuity, not that it is

9. Reading *sāra li-ḥarakah* with **Z** and the Latin's dative of the possessor *motui* for **Y**'s *sāra bi-ḥarakah* (come to be through motion); it is not clear what if any preposition is used in **T**.

سكون. وليس هذا الاتصال علّة لصيرورة الزمان متصلاً؛ بل لإيجاد الزمان، فإنّه ليس الزمان شيئاً يعرض له الاتصال الخاص به، بل هو نفس ذلك الاتصال. فلو كان شيء يجعل للزمان اتصالاً، لا على معنى اتحاد ذات الزمان المتصل، لكان الاتصال عارضاً للزمان لا جوهر الزمان. وكما أنّا نقول إنَّ لوناً كان سبب لونٍ، أو حرارة كانت سبب حرارة، ونعني بذلك أنّها كانت سبباً لوجود اللون أو الحرارة، لا لكون الكيفية حرارة، كذلك نقول إنَّ اتصالاً هو سببٌ لوجود اتصال، لا أنّه سببٌ لصيرورة ذلك الشيء اتصالاً؛ فإنّه اتصال بذاته، كما أن تلك حرارة لذاتها.

(٥) وليس لقائلٍ أنْ يقول إنّا لا نفهم للحركة اتصالاً إلاّ بسبب المسافة أو الزمان، وأنتم أبيتم أنْ يكون الاتصال المسافي سبباً للزمان، ولا يجوز أنْ تقولوا إنَّ الاتصال الزماني هو سبب الزمان، ثم تقولون إنَّ اتصال الحركة سببٌ للزمان، وليس هناك اتصال غير هذين. فإنّا نجيبه ونقول: إنّا نجعل الاتصال المسافي سبباً للزمان ولكن لا مطلقاً بل من حيث صار لحركةٍ فصارت الحركة بها متصلة، واعتبار اتصال المسافة بنفسه شيء، واعتباره مقارناً للحركة شيء. فافهم الآن أن اتصال المسافة من حيث هي للحركة علّة لوجود الزمان، الذي هو بذاته متصل أو اتصال، لا أنه علّة لكون ذات الزمان متصلاً، فذلك أمرٌ لا علّة

a cause of the time itself being something continuous (for that is something that has no cause). So, by this, it turns out that time is something accidental to motion and is not its genus or difference or one of its causes, and yet it is still something necessarily accompanying it that measures every single instance of it.

(6) Recognizing what it is *to be in time* also belongs to a discussion of things temporal. So we say: Something is in time precisely according to the preceding principles—namely, it is understood as having earlier and later[10] parts. Now, the whole of what is understood as having earlier and later parts is either motion or something that undergoes motion. As for motion, that [namely, being earlier and later] belongs to it on account of its substance, whereas that belongs to the mobile on account of motion. Because it may be said of the kinds, parts, and extremities that they are in something, the earlier, later, and present instant, as well as hours and years, are said to be in time. So the present instant is in time as the unit is in number, and the earlier and later are like the even and odd in number, while hours and days are like two, three, four, and ten in number. Motion is in time as things that happen to be ten are in ten-ness, while the mobile is in time like the subject of the ten things that are accidental with respect to the ten-ness. Also, because the estimative faculty might imagine rest either as something continuing to go on forever *or* inasmuch as it accidentally happens to have an earlier and later [aspect]—namely, because of two motions that are [as it were] on either end—[and] since rest is a privation of motion in that which is disposed to being moved, not the privation of motion absolutely, it is not unlikely that there be [a rest] between two motions, and so something like this latter [type] of rest has an *earlier* and a *later,* in a certain way, and so the two ends of the rest enter into time accidentally. Also, changes that resemble local motion in that they advance from one limit to [another] limit (like heating going from one limit to another) enter into time on account of the fact that they have an earlier and later [aspect]. So, when

10. Reading *al-muta*ʾ*khkhir* with **Z**, **T**, and the Latin (*posterius*), which is (inadvertently) omitted in **Y**.

لـه . فبهذا يصح أنَّ الزمان أمر عارض للحركة وليس بجنس ولا فصل لها ولا سبب من أسبابها ، بل أمر لازم لها يقدّر جميعها .

(٦) ومن المباحث في أمر الزمان ، أنْ نعرف كون الشيء في الزمان فنقول إنّما يكون الشيء في الزمان على الأصول التي سلفت ، بأنْ يكون له معنى المتقدّم والمتأخر ، وكل ما له في ذاته معنى المتقدّم والمتأخر ، فهو إمّا حركة وإما ذو حركة . أمّا الحركة فذلك لها من تلقاء جوهرها ، وأمّا المتحرّك فذلك له من تلقاء الحركة . ولأنّه قد يقال لأنواع الشيء ولأجزائه ولنهاياته إنّها شيء في الشيء ؛ فالمتقدّم والمتأخر والآن أيضاً والساعات والسنون يقال إنّها في الزمان . فالآن في الزمان كالوحدة في العدد ، والمتقدّم والمتأخر كالزوج والفرد في العدد ، والساعات والأيام كالإثنين والثلاثة والأربعة والعشرة في العدد ، والحركة في الزمان كالعشرة الأعراض في العشرية ، والمتحرك في الزمان مثل الموضوع للأعراض العشرة في العشرية . ولأنَّ السكون إمّا أنْ يتوهم مستمراً ثابتاً أبداً ؛ وإمّا أنْ يتوهم بحيث يعرض له تقدّمٌ وتأخرٌ بالعرض ؛ وذلك بسبب الحركتين اللّتين تكتنفانه ، إذ السكون عدم حركةٍ فيما من شأنه أنْ يتحرّك ، لا عدم الحركة مطلقاً ؛ فلا يبعد أنْ يكون بين حركتين . فمثل هذا السكون له - بوجهٍ ما - تقدّم وتأخر فهو أدخل وجهي السكون في الزمان دخولاً بالعرض . والتغيّرات التي تشبه الحركة المكانية في أنّها تبتدىء من طرف إلى طرف ، كما يأخذ التسخّن من طرفٍ إلى طرف ، فهي داخلة في الزمان ، لأجل أنَّ لها تقدّماً وتأخّراً .

a certain change wholly and completely overtakes what undergoes the change and then progressively increases or decreases, the continuity it has is only temporal continuity, for it has an earlier and later in time only.[11] Thus,[12] it does not have what makes time, which is the continuity of motion along a distance or something similar to distance; nonetheless, it possesses earlier and later parts and so is dependent upon time. So it exists after the existence of time's cause (that is, the motion with respect to which there is a transition). So these changes and spatial motions share the common feature that they are measured by time, while not sharing in common as their effect that time depends upon them for its existence, for this belongs to spatial [motions] alone. Now you have learned our intention concerning the account of spatial motions.

(7) The things in which there is neither an *earlier* nor a *later* in some way are not in time, even if they are together with time—as, for example, the world, for it is together with a mustard seed but is not in the mustard seed.[13] If, from one perspective, something has an earlier and later [aspect] (as, for example, from a certain perspective it is undergoing motion), whereas it has another perspective that is not susceptible to *earlier* and *later* (for example, from a certain perspective it is a being and substance), then it is not in time from the perspective that it is not susceptible to an *earlier* and a *later,* while it is in time from the other. The thing existing together with time, but not in time, and so existing with the whole of uninterrupted time is the *everlasting,* and every one and the same uninterrupted existence is in the everlasting. I mean by *uninterrupted* that it exists the very same, just as it is, at every single moment continuously. So it is as if the everlasting is a comparison of the permanent to the impermanent, and the relation of this simultaneity to the everlasting is like the relation of that instant of time to time. The relation of some permanent

---

11. Reading with **Z**, **T**, and the corresponding Latin *fa-inna lahu taqadduman wa-taʾkhkhur fī al-zamān faqaṭ* (for it has an earlier and later in time only), which is (inadvertently) omitted in **Y**.

12. Reading *li-dhālika* with **Z**, **T**, and the Latin (*quia*) for **Y**'s *ka-dhālika* (likewise).

13. Cf. Aristotle, *Physics* 4.12.221a21–23 for a similar example.

فـإذا كان تغيّرٌ ما يأخذ المتغيّر كله جملة، فيذهب إلى الاشـتداد أو التنقّص؛ فإنَّ له من الاتصـال الاتصال الزمانـي فقط، فإن له تقدما وتأخرا في الزمـان فقط. ولذلك ليس له فاعل الزمان الذي هو اتصال الحركة في مسافةٍ أو شبه مسافة، وهو مع ذلك ذو متقدّم أو متأخّر فهو متعلّق بالزمان، فوجوده بعد وجود علّة الزمان، وهو الحركة التي فيها انتقال. فهذه التغيّرات تشارك الحركات المسافية في أنّها تتقدر بالزمان، ولا تشاركها في أنَّ الزمان متعلّق الوجود بها معلول لها، فإنَّ هذا للمسافيات وحدها، وقد علمت غرضنا في قولنا الحركات المسافية.

(٧) وأمّا الأمور التي لا تقدّم فيها ولا تأخر بوجهٍ، فإنّها ليست في زمان وإنْ كانت مـع الزمـان؛ كالعالم فإنّه مع الخردلة وليس في الخردلة. وإنْ كان شـيء له من جهةٍ تقدّم وتأخّـر، مثلاً من جهة ما هو متحـرّك، وله جهة أخرى لا تقبل التقدّم والتأخر، مثلاً من جهـة مـا هو ذات وجوهر، فهوّ - من جهةٍ ماّ - لا يقبل تقدّماً وتأخراً ليس في زمان، وهو من الجهة الأخرى في الزمان. والشيء الموجود مع الزمان وليس في الزمان، فوجوده مع اسـتمرار الزمان كله هو الدهر. وكل اسـتمرار وجودٍ واحدٍ فهو في الدهر، وأعني بالاسـتمرار وجوده بعينه، كما هو، مـع كل وقت بعد وقتٍ على الاتصال، فكأن الدهر هو قياس ثبات إلى غير ثبات. ونسبة هذه المعيّة إلى الدهر كنسبة تلك الفينة إلى الزمان،

things to others, and the simultaneity that belongs to them from this perspective, is a notion above the everlasting. It seems more worthy to be called *eternity*. So eternity is a whole uninterrupted existence in the sense of the absolute negation of change without a comparison of one moment after another. How odd is the claim of those who say that the everlasting is the duration of rest, or it is a time not numbered by motion,[14] when no *duration* or *time* is understood that does not, in itself, involve a *before* and an *after*. When there is a *before* and an *after* in it, there must be a renewal of some state (as we said),[15] and so it will not be devoid of a motion. Also, being earlier and later exist with respect to resting, although only in the way we mentioned previously.[16]

(8) Now, time is not a cause of anything; but, when something either comes to exist or ceases to exist with the passing of time and no obvious cause is seen for it,[17] the common man attributes that to time, since he either does not find or is not aware of any other conjoined thing except time. So, if the thing is praiseworthy, he praises time; and if it is blameworthy, he blames it. Still, things that come to exist have, for the most part, obvious causes, while ceasing to be and corruption have hidden causes; for the cause of the building is known, whereas the cause of its decline and dilapidation is, for the most part, unknown. The same will hold if you wish to examine many particular cases inductively. Therefore, it appears that most of what is attributed to time involves cases[18] of ceasing to be and corruption, like neglect, old age, decline, the exhaustion of the material, and the like. That is why the common man came to love blaming and speaking ill of time.

(9) Time has certain accidental properties and features indicated by certain expressions, which we ought to mention and enumerate. Among [the temporal expressions] is *now*, which is sometimes understood to be

---

14. The position would seem to be that of Abū Bakr Muḥammad al-Rāzī, who does distinguish between "duration" (*muddah*, which is the same term that Avicenna uses here), which is eternal, and "time," (*zamān*, which again is the same term that Avicenna uses here), which is numbered motion and created; see al-Rāzī, *Opera philosophica fragmentaque quae supersunt*, 195–215.

15. See 2.1.6.

16. See par. 5.

17. Reading *lam yura lahu* with **T**, the Latin (*non apparet*) and most MSS consulted by **Y** for **Y**'s *tuzilhu* (to cause it to disappear); **Z** has *lam nara lahu* (we do not see it).

18. Secluding **Y**'s suggested *arbaʿa* (four).

ونسبة الأمور الثابتة بعضها إلى بعض، والمعيّة التي لها من هذه الجهة هو معنى فوق الدهر. ويشبه أن يكون أحقّ ما سمي به السرمد. فكل استمرار وجودٍ، بمعنى سلب التغيّر مطلقاً من غير قياسٍ إلى وقتٍ فوقتٍ؛ فهو السرمد. والعجب من قول مَنْ يقول إنَّ الدهر مدّة السكون، أو زمان غير معدودٍ بحركةٍ، ولا تُعقل مدّة ولا زمان ليس في ذاته قبْل ولا بَعْد. وإذا كان فيه قبْل وبَعْد وجب تجدّد حالّ - على ما قلناً - فلم يخل من حركة، والسكون يوجد فيه التقدّم والتأخّر، على نحو ما قلنا سابقاً لا غير.

(٨) والزمان ليس بعلّةٍ لشيءٍ من الأشياء، لكنه إذا كان الشيء مع استمرار الزمان، يوجد أو يعدم، ولم يُر له علّة ظاهرة، نسبَ الناس ذلك إلى الزمان، إذ لم يجدوا هناك مقارناً غير الزمان، أو لم يشعروا به. فإنْ كان الأمر محموداً مدحوا الزمان، وإن كان مذموماً ذمّوه. لكن الأمور الوجودية، في أكثر الأمر، ظاهرة العِلل، والعدم والفساد خفي العِلّة؛ فإن سبب البناء معقول، وسبب الانتقاص والاندراس مجهول في الأكثر. وكذلك إنْ شئت استقريت جزئيات كثيرة، فيعرض لذلك أنْ يكون أكثر ما ينسب إلى الزمان هو من الأمور العدمية الفسادية؛ كالنسيان والهرم والانتقاص وفناء المادة، وغير ذلك، ولذلك صار الناس يولعون بذم الزمان، وهجوه.

(٩) والزمان له عوارض وأمور تدل عليها ألفاظٌ، فحري بنا أنْ نذكرها ونعدّها. فمن ذلك الآن؛ وقد يفهم منه الحدّ المشترك بين الماضي والمستقبل الذي فيه الحديث لا غيره.

the common limiting point between the past and future in which there is nothing but the present. At other times, it is understood to be every common division, regardless of whether it is in the past or future. At still other times, it is understood to be time's limit, even though it does not indicate a common point and, instead, permits the estimative faculty to make it a dividing rather than connecting limit (although it is known from external considerations that it is inevitably something common and cannot be a division—that is, by a kind of reflection other than conceptualizing the meaning of the expression [*now*]). They also might use *now* for a short period of time very close to the present *now*. An independent confirmation of this use is that all time comes to be from [the now], for [time] necessarily has two limiting points that belong to it as two instants posited in the mind, even if we are not aware of it. These two instants are simultaneously in the mind as necessarily present; however, in some cases, the mind is aware that one instant is earlier in existence and the other later, owing to the distance between the two, just as one is aware of the earlier instant of two instants [that delimit] the hour or day. In other cases, the two instants are so close that, as long as the mind does not rely on reflection, it is not immediately aware that there is something in between them. In that case, the mind will perceive the two as if they occur simultaneously and are a single instant, although the mind will deny that on the most rudimentary reflection considering the implications. Still, until the mind thinks the matter over, it is as if the two instants occur simultaneously.

وقد يفهم منه كل فصل مشـــترك ولو في أقســام الماضي وأقســام المستقبل، وقد يفهم منه طرف الزمان وإنْ لم يدل على اشـــتراك، بل كان صالحاً لأنَّ يجعل طرفاً فاصلاً في الوهم غير واصلٍ، وإنْ كان يعلم من خارج المفهوم أنّه لا بُدّ من أنْ يكون مشـــتركاً، ولا يمكن أن يكــون فصلاً وذلك بنوعٍ من النظر غير تصــور معنى لفظه. وقد يقولون الآن لزمانٍ قريبٍ جداً من الآن الحاضر قصير؛ وتحقيق ســبب هذا القول هو أنَّ كل زمان يحدث عنه، فله حدّان لا محالة هما آنان يفترضان في الذهن لهّ – وإنْ لم نشعر به. وهذان الآنان يكونان في الذهن حاضرين معاً لا محالة، لكنه قد يشــعر الذهن، في بعض الأوقات، بتقدم آنٍ في الوجود وتأخّر آن، وذلك لبعد المسافة بينهما، كما نشعر بالآن المتقدّم من آني الساعة واليوم. وفي بعضها يكون الآنان من القرب بحيث لا يشــعر الذهن بما بينهما في أول وهلةٍ ما لم يستند إلى استبصار، فيكون الذهن يشعر بهما كأنهما وقعا معاً؛ وكأنهما آنٌ واحد، وإنْ كان التعقّب والاســتقصاء يمنع الذهن عن ذلك فــي أدنى تأمل، ولكن إلى أنْ يراجع الذهن نفسه يكون الآنان كأنهما وقعا معاً.

(10) Also among the temporal expressions is *suddenly,* which is a relation of a thing to its time occurring in a period of time that is so brief that one is unaware of its measure; and, additionally, the thing was not expected to occur. There is, again, the expression *all at once,* which indicates that something happened at an instant and also frequently indicates the opposite of *gradually* (but we have already commented on that).[19] Another expression is *right away,* which signifies some future instant that is close to the present instant and [such that] the measure of the interval between them is so short as to be negligible. Among them is also *just before,* which indicates a relation to some past instant that is close to the present instant, but not [so close] that one is unaware of the duration between them; and *just after* in the future is the same *mutatis mutandis* as *just before* in the past. *Earlier* is either with respect to the past, in which case it indicates what is farther away from the present instant (and *later* indicates its opposite), or it is with respect to the future, in which case it indicates what is closer to the present (and, again, *later* indicates its opposite). When they are taken absolutely, what is earlier is the past, and what is later is the future. The *ancient*[20] is a time that, between it and now, is considered to be extremely long relative to the time's opposing[21] limits. Moreover, there is also the ancient in time absolutely, which, in fact, is that whose time has no beginning.[22]

◆

19. See 2.12.3.

20. The translation "ancient" for the Arabic *qadīm* is not an entirely happy one, for, while it perfectly captures one sense of that Arabic term, it fails to do justice to the philosophical connotation attached to that term that expresses the (purported) pre-eternity or everlasting nature of the cosmos.

21. Reading *mutaqābilah* with **Y** and the Latin (*oppositorum*); **Z** has *mutaᶜālamah* (known), and **T** is not completely clear, although it seems to be *mutaqābilah.*

22. Secluding **Y**'s *Allah aᶜlam* (God knows best), since it is does not appear in the majority of the MSS (in fact none of the MSS consulted by **Z**), nor is it found in **T** or the Latin.

(١٠) ومـن الألفاظ الزمانية قولهم بغتة؛ وبغتة هو نسبة الأمر الواقع في زمانٍ غير مشعورٍ بمقداره قصراً إلى زمانه، بعد أنْ لا يكون الأمر منتظراً متوقعاً. ومن هذه الألفاظ قولهم دفعة؛ وهو يدل على حصول شيء في آنٍ، وقد يدل على مقابل قولنا قليلا قليلا، وقد شرحنا ذلك. ومن هذه الالفاظ قولهم هو ذا وهو يدل على آنٍ قريبٍ في المستقبل من الآن الحاضر، لا يشعر بمقدار البُعْد بينهما قصراً شعوراً يعتد به. ومن ذلك قولهم قُبيل وهو يدل على نسبةٍ إلى آنٍ في الماضي قريبٍ من الآن الحاضر، إلاّ أنَّ المدة بينهما غير مشعور بها، وبُعيد في المستقبل نظير قُبيل في الماضي. والمتقدم، أما في الماضي فيدل على ما هو أبعد من الآن الحاضر، والمتأخر على مقابله، وأما في المستقبل فيدل على ما هو أقرب من الحاضر، والمتأخر على مقابله. وأما إذا أُخذ مطلقاً فالمتقدم هو الماضي، والمتأخّر هو المستقبل. والقديمُ زمانٌ يستطال ما بينه وبين الآن بالقياس إلى الحدود المتقابلة للزمان. وأيضاً القديم في الزمان مطلقاً وبالحقيقة هو الذي ليس لزمانه ابتداء.